MODERN DICTIONARY
BOTANY

MODERN DICTIONARY

BOTANY

Inderjeet Sethi

1990

Discovery Publishing House

New Delhi

ISBN: 81-7141-065-0

1990

Published by:

Discovery Publishing House 4594/9, Darya Ganj
New Delhi-110002

Printed in India
at
Taruna Enterprises Anup Market Maujpur Delhi-110053

Preface

The aim of this dictionary is to provide, within a single volume of moderate size, a comprehensive collection of terms frequently used in botany and allied disciplines. It has been difficult to decide the limit of botany as a subject, as the scope of each science is broadening and they are becoming integrated. This compilation will open up a vast new literature of specialized terms which although relevant may be quite unfamiliar to the reader as they have their origins in other disciplines. This dictionary should be an invaluable aid in the day-to-day learning for students of honours course in Indian University and will also be useful for preparing them for various enterance examinations. I hope teachers as well will find parts of the book useful.

This dictionary provides simple and direct explanation, while still maintaining a high academic standard. It includes a broad range of terms and concepts used in botany from simple to advanced levels. Besides full coverage of botany, the dictionary includes terms used in Agriculture, Cell Biology, Microbiology, Mycology, Phytochemistry, Ecology and Systematics. Commonly used abbreviations are also included. Cross references for additional information appear at the end of an entry. Separate definitions have been given for the terms which are in diverse use. Synonyms and adjectives have also been included along with the entry. SI units and prefixes used in botanical words have been listed in the Appendices.

Several colleagues provided comments and suggestions on the manuscript, I and obliged to them all. I gratefully acknowledge the help of my colleague Dr. Gurcharan Singh Department of Botany, SGTB Khalsa College (Delhi University) for his comments and suggestions and for going through the manuscript And above all, I am indebted to my husband Dr. M.S. Sethi, Department of Chemistry, ARSD College (Delhi University), for his unstinting help throughout the compilation of this manuscript. It was he, who encouraged me to take up this project. Without his help and encouragement it would not have been possible to complete this project.

INDBRJEET SETHI

A

abaca. *Musa textilis* fam. Musaceae also known as Manila hemp. A plant native to Borneo and Phillippines valuable for its hard fibers obtained from outer portions of the leaf stalks.

abaxial. On the underside of a leaf, that is, pointing away from the stem.

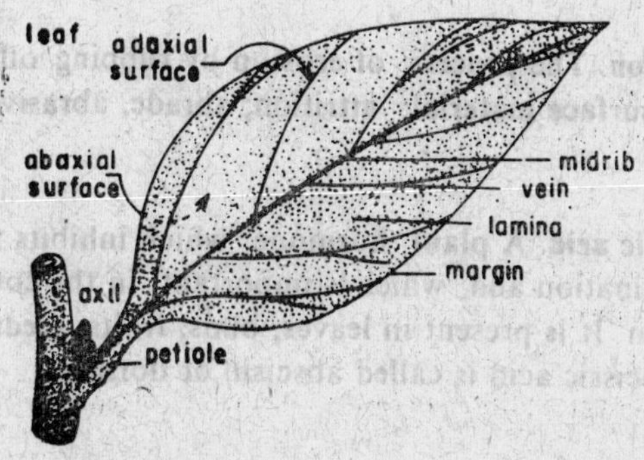

aberration. An unusual phenotype resulting from genetic abnormality or mutation; aberrant.

abiogenesis. Spontaneous generation; the concept that life can arise spontaneously from non-living matter by natural processes without the intervention of supernatural powers; archebiosis; archegenesis; autogenesis; nomogenesis; xenogenesis; *cf.* biogenesis.

abiotic environment. The non-living component of an ecosystem; the physical and chemical factors of the environment; *cf.* biotic environment.

abjection. The separation of a spore from a sporophore or sterigma by an act of the fungus.

aboospore. An oospore produced from an unfertilized female gamete; azygospore; parthenospore.

aborigine. The original or indigenous biota of a geographical region; aboriginal.

abort. To arrest development; abortion, abortive.

abortive transduction. Transduction in which the genetic material is not integrated or replicated but is otherwise functional.

abrasion. The process of erosion by rubbing off or wearing away of surface material; attrition; abrade, abrasive, abrasiveness.

abscisic acid. A plant hormone which inhibits root growth and germination and which is important in the control of leaf abscission. It is present in leaves, buds, fruits, seeds, etc. Sometimes abscissic acid is called abscisin or dormin.

abscisic acid

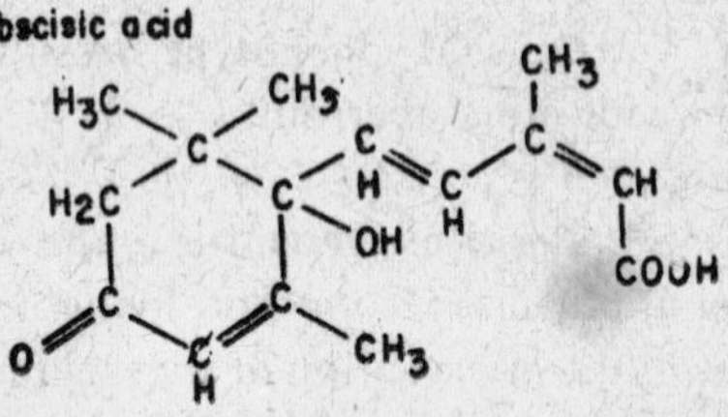

abscission. The shedding of leaves or fruit as a result of cambial activity of cells in a zone (abscission layer) that extends across the base of the petiole; premature formation of abscission layers (proleptic abscission) is a hyperplastic symptom of some plant diseases.

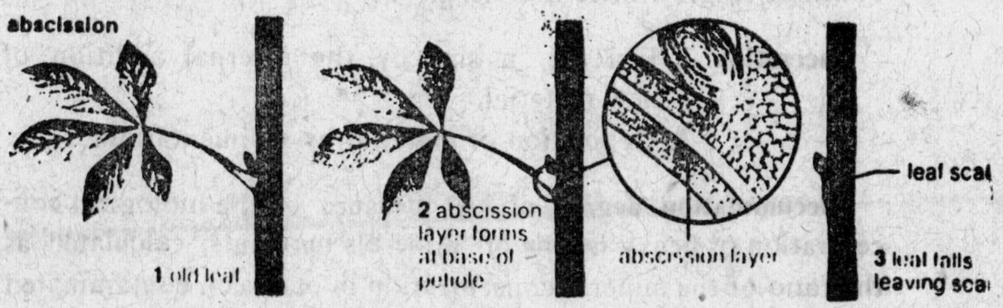

abscission layer. A zone of delicate, thin walled cells surrounding a lesion on a leaf, the breakdown of which disjoins the affected area from the rest of the leaf.

absorption spectrum (*pl.* **spectra**). The wavelengths or colours of light which are absorbed by a pigment. Plants appear green because chlorophyll absorbs red and blue light and reflects green light (See Fig. action spectrum).

acarophytism. Symbiosis between plants and mites; acarophytium.

acaudate. Not having a tail.

accessory pigment. A pigment which is involved in photosynthesis but not directly in the capture of solar energy, e g. carotenoids.

accessory species. A plant species with a moderate degree of fidelity in a given community or association.

accidental. 1: Not normally occurring in a particular community or habitat.

2: Used of a plant species with a low degree of fidelity in a given community or association.

accidental parasite. A parasite found associated with an organism which is not its normal host.

acclivous. Having a gentle upward slope.

accrescent. Increasing in size with age; used of plants that continue to grow after flowering.

accretion. 1: Increase in size by the external addition of new material.
2: Deposition of material by sedimentation.

accumulation, degree of. A measure of the biological concentration of heavy metals or minerals in plants, calculated as the ratio of the mineral concentration in plants on contaminated soils to that of plants on normal soils expressed as a percentage.

accumulation, zone of. The B-horizon of the soil profile, the upper subsoil horizon.

accumulator organism. Any organism that actively concentrates a particular element or compound in its tissues.

acellular. Not made of cells, like the multinucleate plasmodium of a myxomycete slime mould.

acentric. Used of a chromosome or chromosome fragment lacking a centromere; akinetic; *cf.* centric.

acephalous. Not having a head.

acerose. Needle-like and stiff like a pine needle.

acervulus (*pl.* acervuli). A fruiting body of certain imperfect fungi: a shallow, saucer-shaped structure with a layer of conidiophores that bear conidia, as the acervulus expands to form a cushion-like mass, it ruptures the epidermal layer of the substrate, exposing the spore-bearing surface.

acetabuliform. Saucer-like in form.

aceto-orcein. A stain which is used in the preparation of root tip or anther squashes for chromosome number.

achene. A dry indehiscent fruit with one seed the product of one carpel.

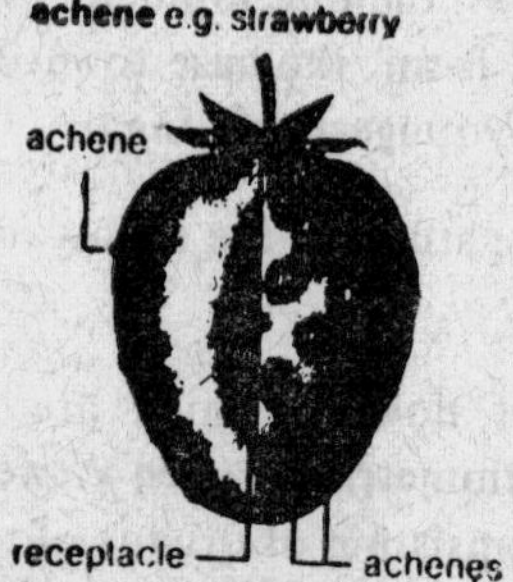

achlorophyllous. Lacking chlorophyll.

achroous: Without colour, unpigmented; achromatic.

acidobiontic. Living in an acidic habitat.

acidophilic. Thriving in an acidic environment; specially of lichens on peaty soils; acidophilous; aciduric; oxyphilous; acidophile; acidophily; *cf.* acidophobic.

acidophobic Intolerant of acidic environments. acidophobe; acidophoby; *cf.* acidophilic.

acidotrophic. Feeding on acidic food or acidic substrates.

aciduric. See acidophilic.

acme. A period of maximum vigour; the highest point attained in phylogenetic or ontogenetic development.

acrocarpous. Of mosses which have an upright stem, with the reproductive organs at the apex.

acronematic. A smooth flagellum which lacks hair.

acropetal. Describes the development of structures (such as spores) in succession from the base towards the apex, i.e., the apical member is the youngest; basifugal.

actinobiology. The study of the effects of radiation on living organisms; actinology.

actinomorphic. Of flowers which are symmetrical in all directions (radially symmetrical) when viewed from above, that is, with each whorl consisting of organs of the same size.

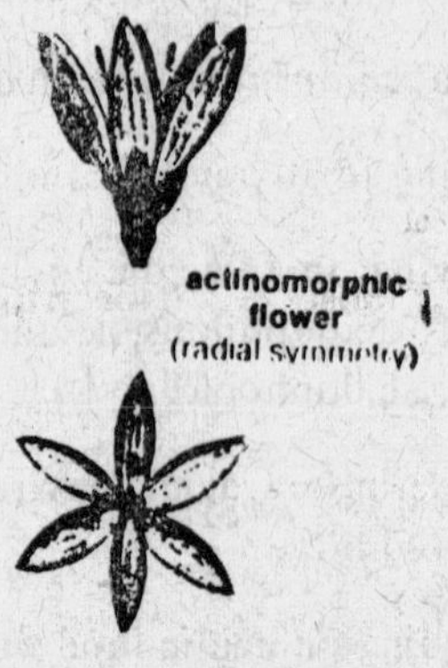

actinomycetes. Organisms characterized by fine hyphae, usually less than 1.0 μm in diameter, that readily break into fragments resembling bacterial cells. They are classified as bacteria by some, as fungi by others. Actinomycetes are typically saprobes (esp. in soil) but a few are pathogenic to man, animals and plants; some are important sources of antibiotics e.g., amphotericin, cycloheximide, streptomycin etc.

actinomycin D. An antibiotic that inhibits the elongation of RNA molecules.

actinostele. A protostele which is stellate in cross-section, having the protoxylem at the tips of the star. A primitive type of stele occurred in the early Pteridophytes.

action spectrum. The wavelength or wavelengths of light which activate a biochemical process. Blue and red light are needed for photosynthesis.

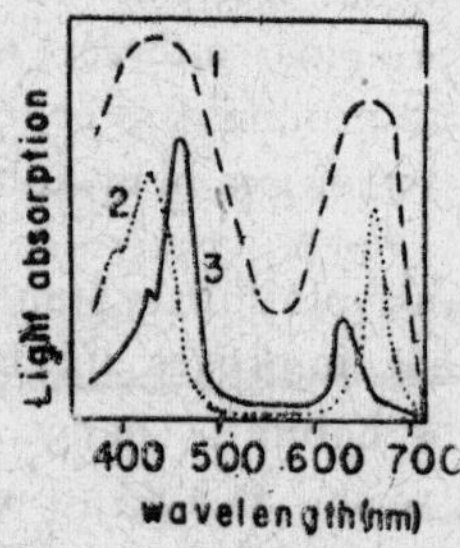

1. action spectrum of photosynthesis
2. absorption spectrum of chlorophyll a
3. absorption spectrum of chlorophyll b

activation energy. Energy required to initiate a chemical reaction.

active site. The part of an enzyme molecule to which substrate molecule become attached and where catalysis takes place.

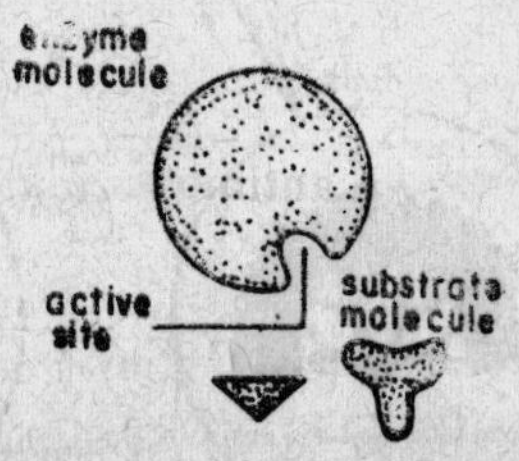

active transport. The movement of substances across membranes using energy. This is necessary when a substance is being transported from the side of the membrane where it is less concentrated to the side where it is more concentrated.

actomyocin. A term that describes the complex of actin and myosin when these proteins interact to form a contractile system in the cell. These proteins are believed to be responsible for the movement of plasmodium in myxomycetes (slime moulds).

adaptation. From an evolutionary standpoint, a characteristic of a living organism that improves its chances for survival in the environment of its habitat; change brought about in a population of an organism as a result of exposure to a particular set of environmental conditions, the change enabling the organism to adjust to the environmental conditions in question; adapt; adaptive.

adaptive radiation. A process of evolution from one primitive species to more than one advanced species, each adapated to a particular niche or habitat.

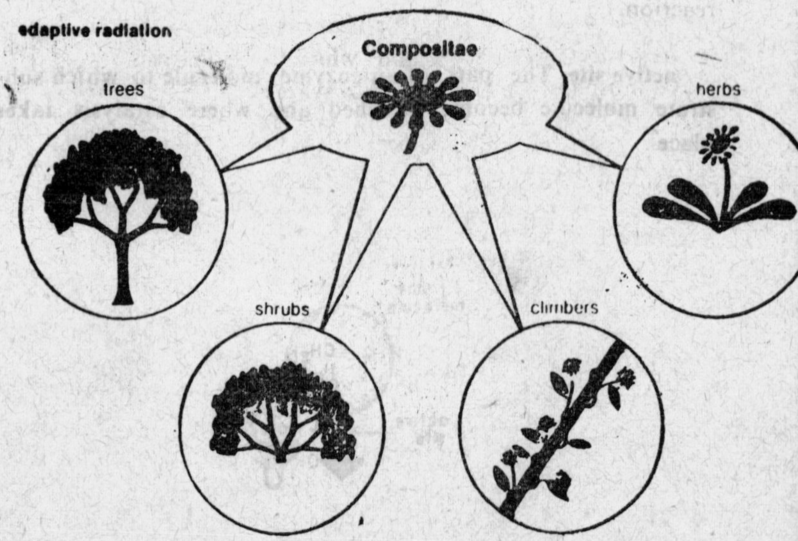

adaxial. On the top side of a leaf, that is, pointing towards the stem. (See Fig. abaxial).

adeloparasite. Parasitism between two closely related plants or animals.

adelphogamy. 1: Fertilization between two different individuals derived vegetatively from the same parent plant; sib mating; sib pollination.

2: pseudomictic copulation of mother and daughter cells, as in some yeasts.

adenine. A purine base which pairs with thymine in DNA and with uracil in RNA.

Adenine
(6-aminopurine)

adenosine diphosphate (ADP). Then ucleotide which results from the hydrolysis of ATP and which with the addition of an orthophosphate group, is used to synthesize ATP.

high energy bond

ADP
(two phosphate groups)

adenosine triphosphatase (ATPase). An enzyme that hydrolyzes ATP to form ADP and inorganic phosphate.

adenylcyclase. An enzyme that catalyzes production of cyclic AMP from ATP; adenylate cyclase.

adichogamy. Simultaneous maturation of male and female reproductive organs of a flower or hermaphroditic organism; homogamy.

adnate. 1: Pertaining to lamellae, widely joined to the stipe.
2: Pertaining to pellicle, scales etc., tightly fixed to the surface.

adnexed. Of lamellae or tubes narrowly joined to the stipe

aduncate. Bent; hooked; crooked.

adventive. Not native; an organism transported into a new habitat, whether by natural means or by the agency of man.

adynamandrous. 1: Having non-functioning male reproductive organs; adynamandry; *cf*. adynamogynous.
2: Incapable of self-fertilization.

adynamogynous. Having non-functioning female reproductive organs; apogynous; adynamogyny; *cf*. adynamandous.

adventitious embryony. The production of an embryonic sporophyte by mitotic divisions from tissues of another sporophyte without an intervening gametophyte generation; seed production without a sexual process.

adventitious root. Any root which grows from a tissue other than the pericycle or endodermis of an older root. These roots appear in an unusual place or position, e.g., on the stem in tulips and onion. Some plants may even produce adventitious roots from leaves.

aecidium (*pl.* aecidia). A cup-shaped aecium that contains a peridium. Aecidia usually are brightly colored (orange) and occur in groups that make up the cluster-cup stage of certain rust fungi.

aeciospore. A binucleate rust spore produced in an aecium.

aecium (*pl.* aecia). The fruiting body of certain rust fungi produced immediately after the pycnium; the aecium produces dikaryotic spores in chains; spores may infect the suscept in which they are formed, but usually infect an unrelated alternate suscept. Aecia may be of the aecidium, caeoma, peridermium, or roestelium type.

aerenchyma. Tissue with air-filled spaces between its cells, common in aquatic plants.

aerial root. A root growing from a part of a plant which is above the ground.

aerobic. Of respiration using molecular oxygen and involving oxidative processes. Also, of organisms which respire aerobically; aerobe.

aerohygrophilous. Thriving in high atmospheric humidity; aerohygrophile, aerohygrophily; *cf.* aerohygrophobous.

aerohygrophobous. Intolerant of high atmospheric humidity; aerohygrophobe, aerohygrophoby; *cf.* aerohygrophilous.

aestivation. 1: Passing the summer or dry season in a dormant or torpid state; **estivation**; **aestivate**; *cf.* hibernation.
2: The manner in which flower parts i.e. calyx, corolla, stamens and pistil are folded prior to expansion or opening.

aethalium (*pl.* **aethalia**). A rather large, sometimes massive, generally cushion-shaped fructification of some Myxomycetes.

aetiology. 1: The branch of science dealing with the study of origins or causes.
2: The demonstrated cause of a disease or trait; causation; aetiological; etiology.

affinity chromatography. A technique for separating macromolecules based on the biological affinity of the molecule for a matrix-bound ligand.

afforestation. The process of establishing a forest in a non-forested area; *cf.* reforestation.

aflatoxin. A mycotoxin produced by *Aspergillus flavus* and other species of this fungus; the cause of a fatal disorder (aflatoxicosis) in ducklings and young turkeys and carcinogenic for rats and possibly man.

agamandroecious. Used of a plant having male and neuter flowers in the same inflorescence; *cf.* agamogynoecious, agamohermaphrodite

agameon. A species comprising only non-sexually reproducing individuals; agamospecies; binom.

agamogynoecious. Used of a plant having female and neuter flowers in the same inflorescence *cf.* agamandroecious, agamohermaphrodite.

agamohermaphrodite. Used of a plant having hermaphrodite and neuter flowers in the same inflorescence; *cf.* agamandroecious, agamogynoecious.

agamont. The asexual individual or generation producing agametes.

agamospecies. A species or population comprising only asexually reproducing (apomictic) individuals; agameon.

agamospermy. Apomixis in which embryos and seeds are formed asexually, but not including vegetative reproduction; agamospermous.

agamotropic. Used of flowers that do not close again once they have opened; *cf.* gamotropic, hemigamotropic.

agar. A polysaccharide from certain red algae e.g. *Gelidium* and *Gracillaria* (seaweeds): agar forms a gel with water, and is used at a concentration of 1.5 to 2.0 per cent to solidify media used for culturing microorganisms.

agaricolous. Living on mushrooms and toadstools; agaricole.

agaritine. An amino acid from *Agaricus bisporus*

agent of inoculation. That which transports inoculum from its source to or into the infection court. Examples: wind, splashing rain, running water, insects, man, other animals.

agglutination. A serological test in which viruses or bacteria suspended in a liquid collect into clumps whenever the suspension is treated with antiserum containing antibodies specific against these viruses or bacteria.

α-granules. Glycogen-rich granules found in the cells of blue-green algae.

agrad. A cultivated plant.

agrarian. Pertaining to cultivation or cultivated plants.

agrology The branch of agriculture dealing with the study of soils.

agronomy. The theory and practice of agricultural management, crop production and husbandry.

agrophilous. Thriving in cultivated soils; agrophile, agrophily.

agrostology. The study of grasses; graminology.

agrotype. An agricultural variety or race.

aigialophyte. A beach plant.

aigicolous. Living in beach habitats; **aigicole.**

aiphyllophilus. Thriving in evergreen woodland; aiphyllophile; aiphyllophily

air layering. A way of causing the formation of roots from nodes on a shoot. Wet moss is wrapped around the shoot, when roots have formed, the shoot can be cut from the plant and be grown as a new individual.

akaryote. A phase in the life cycle of the Plasmodiophorales during which the nucleoplasm loses its affinity for stains thus little or no chromatin can be seen in the nucleus.

akinete. A non-motile thick-walled resting spore derived from a vegetative cell. Thus it is a modified vegetative cell with concentrated food.

albidin. A red quinone pigment from *Penicillium albidum*; antifungal and weakly antibacterial.

albumen. The endosperm of a seed; albuminous

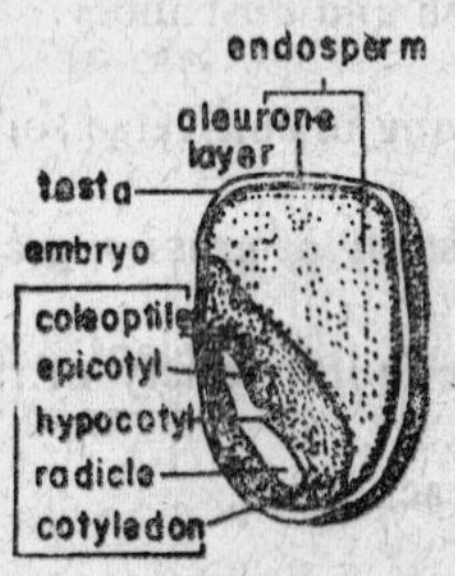

albuminous seed where most food is stored in endosperm e.g. maize.

alburnum. See sap wood.

aldose. Any monosaccharide in which one carbon atom is an aldehyde (—CHO) group e.g., glucose, glyceraldehyde.

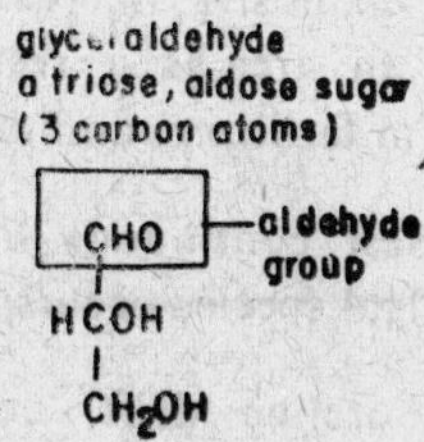

aleppo galls. Galls which form on the twigs of aleppo oak; it has gallic acid and gallitannic acid, used in tanning leather.

aletophilous. Thriving on roadside verges and beside railway tracks; aletophiie, aletophily.

aletophyte. A plant growing in a mesic habitat or on roadside verges.

aleurone layer. The outer layer of thick-walled cells in the endosperm of the seeds of many grasses. The aleurone layer is rich in protein (See Fig. albumen).

aleuroplast. See leucoplast.

alfalfa. A leguminous forage plant (*Medicago sativa*) also known as lucerne. A herbaceous perennial legume characterized by a deep tap root.

alga (*pl.* algae). Aquatic plants or plants of damp habitats, unicellular or multicellular plants with filamentous thalli; distinguished from the fungi by the presence of chlorophyll. The most common types are the green, blue-green, brown and red algae, all of which contain chlorophyll as well as other pigments.

algal bloom. An explosive increase in the density of phytoplankton within an area as a result of high phosphate concentration from farm fertilizers and detergents.

algal line. The highest continuous line on the shore along which any particular algal species occurs; *cf.* tang line.

algal wash. A shoreline drift, comprising mainly of filamentous algae.

algicide. A substance toxic to algae.

algin. A sodium salt of alginic acid found commonly in the cell walls of Phaeophyceae. (brown algae), widely used for its water-binding, thickening, and emulsifying properties.

alginic acid. A polysaccharide consisting of variable mixture of β-1, 4-linked *d*-mannuronic acid and 1, 4-linked *l*-gluronic acid units found in the cell walls of brown algae, *Fucus*, *Laminaria*, *Macrocystis* etc., used as an emulsifying agents in preparation of polishes and paints; can stop bleeding effectively and is employed as a highly efficient gauze in internal operations; derivatives of alginic acid used in preparation of soups. sauces and creams and also antibiotic capsules.

algology. Study of algae.

algophagous. Feeding on algae; algophage; algophagy

alien. Non-native; a species occurring in an area to which it is not native.

aliform. Wing like in form.

alkaloid. An organic nitrogen-containing compound, produced by many plants. Alkaloids have alkaline properties, are mostly toxic and often defend plants against attack by herbivores.

allantoid. Pertaining to fungal spores, slightly curved with rounded ends; sausage shaped.

alleles, Two genes each occupying the same position or locus on two homologous chromosomes. Alleles may have small differences in the sequence of bases in their DNA; allelomorph.

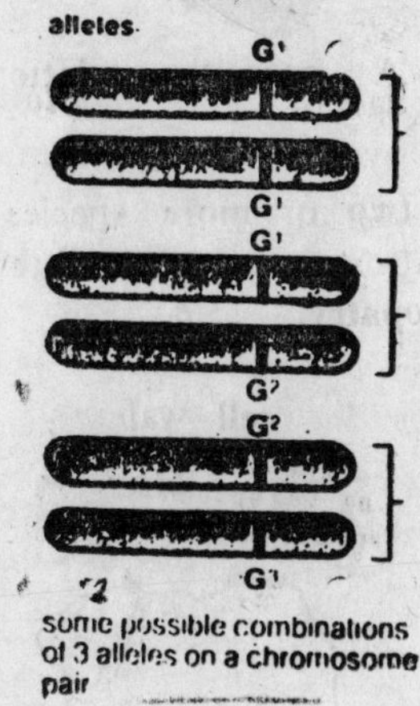

some possible combinations of 3 alleles on a chromosome pair

allelopathy. Discouragement by one plant of the growth of other plants around it. e.g. by toxins contained in the fallen leaves; **allelopathic**.

allochemic. Any secondary compound produced by plants as part of their defence mechanism against herbivores; acting either as a toxin or digestibility reducer.

allochronic speciation. 1: Speciation without geographical separation through the acquisition of different breeding seasons or patterns. 2: Speciation occurring by the sequential replacement of species through time.

allodiploid. A hybrid diploid in which one or more chromosome pairs are derived from different species.

allogamy. The production of zygotes by cross-fertilization; **allogamous.**

allometric growth. Used for describing growth rate of a part of an organism to that of whole organism. All parts of an organism rarely grow with same rate of growth. Parts growing faster than the whole organism are said to exhibit *positive allometry.* Parts growing slower than the whole organism are said to exhibit *negative allometry.*

alloparasitism. A parasitic relationship between two unrelated organisms.

allopatric. Of two or more species living in separate place, or of speciation in which different species evolve in different places; **allopatry.**

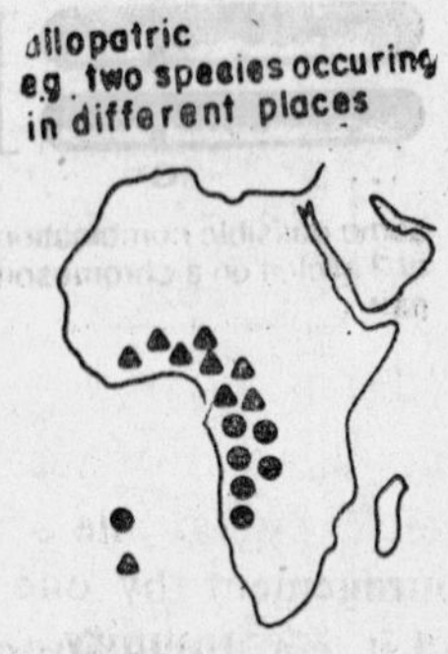

allopolyploid. A polyploid species with sets of chromosomes from two or more different species. This can be the result of hybridization between species.

allosome. Heterochromosome; any chromosome or chromosome fragment other than a normal A-chromosome.

allosteric effectors. Small molecules, usually metabolites, that bind to allosteric proteins at a site other than the active site and cause a change in protein shape and activity.

allosteric enzymes. Enzymes whose activity is modulated by the binding of allosteric effectors at sites other than the active site, allosteric proteins.

allotriploid. A hybrid triploid; a type of allopolyploid.

allotype. A paratype of opposite sex to the holotype and originally designated by the author.

alsocolous. Living in woody groves; alsocole.

alsophilus. Thriving in woody grove habitats; alsophile, alsophily.

alsophyte. A grove plant; alsad.

alternate host. One of two species of plants required as a host by heteroecious rust fungi for the completion of their developmental cycles.

alternation of generations. The life cycle which consists of a haploid gametophyte producing gametes followed by a diploid sporophyte producing spores.

alternative host. A plant other than the main host that is fed upon by a parasite; alternative hosts are not required for completion of the developmental cycle of the parasite.

altherbosa. A tall herb community, often found in areas of forest clearance.

ambisexual. 1: Having separate male and female flowers on the same plant; ambosexual; hermaphrodite; monoecious. 2: Pertaining to both sexes.

ameiosis. Nuclear division in which no reduction in chromosome number occurs.

ametoecious. 1: Used of a parasite having only a single host during its life cycle; autoecious; autoxenous; *cf.* heteroecious. 2: Used of a parasitic species that is highly host specific; **ametoecius**; *cf.* metoecious.

ameiotic parthenogenesis. Parthenogenesis in which meiosis has been suppressed so that neither chromosome reduction nor any corresponding phenomenon occurs.

amensalism. Used for describing association betweet two different types of organisms in which one organism harms the other by its activities but does not get any benefit by such activity.

amictic. 1: Pertaining to females that produce unfertilized eggs that develop into offspring of one sex only. 2: Used of a lake that has no period of overturn because it is perennially frozen; *cf.* mictic.

amino acid. Any one of a class of organic compounds with a carboxyl group (—COOH), an amino group (NH_2) and a 'side-group' all attached to a central carbon atom. Different amino acids have different sidegroups. There are about 20 different amino acids found in proteins which they form when linked together in a chain or polymer.

aminoacyl adenylate. In protein synthesis, an intermediate in the formation of a covalent bond between an amino acid and its tRNA.

amitosis A process of nuclear division by simple constriction into two portions, often forming dissimilar daughter nuclei; holoschisis.

amixis. 1: Absence or failure of interbreeding; amlxia
2: A process of pseudosexual reproduction in certain haploid fnngi, in which both meiosis and karyogamy are absent.

ammochthad. A plant living on sand banks; ammochthophyte.

amphicarpogean. Producing fruits above ground that are subsequently buried; **amphicarpogenous;** *cf.* hypocarpogean.

amphicryptophyte. A marsh plant with amphibious vegetative parts.

amphigean. Used of a plant having underground as well as aerial flowers.

amphigeic. Pertaining to both the Old World and the New World; *cf.* gerontogeic, neontogeic.

amphimixis. True sexual reproduction: the union of male and female gametes; may be either autogamy (inbreeding) or allogamy (outbreeding); **amphimict, amphimictic;** *cf.* apomixis.

amphinereid. An amphibious plant.

amphiospore, Urediniospore with thickened walls and capable of hibernating; amphispore.

amphipathic. Describing a molecule possessing both a hydrophilic and a hydrophobic region.

amphiphyte. A plant able to live either rooted in damp soil above the water level or completely submerged.

amphitoky. Parthenogenesis in which both male and female offspring are produced; ampherotoky; amphoterotoky; deuterotoky; gametotoky; *cf.* thelytoky.

amphitrophic. Used of an organism that lives as a phototroph during daylight and as a chemotroph in darkness; amphitroph.

amplification. The production of additional copies of a gene.

amoeboid. Amoeba-like; moving by temporary processes (pseudopodia) from the surface of the vegetative body.

amygdaliform. Almond-shaped.

Amylase. An enzyme which catalyzes the breakdown of starch into monosaccharide units. Amylases are widely distributed in plants and animals e.g. in germinating grains of barley and in saliva.

amylopectin. A form of starch in which the glucose molecules are in branched chains which in solution forms a paste and reacts with iodine to give the characteristic red-violet colour; *cf.* amylose.

amyloplast. A plastid in the cortical cells of roots in many plants. The function of amyloplasts is to store starch; see leucoplast.

amylose. A form of starch made of straight chains of glucose monomers that reacts with iodine to give the characteristic blue-black colour; *cf.* amylopectin.

anabolism. Energy-requiring synthesis of more complex molecules from simpler ones involving the taking up and storing of energy; *cf.* catabolism.

anaerobic. Of respiration in which molecular oxygen is not used, that is, in glycolysis and fermentation. Also, of organisms which can live without molecular oxygen, e.g. the bacteria that live in mud or in the gut of animals; these organisms are sometimes called anaerobes.

anaphase. The third stage of cell division following metaphase in which the chromatids separate from each other and move to the ends, or poles, of the spindle.

anaphase chromatids separate at centromeres Sister chromatids drawn to opposite poles of the spindle

anastomosis (pl. anastomoses). The union of a hypha or vessel with another resulting in intercommunication of their contents, resulting into a network; **anastomosing: running** together irregularly to give a vein-like network.

anatomy. The study of the way in which tissues and organs are arranged in organisms; anatomical.

anatropous. Of ovules with the funicle bent back on itself and the micropyle facing the placenta.

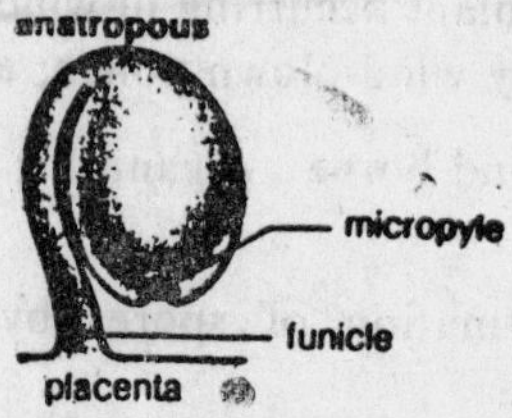

ancophyte. A plant inhabiting canyon forests; **anchophyte.**

androecium. The male part of a flower, consisting of stamens. The function of the androecium is to produce the male gametes contained in pollen.

andromonoecious. Having male and hermaphrodite flowers on the same individual; andromonoecy.

androphages These are phages (viruses) which infect only male strains of bacteria; they may be filamentous and contain single-stranded DNA (e.g. f^1 and fd) or are spherical and contain single stranded RNA (e.g. f_2).

androsporangium. An androspore-producing sporangium.

androspore. 1: A multiflagellate, zoospore-like unicellular spore of nannandrous species of *Oedogonium*. It germinates into dwarf male.

2: A male spore; pollen grain.

anemochorous. Having seeds, spores or other propagules dispersed by wind; **anemochoric, anemochore, anemochory.**

anemophilous. 1: Pollinated by wind-blown pollen; anemogamous, **anemophile**, **anemophily**. 2: Thriving in sand draws; anemophilus.

anemophily. Pollination by wind. Plants which are pollinated by wind produce large amounts of pollen. They are not usually scented, do not produce nectar, and are sometimes apetalous; anemophilous.

anemophobic. Intolerant of exposure to wind; anemophobe, anemophoby.

anemophyte. 1: A plant occurring in wind swept situations. 2: A plant fertilized by wind-blown pollen; anemophile.

anemoplankton. Wind borne organisms; aerial planktons aeroplankton.

aneuspory. The formation of spores by irregular meiotic divisions; aneusporous.

angiosperm. A spermatophyte of the subdivision Angiospermae. They differ from gymnosperms in that their ovules are protected in the ovary and in their possession of vessels in the xylem. They also have double fertilization of the egg and the endosperm. There are more than 200 families and 2,50,000 species of angiosperms. They are divided into two classes, the Monocotyledones and the larger Dicotyledones.

anisogamous. Of plants which produce gametes of different sizes, sometimes known as microgametes (male) and megagamates (female). All plants growing on land are anisogamous; anisogamy.

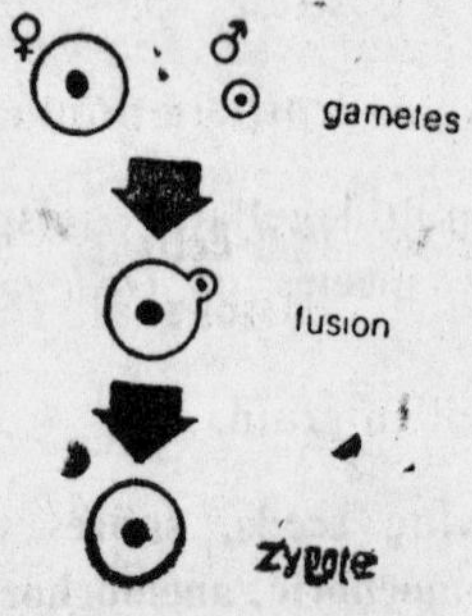

anisogamous planogametes. Motile gametes which are morphologically similar but which differ in size.

anisosymbiosis. Symbiosis in which one partner (the macrosymbiont) is larger than the other (the microsymbiont).

annual. Of plants which complete their entire life cycle, from seed to reproduction to death, in one year. e.g. *Nicotiana* (tobacco).

annual ring. The growth of secondary xylem (wood) in the stem of a woody plant in a temperate climate during one year. Annual rings appear as a series of concentric lines (rings) in a cross-section of a stem. One light-ring and one darker ring are produced each year. From the number of annual rings the approximate age of the plant may be determined.

annulus (pl. annuli). 1: The ring found on the stem of certain species of mushrooms. Remnant of the inner veil.

2: The ring of tissue which is found around the inside of the top of a moss capsule. The cells below it dry out, causing the lid to fall off, and allowing the spores to get dispersed.

3: An arc of cells which is found around the sporangium of a fern. The outer walls of the cells are not thickened, so that when the cells dry out, tensions are set-up which burst open the sporangium.

antagonistic symbiosis. 1: One organism of an association benefits at the expense of the other.

2: Parasitism.

antapical. Away from the apex.

anther. The part of a stamen in which pollen is produced. The anther is attached to the receptacle by the filament. Anthers are hollow organs which dehisce along one side to release pollen.

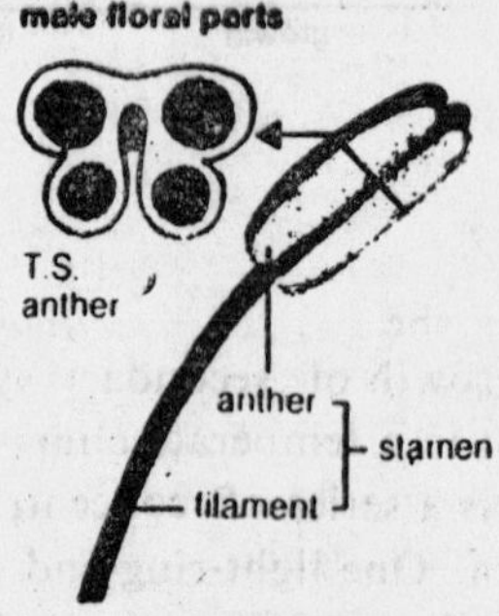

antheridium (*pl.* **antheridia**). 1: The organ producing male gametes in bryophytes and ferns.

2: The finger-like male sexual organ of the Oomycetes.

antherozoid. A flagellate motile male gamete of bryophytes and some ferns. Antherozoids are produced in antheridia.

anthesis. Flower opening or flowering period.

anthocyanescence The condition of being anthocyanescent; reddish-purple color in tissues that are normally green—often a plesionecrotic symptom of plant disease caused by an infectious agent, usually appearing in the margins around holonecrotic spots in green leaves.

anthracnose. Common name of plant diseases usually characterized by development of ulcerous lesions on stems, leaves, and fruit; often caused by certain imperfect fungi that produce conidia in acervuli (examples; *Colletotrichum*, *Gloeosporium*, *Kabatiella*).

anthropophyte. A plant introduced unintentionally into an area by man during cultivation.

antiauxin. These are chemicals which are strong inhibitors of polar transpart of auxin e.g. 2,3,5-triiodobenzoic acid (TIBA) and naphthylphthalamic acid (NPA).

antibiosis. An association between two organisms that is detrimental to the vital activities of one of them.

antibiotic. A chemical compound produced by one microorganism which inhibits or kills other microorganism. Antibiotics are produced by many fungi e.g. penicillin is produced by several species of the fungus *Penicillium*.

antibody. A new or altered protein produced in animal in response to activities of a normally foreign substance (antigen) that gains access to its tissues; it combines chemically with the antigen and renders it harmless.

anticlinal. Of cell divisions which occur at right angles to the surface of the plant.

anticodon. A sequence of three nitrogen-containing bases in a molecule of transfer RNA, which pairs with a codon on a molecule of messenger RNA during translation. Each molecule of transfer RNA has only one anticodon. Corresponding to the particular amino acid to which it is attached during protein synthesis e.g. one of the anticodons for the amino acid serine is a sequence consisting of uracil, cytosine and guanine or UCG.

antigen. Foreign proteins, and occasionally complex lipids and carbohydrates, which upon injection into an animal induce the production of antibodies.

antigibberellins. One of a group of organic compounds which cause plants to grow with short, thick stems; when applied to grass they retard its growth.

antipodal cells. The three cells at the other end of the embryo sac from the ovum in the female gametophyte of an angiosperm.

antiserum. The blood serum of a warm-blooded animal that contains antibodies.

apetalous. Of flowers without petals. Apetalous flowers are often pollinated by wind.

aphanoplasmodium (pl. aphanoplasmodia). A plasmodium consisting, in its early stages, of a network of very fine trans parent strands which are not conspicuously differentiated into ectoplasm and endoplasm, and in which the protoplasm is not coarsely granular. Characteristic of *Stemonitis*

aphid. A plant-louse of great importance to agriculture. It damages crops by sucking the cell-sap and by transmitting certain virus diseases. The commonest aphid is the Myzus persicae.

apical dominance. The inhibition of the development of lateral buds by hormones produced in the shoot apex. If the apex is cut off, the buds develop.

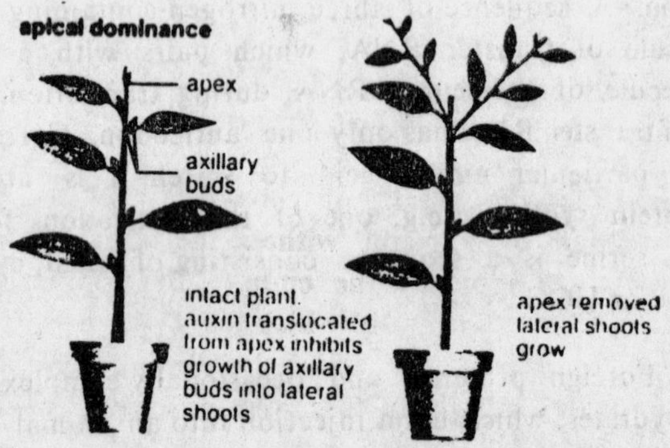

aplanosporangium. Sporangium containing aplanospores.

aplanospore. A non-motile spore

apocarpous. Of ovaries in separate carpels not joined together at their margins. This is characterstic of many primitive flowers.

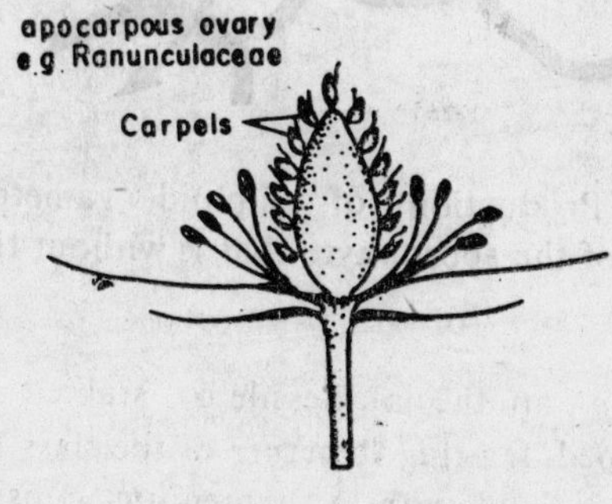

apogamy. Asexual reproduction in which embryos and propagules are produced without meiosis occurring.

apomixis. The production of propagules by the female reproductive organs of a plant, without the sexual fusion of cells. In one type of apomixis, the embryo develops from the unfertilized haploid egg-cell. In which case the offspring are usually sterile. In others, the embryo develops from diploid tissue in the ovule, in which case the offspring are fertile; apomictic.

apoplast. The non-living parts of a plant, i.e. the xylem, the cellulose cell walls and the intercellular spaces.

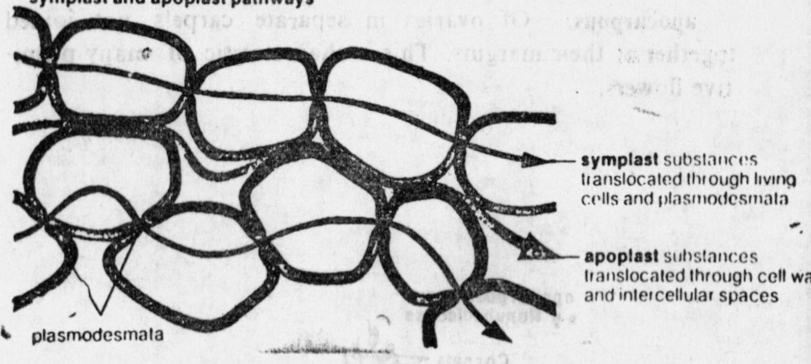

apospory. Production of diploid gametophyte from vegetative cells of the sporophyte that is without the production of spores.

apothecium (*pl.* apothecia). Sessile or stalked saucer-shaped or open cup-shaped, fruiting structure of the class Discomycetes; the inner surace is lined with a hymenium consisting of asci and paraphyses (sterile cells).

apposition. In plant cells the walls become thickened by the deposition of successive layers of cellulose.

appressorium (*pl.* appressoria). A bulbous or lobed swelling of a hyphal tip; it is often held to the surface of the substrate by a gelatinous secretion. Suscept tissue is penetrated by a slender peg that forms at the center of the appressorium.

arboreal. Living in trees; adapted for life in trees; arboricolous.

arboreous desert. An area of sparsely scattered trees with little or no vegetation between; desert forest.

arboretum. A botanic garden or parkland dominated by trees.

arboricide. A chemical that kills trees.

arboricolous. Living predominantly in trees or large woody shrubs; dendricolous; arboreal; arboricole.

arboriculture. The cultivation of trees.

archegonium. The flask shaped female organ of bryophytes, pteridophytes and gymnosperms. The archegonium consists of a hollow neck, whose wall is one cell thick, and a swollen base containing the ovum. The antherozoid swims down the neck to reach the ovum.

archicarp. An ascogonium with supporting cells a female complex formed in Ascomycetes around, protocarp.

archicleistogamic. Having reduced reproductive organs with in permanently closed flowers; *cf*. archocleistogamic.

archocleistogamic. Having mature reproductive organs within permanently closed flowers; *cf*. archicleistogamic.

areolae. Cavity-like depressions of diatoms covered with perforated membrane.

areolate. Surfaces of cap and stem showing cracks or crevices in some places.

argillophilous. Thriving in clay or mud; **argillophile**, argillophily.

aril An extra seed envelope, often coloured and fleshy found in some angiosperms. The aril is produced from the tissues of the funicle or base of the ovule; **arillate**.

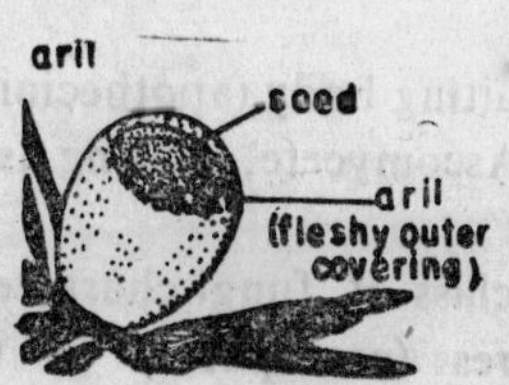

armed. Having thorns or spines.

aromatic. Of organic compounds in which carbon atoms are arranged in rings of six

arthrospore. 1: A spore resulting from the fragmentation of a hypha in fungi. Also called oidium.
2: From filament segmentation in some blue green algae.

articulated. Jointed.

artificial classification. A classification structured for convenience, using easiiy observed phenotypic characters and not necessarily indicating phylogenetic relationships; key classsfication; *cf.* natural classification.

artificial key. A way of identifying plants by steps. Each step involves a choice between at least two different characters, each of which leads to another choice of two characters, finally leading to the right identification.

artificial selection. The process by which man selects plant varieties or individuals with useful characters or traits e.g. wheat plants with large seeds, in order to breed them for his own purposes. Because of this, many cultivated plants appear to be very different from their wild ancestors.

asaccharolytic Used of microorganisms that are unable to break down carbohydrates, alcohols or polysaccharides.

ascigerous. Ascus-bearing (perfect) stage of development of an Ascomycete.

ascocarp. The fruiting body (apothecium, cleistothecium, or perithecium) of an Ascomycete, bearing asci.

Ascomycetes. A class of fungi characterized by endogenous production of spores (**ascospores**) in the organ of meiosis (**ascus**).

ascospore. The haploid spore of an ascomycete formed by meiosis immediately after nuclear fusion. Ascospores are contained within asci from which they are violently ejected when ripe.

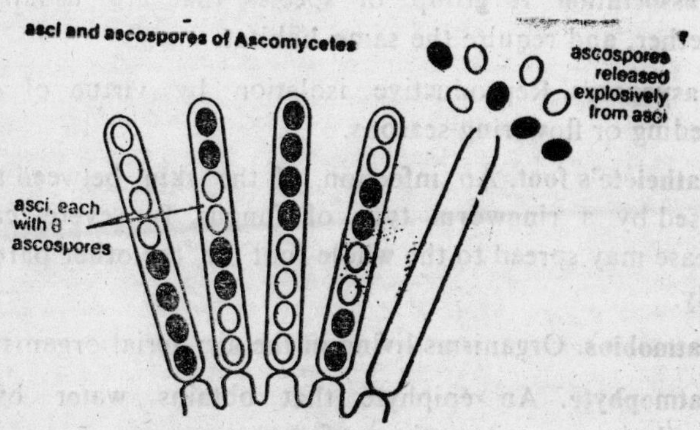

ascostroma. A stromatic ascocarp bearing asci directly in locules within the stroma.

ascus. The reproductive organ of an ascomycete usually containing 8 ascospores. Asci (*pl.*) are usually long and thin, with the ascospores arranged in a row.

ascus mother cell. The binucleate crook cell in the Ascomycetes in which karyogamy occurs and which develops into the ascus.

aseptate. Lacking cross-walls.

asexual. Of reproduction from one individual without the fusion of sex cells from two different parents. Asexual reproduction is common in the plant kingdom. (For example, reproduction by conidia in the imperfect Fungi). Many plant sepecies can reproduce both sexually and asexually.

aspergillosis. One of a group of diseases of animals and human beings caused by various species of *Aspergillus*. (usually *A. fumigatus*).

asporogenic. Not producing, or developing from spores; asporogenous.

assimilator. Leaf-like upright part of some Chlorophycean members.

association. A group of species that are usually found together, and require the same habitat.

asyngamy. Reproductive isolation by virtue of different breeding or flowering seasons.

athelete's foot. An infection of the skin between the toes, caused by a ringworm type of fungus. In severe cases the disease may spread to the whole foot or the other parts of the body.

atmobios. Organisms living in the air; aerial organisms.

atmophyte. An epiphyte that obtains water by aerial assimilation over its entire surface.

atokous. Without offspring; non-reproductive; vegetative; *cf.* epitokous.

ATP, adenosine triphosphate. The nucleotide which stores energy in the bonds between its three phosphate groups. This energy is released by hydrolysis to drive synthetic reactions in the cell.

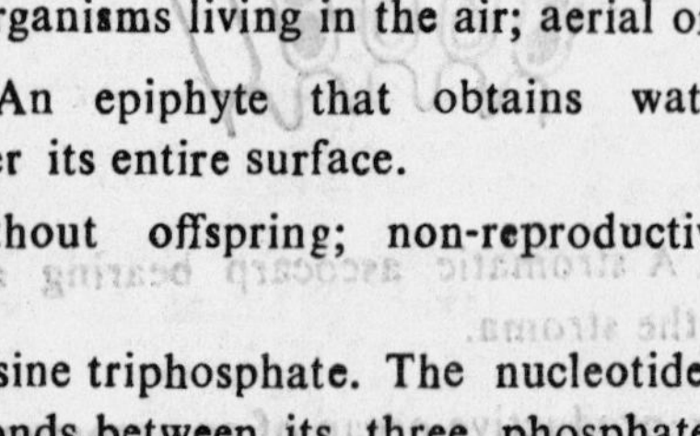

ATP

atropine. An alkaloid which can be extracted from the plant *Atropa belladona*, also called deadly nightshade plant. It is found in a number of other plants of the family Solanaceae and is also available as a synthetic product. Atropine causes dilation of pupil and is used as an aid in the examination of the retina. It is also used in the treatment of inflammation in the cornea It also reduces the secretion of tears, sweat, saliva and digestive juices. As atropine inhibits the secretion of the bronchial glands, it is sometimes used in the treatment of asthma.

aulophyte. A non parasitic plant living within a hollow cavity in another plant.

auricle. A small outgrowth at the side of the base of the leaf in some grasses.

autecology. The ecology of a single species in a habitat.

authority. The name of the author who was the first to give a species or other taxon its name. In the case of species, the authority is given after the binomial.

autoallogamy. The condition of a species in which some individuals are adapted to cross fertilization and others to self-fertilization; homodichogamy; allautogamy; autallogamia, autallogamy.

autoantibiosis. Self-inhibition of growth; growth retardation due to nutrient depletion by the same organism.

autochorus. Having motile spores or propagules disseminated by the action of the parent plant; autochore, autochory.

autocolony. A daughter colony of many Volvocales formed within the parent colony.

autoecious. Having all different spore forms (rust fungi) produced on only one species of host plant. It is a parasitic fungus that can complete its entire life cycle on the same host, e.g. *Puccinia menthae*, mint rust; *cf.* heteroecious.

autogamy. 1: The process of self-fertilization or self-pollination; orthogamy.
2: The fusion of two reproductive nuclei within a single cell, derived from a single parent.
3: Reproduction in which a single cell undergoes reduction division producing two autogametes which subsequently fuse.
4: In protistans, the division of the micronuclei to produce eight or more nuclei, of which two fuse and give rise to a new macronucleus.

autonereid. An autotrophic water plant.

autoorthotropism. The tendency of a plant to grow in a straight line; autoorthotropic.

autoparasite. A parasite living on another parasite; hyperparasite; superparasite.

autoparthenogenesis. The development of an unfertilized egg that has been activated by a chemical or physical stimulus.

autophagy. The intracellular digestion of endogenous material of the cell involving lysosomes.

autopolyploid. A polyploid species with all sets of chromosomes coming from the same species.

autoradiography. A method used to locate radioactive materials in cells by microscopy of exposed photographic film developed after a period of contact with labeled cells.

autosome. A chromosome which is not a sex chromosome.

autosporangium. A sporangium containing autospores.

autospore. A non-motile spore resembling in shape and structure the parent cell.

autotrophic. Able to synthesize food from simple chemical compounds, using energy from light or chemical reaction. Most plants are autotrophic; autotroph.

auxanometer. An instrument used to measure plant growth.

auxiliary cell. A cell of female gametophytic plants of some red algae that receives nucleus from zygote. It gives rise to carposporophyte.

auxin. General name for an important group of plant hormones. Indole acetic acid, IAA, is the most common auxin. Auxins affect many processes, e.g. tropisms, fruit growth, apical dominance, stem growth. They can also inhibit root growth. Very low concentration of auxin are sufficient to show an appreciable effect.

auxin e.g. Indole acetic acid (IAA)

auxoautotrophic. Used of an autotrophic organism requiring certain specific additional nutrients for growth; auxoautotroph.

auxoheterotrophic. Used of an organism that is unable to synthesize all substances necessary for growth and development.

auxospore. A spore of diatoms resulting from syngamy.

auxotrophic. Used of a microoganism having a biochemical deficiency and requiring supplementary growth factors not needed by the wild type (prototroph); auxotroph, auxotrophy.

available water. That part of the soil water that can be readily taken up by plants; *cf.* holding capacity.

avoidance. Principle of plant-disease control marked by deliberate actions to take advantage of environmental factors and time.

axenic. A culture without another organism being present; pure culture.

axil. The point where the upper side of the petiole of a leaf joins the stem; axillary.

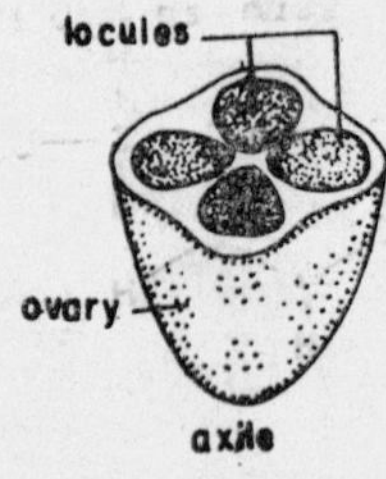

axile. Of a kind of placentation in which the margins of the carpels grow inwards to the centre of the ovary forming several locules so that the ovules are arranged in a divided central column.

axoneme. The microtubular system and associated structures that make up the interior of a cilium or flagellum in eucaryotes.

azotification. Nitrogen fixation.

Azotobacter. A genus of bacteria which can fix atmospheric nitrogen, N_2, to form nitrogenous organic matter; see also nitrogen fixation.

azygospore. A spore produced from an unfertilized female gamete; aboospore; parthenospore.

azygote. An organism produced by haploid parthenogensis.

B

baccate. 1: Bearing berries.

2: Having pulp like a berry.

baccatin. A wilt toxin from *Gibberella baccata.* This is antibacterial.

baccillary. 1: Rod-shaped; composed of rod-shaped structures

2: Produced by, pertaining to or resembling bacilli.

bacciferous. Producing berries.

bacciform. Shaped like a berry.

bacillus (pl. bacilli). Any rod-shaped, spore producing bacterium.

baccivorous. Feeding principally upon berries.

bacitracins. These are peptide antibiotics produced by *Bacillus* spp. They are effective against gram + bacteria and inhibit its cell wall growth.

back bulb. A pseudobulb in certain orchid plants that remains on the plant after removal of the terminal growth, and that is used for propagation.

backcross. A cross between a hybrid offspring and one of its parents; the notation for backcross generations is B_1, B_2. . .

back mutation. A mutation of a gene so that it assumes a function lost by an earlier mutation; reverse mutation; *cf.* forward mutation.

bacteria (sing. bacterium). A division of unicellular motile or non-motile prokaryotic organisms, most of which are heterotrophic and reproduce by simple fission. Bacterial cells are usually 0.5-2μm across. Bacteria are important in the decay of organic matter in the soil. Many are parasitic on other organisms, often causing disease. They are of different shapes like:

1. bacilli (straight rods)
2. cocci (spherical)
3. vibrios (curved rods; comma shaped)
4. spirochaetes (spirals);

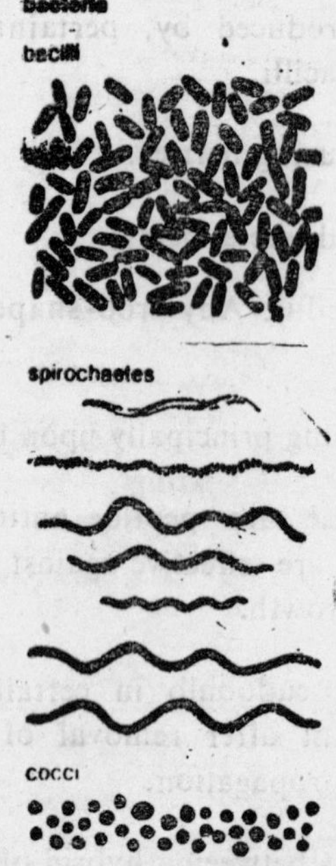

bacteria capsule. A thick, mucous envelope, composed of polypeptides or carbohydrates, surrounding some bacteria.

bacterial coenzyme. Organic molecules that participate directly in a bacterial enzymatic reaction and may be chemically altered during the reaction.

bacterial competence. The ability of cells in a bacterial culture to accept and be transformed by a molecule of transforming deoxyribonucleic acid.

bacterial endoenzyme. An enzyme produced and active within the bacterial cell.

bacterial endospore. A body, resistant to extremes of temperature and to dehydration, produced within the cells of gram-positive, spore forming rods of *Bacillus* and *Clostridium* and by the coccus *Sporosarcina.*

bacterial luminescence. A light-producing phenomenon exhibited by certain bacteria.

bactericidal. Lethal to bacteria.

bactericidin. A substance that kills bacteria without breaking down or dissolving its cells.

bacteriochlorophyll. $C_{52}H_{70}O_6N_4Mg$. A tetrahydroporphyrin chlorophyll compound occurring in the forms *a* and *b* in photosynthetic bacteria; there is no evidence that *b* has the empirical formula given.

bacteriocins. Non-replicating, bactericidal protein-containing substances produced by certain strains of bacteria and active against some other strains of the same or closely related species.

bacteriods. Irregular, enlarged forms of rod-shaped bacteria especially of the root-nodule forming species of *Rhizobium.* These are ultimately absorbed by the cells of the root-nodule.

bacteriological code. International Code of Nomenclature of Bacteria.

bacteriolysin. An antibody that is active against and causes lysis of specific bacterial cells.

bacteriophage. A kind of virus which attacks the cells of bacteria. Many bacteriophages have a head consisting of a protein coat containing nucleic acid-DNA and a 'tail' of protein through which the nucleic acid is injected into a bacterial cell.

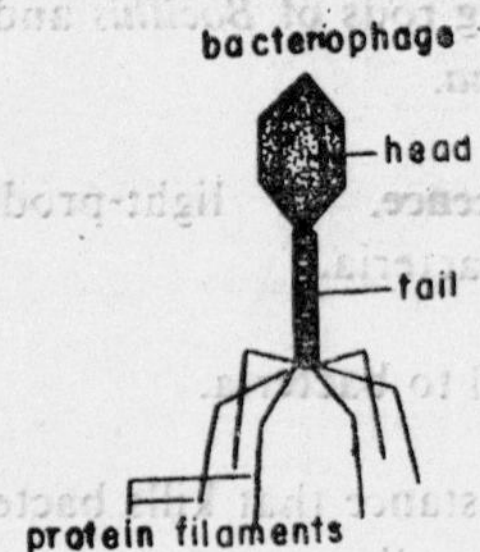

bacteriostatic. A chemical or physical agent that prevents multiplication of bacteria without killing them.

balbiani rings. Localized swellings of a polytene chromosome.

ballistic. Used of dehiscence in which the seeds are catapulted from the fruit.

balsa. *Ochroma lagopus.* A tropical American tree in the order Malvales; its wood is strong and lighter than cork.

banner. The fifth or posterior petal of a butterfly-shaped (papilionaceous) flower.

barachorous. Having propagules dispersed by their own weight; barachore, barachory.

bark. Tissue usually made of dead cork cells and phloem on the outside of woody stems. Its function is to protect the stem.

bark graft. A graft made by slipping the scion beneath a slit in the bark of the stock.

barm 1: The froth on the surface of fermenting malt liquids
2: Baker's yeast.

barrage. The space between two mycelia which have an aversion for one another.

barrier zone. In a tree, new tissue formed by the cambium after it has been wounded; serves as both an anatomical and a chemical wall.

basal body. An organelle located at the base of cilia and believed to be involved in the organization of ciliary microtubules.

base. 1: any substance which can accept protons from water molecules; basic.

2: An alkaline, usually nitrogenous organic compound, used particularly for the purine and pyrimidine moieties of the nucleic acids of cells and viruses.

base pairing. The hydrogen bonding of complementary purine and pyrimidine bases—adenine with thymine, guanine with cytosine—in double-stranded deoxyribonucleic acids (DNA) or ribonucleic acids (RNA) or in DNA/RNA hybrid molecules.

basic number. The number of chromosomes in a basic chromosome set, denoted by x, and representing the lowest monoploid number in a polyploid series.

basidia. Typically club-shaped organs of meiosis of Basidiomycetes; basidiospores (usually four) are borne externally on stalks (sterigmata) that extend from the apex of the basidium.

basidiole. A basidium like hymenial element that lacks sterigmata because it is either young or permanently sterile.

basidiograph. The straight line graph obtained by plotting the ratio of the length (l) to the width (w) against the length of the basidia of a species of agaric.

Basidiomycetes. A class of fungi recognized by their production of spores (basidiospores) externally on a club shaped organ—basidium. Mushrooms and toadstools are the fruiting bodies of basidiomycetes.

basidiospore. A sexually produced haploid spore of a basidiomycete produced on a basidium exogenously.

basifixed. Of an organ which is attached to another organ by its base. This is one way in which anthers are attached to filaments.

basifuge. A plant unable to tolerate basic soils; calcifuge.

basikaryotype. See Basic number.

basionym. The original name of a taxon subsequently replaced by another using the same stem, as a result of a change in rank or position of the taxon; basinym.

basipetally. Successively from morphological apex to morphological base.

basts. The inner fibrous bark of certain trees. Consisting of pericycle and phloem. This yields fibers known as bast fibers, e.g., *Cannabis*, *Linum* (Alsi or flax), *Corchorus* etc.

bathyplankton. Planktonic organisms which undergo diurnal vertical migration, moving up towards the surface at dusk and down away from the surface at dawn; bathyplanktonic; *cf.* epiplankton.

beard moss. Species of lichens belonging to genera *Alectoria* and *Usnea*.

beef-steak fungus. *Fistulina hepatica.* A member of Fistulinaceae (Basidiomycotina) parasitises *Quercus* and *Castanea.* The fruiting bodies are edible.

benthic. Bottom-loving algae of lakes or sea.

benthophyte. A plant living at the bottom of a water body or on the bed of a river.

benthopleustophyte. Any large plant resting freely on the floor of a lake but capable of drifting slowly with the currents.

berry. A succulent or juicy fruit with many, usually small, seeds. The pericarp consists of the inner soft membranous portion (the endocarp), the middle fleshy region (the mesocarp) and the outer skin (the epicarp) e.g. banana, tomato, orange etc.

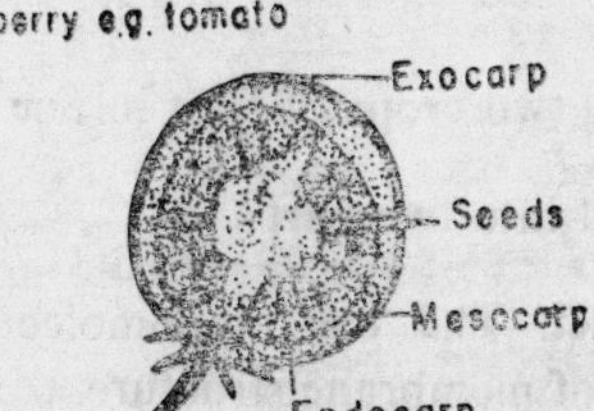

beta carotone $C_{40}H_{56}$. A carotenoid hydrocarbon pigment found widely in nature, always associated with chlorophylls; converted to vitamin A in the liver of many animals.

beta-galactosidase. An enzyme that catalyzes the hydrolysis of lactose, producing glucose and galactose.

betacyanin. A group of purple plant pigments found in leaves, flowers, and roots of members of the order Caryophyllales.

betalain. The name for a group of 35 red or yellow compounds found only in plants of the order Caryophyllales, including red beets, red chard, and cactus fruits.

betanin. An anthocyanin that contains nitrogen and constitutes the principal pigment of garden beets.

betaxanthin. The name given to any of the yellow pigments found only in plants of the order Caryophyllales; they always occur with betacyanins.

betel nut. A dried, ripe seed of the palm tree *Areca catechu* in the family Arecaceae; contains a narcotic.

bhang. See *Cannabis*.

bicostate. Of a leaf, having two principal longitudinal ribs.

biennial. Of plants which complete their life cycle in two years, growing in the first, reproducing and dying in the second.

biferous. Bearing two crops of fruit in one season.

bifurcate. Forked into two parts.

bimolecular leaflet. The two monomolecular layers of lipid that form the basis of membrane structure.

binary fission. The division of a cell into two identical cells. A form of asexual reproduction.

binomial nomenclature. The method of scientifically naming plants and animals in descriptive Latin terms; the first term identifies the genus, the second the species to which an organism belongs. The first letter of the generic name is capitalized, and both names are italicized. The name (often abbreviated) of the author responsible for naming the organism may follow the Latin binomial. When another author transfers a species to another genus, the name of the first author is placed in parentheses and the name of the second author follows. Thus, the scientific name of the fungus that causes brown rot of peach is written *Monilinia fructicola* (Wint.) Honey.

bioassay. Biological assay. The use of an organism for assay purposes; any quantitative biological analysis.

bioenergetics. Thermodynamics as applied to living cells or whole organisms.

bioerosion. Erosion resulting from the direct action of living organisms.

bioflavonoid. A group of compounds obtained from the rinds of citrus fruits and involved with the homeostasis of the walls of small blood vessels; in guinea pigs a marked reduction of bioflavonoids results in increased fragility and permeability of the capillaries; used to decrease permeability and fragility in capillaries in certain conditions. Also known as citrus flavonoid compound; vitamin P complex.

biogenesis. The principle that all living organisms have derived from previously existing living organisms; **biogenetic**; *cf.* abiogenesis.

biogenic. Produced by the action of living organisms; biogenetic.

biogenetic. See biogenic.

biogenetic law. The theory that an individual during its development passes through stages that resemble adult form of its successive ancestors so that the ontogeny of an individual recapitulates the phylogeny of its group; Haeckel's law; recapitulation.

biogenous. Living on or in other organisms; parasitic; symbiotic.

biological clock. An inherent physiological mechanism for measuring time or maintaining endogenous rhythms; escapement clock.

biological control. The control of a pest by the introduction, preservation or facilitation of natural predators, parasites or other enemies, by sterilization techniques, by the use of inhibitory hormones or by other biological means; biocontrol.

biological indicator. A microorganism which is used as an indicator of chemical activity in, or the chemical composition of, a natural system.

biological spectrum. A list of plant species in a given region or community together with percentage abundances of individual species.

biology. The science of life; the study of living organisms and systems; biological.

bioluminescence. Light produced by living organisms and the emission of such biologically produced light; bioluminescent; *cf.* phosphorescence.

biolysis. Death and subsequent tissue degradation of an organism.

biomass. The weight of a living organism, or of all the organisms in an ecosystem or in a habitat.

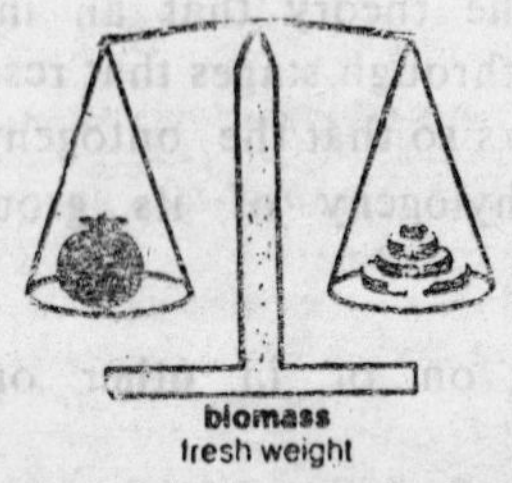

bionomics. Ecology; the study of organisms in relation to environment; bionomy.

biont. An individual organism; bion.

-bionta. The ending of the name of a subkingdom in botanical nomenclature.

biophagous. Consuming or destroying other living organisms; biophage, biophagy; *cf*. saprophagous.

biophilous. Thriving on other living organisms, often used specifically of plant parasites; biophile, biophily.

biophotogenesis. See bioluminescence.

biophysics. The application of physics to the study of living organisms and systems.

biophyte. A plant that feeds on other living organisms; a parasitic or predatory plant.

biosphere. The parts of the earth in which organisms live, including the oceans, the land, the soil, and the atmosphere.

biosystematics. The study of evolution and the classification of living organisms, based on biological information at population level such as genetic variability, hybridization, breeding strategies, competition and local adaptations; biosystematy; genonomy.

biotic factors. The effects of living organisms on an ecosystem and on each other, e.g., herbivores eating plants, or trees casting shade.

biotrophic. A parasite, that derives its food from the living cells of the host; biotroph, biotrophy.

biotrophic symbiosis. A symbiosis in which one symbiont uses its living partner as a food source; *cf*. necrotrophic symbiosis.

biotype. A subgroup within a species usually characterized by the possession of a single or a few characters in common. This subgroup show identical genetic features but is different physiologically; a subdivision of a pathologic race.

biphasic. Possessing both a sporophytic and a gametophytic generation in the life cycle.

bipinnate. Of pinnate leaves with their pinnae divided into pinnules as in many ferns.

bipolarity. A condition of sexual compatibility in certain Basidiomycetes in which two basidiospores of each basidium are of one strain, and two are of another.

birds's nest fungi. These are fungi belonging to order Nidulariales of the class Gasteromycetes; common examples are *Cyathus*, *Crucibulum.* The fruiting bodies are funnel-shaped and the gleba is differentiated into one or more spherical to lens shaped peridioles which contain basidiospores. These peridioles appear like bird's eggs in a nest and the fungi so named. The fruiting bodies grow on cereal stubble, old stumps and twigs in autumn.

birthwort. A species of low herbs or twining shrubs of the family Aristolochiaceae, aspecially those of the genus *Aristolochia* whose aromatic roots were thought to aid in childbirth.

bisbifid. A two-lobed leaf in which each lobe is also divided into two equal parts.

biseriate. Arranged in two rows.

bisexual. 1: Used of a population or generation or composed of functional males and females; gonochoristic.

2; Used of an individual possessing both male and female functional reproductive organs; hermaphrodite; totipotent.

3: Used of a flower possessing both male and female reproductive organs; perfect; *cf.* unisexual.

bisporangiate. Having two different types of sporangia.

bistratose. Two cell layers in thickness.

bitegmic. Having two integuments, especially in reference to ovules.

bitunicate. An ascus in which the inner wall is elastic and expands greatly beyond the outer wall at the time of spore liberation.

bitypic. Used of a taxon comprising only two immediately subordinate principal taxa, as in a family or genus comprising two genera or species respectively; *cf.* monotypic, polytypic.

bivalent. A pair of joined homologous chromosomes during the first meiotic prophase.

blackfellow's bread. The sclerotium of the Australian *Polyporus mylittae* which can reach the size of a man's head. There are similar sclerotia in India ('little man's bread') and China.

bladder. An expanded, gas containing structure on the blade of various marine algae; a pneumatocyst.

bladderwort. An insectivorous plant of the genus *Utricularia.* It is an aquatic plant.

blade Either the whole leaf, excluding the petiole, or all the parts of the leaf except the midrib, i.e., the lamina.

blasting. A symptom of plant disease characterized by shedding of unopened buds; classically, the failure to produce fruit or seed.

blastocarpous. Germinating in the pericarp.

blastogenesis. Asexual reproduction by budding or gemmation.

blastospore. An asexual spore formed by budding.

blematogen. The undifferentiated tissue which becomes the universal veil in agarics blematogea layer.

blepharoplast. A cytoplasmic granule from which a flagellum originates.

blight. A nonrestricted necrotic symptom characterized by ultimate death of tissues throughout entire organs, such as leaves or flowers.

bloom. 1: An individual flower. Also known as blossom.

2: To yield blossoms.

3: The waxy coating that appeaas as a powder on certain fruits, such as plums, and leaves, such as cabbage.

4: A dense growth of planktonic algae in water giving it a definite colour.

blossom. *See* bloom.

blotch. A necrotic spot with fibrillose margins of visible mycelial strands. These spots are irregular in shape and are found on leaves, shoots and stems.

blowball. A fluffy seed ball, as of the dandelion.

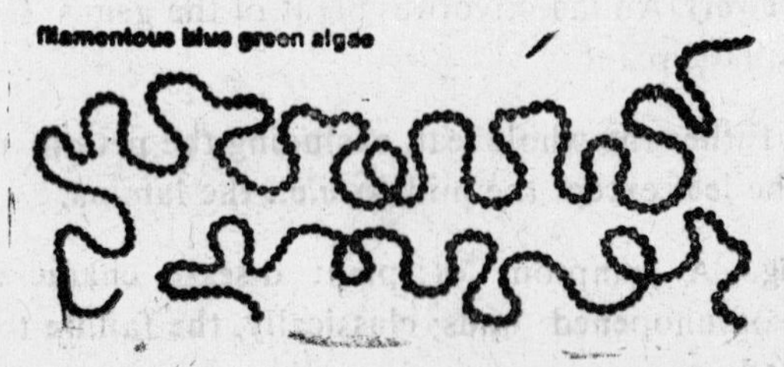

blue-green algae. Algae of the division Cyanophyta with prokaryotic cells. They are unicellular or multicellular, lack flagella and have their own carotenoid photosynthetic pigments; Cyanobacteria.

bog. A plant community that develops and grows in areas with permanently water-logged peat substrates; moor; quagmire.

bog. moss Any moss which grows in bogs, especially those of the genus *Sphagnum*.

bole. The trunk of a tree; a caudex.

boll. A pod or capsule (pericarp), as of cotton and flax.

bolting. A process leading to the rapid elongation of a flowering shoot in the second spring in biennials which remain rossette plants the first season.

bordered pit. A wood-cell pit having the secondary cell wall arched over the cavity of the pit.

botany. The study of plants. This word has originated from a Greek word botane meaning herb; 'phytology; botanical'.

botrylogy. The science of organizing items or concepts into groups or clusters.

bottle graft. A plant graft in which the scion is a detached branch and is protected from wilting by keeping the base of the branch in a bottle of water until union with the stock.

bottom break. A branch that arises from the base of a plant stem.

botulin. The neurogenic toxin which is produced by *Clostridium botulinum* and *C. parabotullnum* and causes botulism; botulinus toxin.

brachiate. Possessing widely divergent branches.

brachyblast. A short shoot often bearing clusters of leaves.

brachymeiosis. A simplified form of meiosis completed in a single division by suppression of the second division.

brachysclereid. A sclereid that is more or less isodiametric and is found in certain fruits and in the pith, cortex, and bark of many stems; stone cell.

brachysm. Plant dwarfing in which there is shortening of the internodes only.

bracketed key. A dichotomous key in which contrasting parts of a couplet are numbered and presented together, without intervening couplets although the brackets joining each couplet (from which the key derives its name) are now omitted; bracket key; parallel key; *cf.* indented key.

bracket fungus. A Basidiomycetes fungus growing in the wood of living or dead trees, producing large flat-topped bracket-like fruiting bodies on the side of the host.

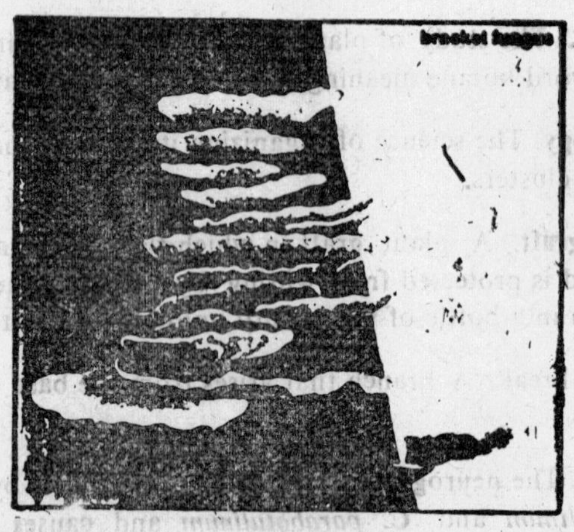

brackish water. Water in which the concentration of dissolved ions is more than in freshwater habitats but less than in seawater.

bract. 1: A modified small leaf associated with a flower or part of an inflorescence growing from its axil.
2: A modified, scale-like leaf of a cone or strobilus.

bracteole. A small bract, especially if on the floral axis; bractlet.

bracteolate. Having bracteoles.

bradytelic. Used in chronistics, to describe a species that has remained more or less unchanged for millions of years. e.g. certain gymnosperms like *Podocarpus*, *Ginkgo* and *Welwitchia*. Evidence of such nature of these plants comes from studies of fossils and karyotypes of these plants.

bramble. 1: A plant of the genus *Rubus*.
2: A rough, prickly vine or shrub.

brassins. These are recently discovered steroid growth-promoters first isolated from pollen grains of rape plants but now known to occur in tea, bean and rice; brassinosteroids.

Braun-Blanquet cover scale. A scale for estimating cover of a plant species, comprising six categories; + (under 1%), 1 (1-5%), 2 (6--25%), 3(26--50%), 4 (51--75%), and 5 (76--100%).

Brazil nut. *Bertholettia excelsa.* A large broad-leafed evergreen tree of the order Lecythedales; an edible seed is produced by the tree fruit.

breadfruit. *Artocarpus altilis.* An Indo-Malaysian tree, a species of the mulberry family (Moraceae). The tree produces a multiple fruit which is edible.

brevicollate. Short-necked.

bridge graft. A plant graft in which each of several scions is grafted in two positions on the stock, one above and the other below an injury.

brittlewort. Common name of calcareous Charales; an order of algae.

brotochorous. Having propagules dispersed by the agency of man; brotochore, brotochory

brown algae. Algae of the division Phaeophyceae, including many of the larger seaweeds Brown algae contain brown accessory pigments.

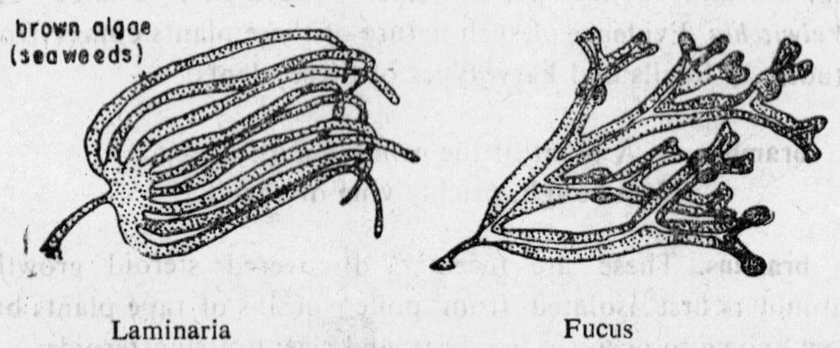

brussels sprouts. *Brassica oleracea* var. *gemmifera* A biennial crucifer (family Cruciferae) cultivated for its small, edible, headlike buds.

bryocolous. Living on or in moss; bryocole.

bryokinin. It is a cytokinin-like substance obtained from the callus cultures of sporophyte of the hybrid *Physcomitrium piriforme* X *Funaria hygrometrica*. It is known to induce buds in large numbers in the normally bud-lacking strain of *Funaria*.

bryology. The study of mosses and liverworts; hepaticology; muscology; bryological.

bryophilous. Thriving in habitats rich in mosses and liverworts; bryophile, bryophily.

Bryophyta. A small division of the plant kingdom, including mosses, liverworts, and hornworts, characterized by the lack of true roots, stems, and leaves and live in wet shady places.

Bryopsida. The mosses, a class of small green plants in the division Bryophyta. Also known as Musci.

bud. An embryonic shoot containing the growing stem tip surrounded by young leaves or flowers or both and frequently enclosed by bud scales.

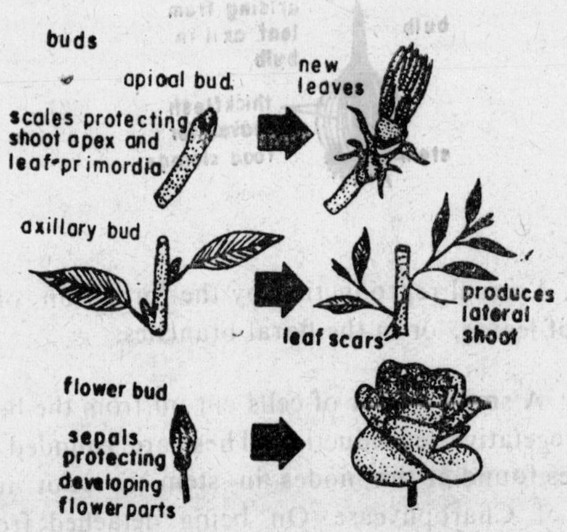

budding. 1: the production of a small outgrowth (bud) from a parent cell. A method of asexual reproduction, e.g., yeast.

2: A method of artificial propagation in which a bud from one plant is inserted under the bark of another.

budbreak. Initiation of growth from a bud.

budling. The shoot that develops from the bud which was the scion of a bud graft.

bud scale. One of the modified leaves enclosing and protecting buds in perennial plants.

buffer species. A plant or animal acting as an alternative food supply for another organism and thereby buffering the effect of the predator on its normal prey.

bulb. An organ of perennation and vegetative reproduction in many monocotyledons. Bulbs are usually underground, and consist of a short axis with many overlapping thick leaves. These leaves generally lack chlorophyll and contain stored food.

bulbifery. Asexual reproduction by the formation of bulbils in the axils of leaves, or in the floral branches.

bulbils. 1: A small cluster of cells cut off from the body. It is a means of vegetative reproduction. These are rounded or star shaped bodies found at the nodes in stem or root in *Chara* (a member of Charophycase. On being detached from plant the bulbils germinate into new plants.

2: A small bulb.

3: A small selerotium-like structure made up of a small number of cells produced in some fungi.

4: Out growths on the stem of some species of *Lycopodium* which are functioning for vegetative reproduction. They are probably modified leaves.

bulbillosis. In Agaricales the condition in which basidiocarp sporulation is suppressed and the basidial function is taken up by bulbils.

bullate. Appearing blistered or puckered, especially of certain leaves.

Buller phenomenon. The dikaryotization of a homokaryon by a dikaryon in Basidiomycetes and Ascomycetes.

bulliform cell. One of the large, highly vacuolated cells occurring in the epidermis of grass leaves, e.g. in *Poa pratensis*, *Ammophilla arenaria*, maize etc. Also known as motor cell. These cells account for rolling of leaves in dry weather.

bundle scar. A mark within a leaf scar that shows the point of an abscised vascular bundle.

bundle sheath. A layer of cells around the vascular bundle in a leaf.

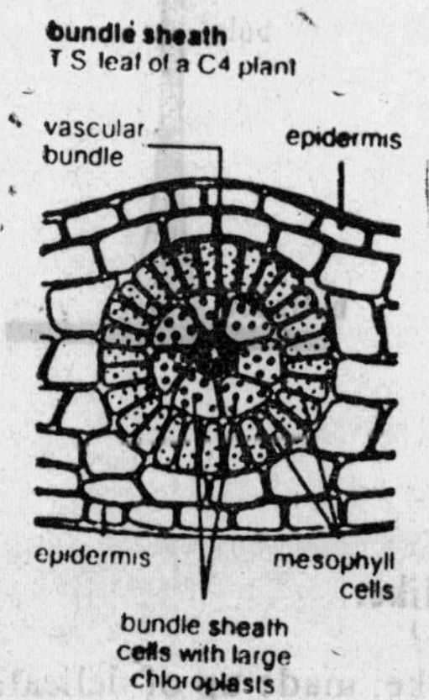

bunt. A fungus disease of wheat caused by two *Tilletia* species and characterized by grain replacement with fishy-smelling smut spores.

burr. 1: A rough or prickly envelope on a fruit.
2: A fruit so characterized.

bursicle. A purse or pouch-like receptacle; bursiculate, bursiform.

burster. An abnormally double flower having the calyx split or fragmented.

butteress. A large, flattened woody structure, growing outwards and downwards from near the base of the trunk of a tree. Buttresses are found especially in very large trees in tropical rain forests.

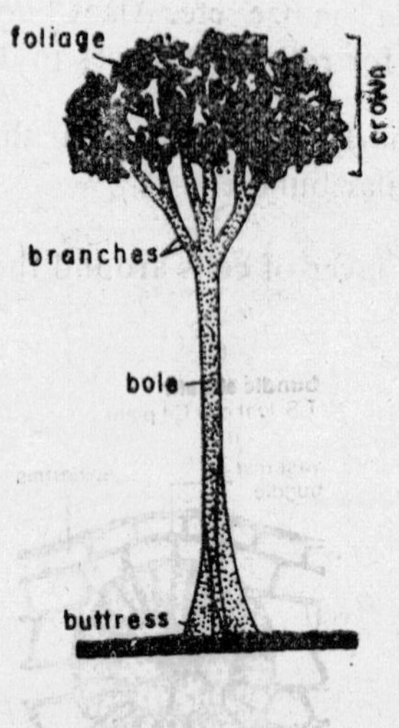

butyrous. Butter-like.

byssoid. Cotton-like, made up of delicate threads; floccose.

C

C_3 pathway. The fixation of CO_2 by ribulose diphosphate to produce two molecules of a compound with three carbon atoms (PGA). Most plants use this pathway, and are called C_3 plants. It is also known as the reductive pentose pathway and the Calvin cycle.

C_4 pathway. Kind of CO_2 fixation found especially in tropical monocotyledons. In this pathway CO_2 is fixed by a compound with three carbon atoms (phosphoenolpyruvate) to produce a molecule with four carbon atoms (malate). This occurs in mesophyll cells in the leaf. The malate is then transported to bundle sheath cells, where the CO_2 is released and fixed by ribulose-diphosphate in the normal way. Plants which do this are called C_4 plants.

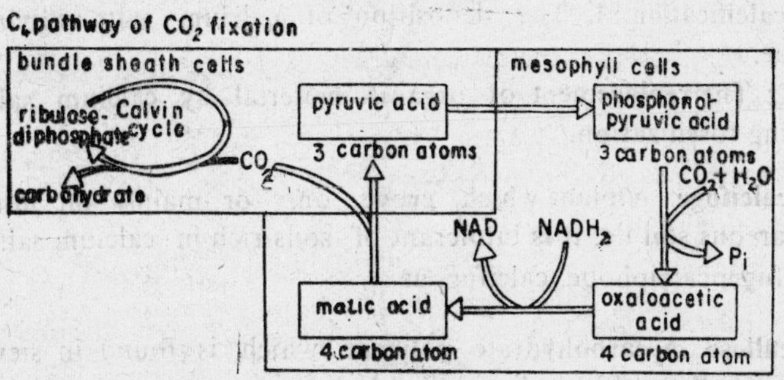

cacao. A South American tree *Theobroma cacoa* family sterculiaceae extensively cultivated for its seeds, cocoa beans, the source of commercial cocoa and chocolate.

caducous. Lasting on a plant only a short time before falling off; evanescent; fugacious.

caeoma. An aecium that is surrounded by fungal filaments but that has no peridium.

caffeine. An alkaloid, $C_8 H_{10} N_4 O_2$, present in tea and coffee, kola and other plants. It stimulates central nervous system, cardiac and respiratory systems and is diuretic.

Cainozoic. The geological era which began 65 million years ago, after the Mesozoic and is still continuing. Angiosperms have been the daminant plants on land throughout this time; Cenozoic.

calcarate. Having spurs.

calcareous. Substrates rich in calcium salts; pertaining to limestone or chalk; growing on or having an affinity for chalky soil.

calcicole. A plant which grows only or mainly on calcareous soil; calcicolous; calciphyte; gypsophyte; calcipete; *cf.* calcifuge.

calcification. 1: The deposition of calcium salts in living tissue.

2: The replacement of organic material by calcium salts during fossilization.

calcifuge. A plant which grows only or mainly on non-calcareous soil i.e; it is intolerant of soils rich in calcium salts; basifuge; calciphobe, calcifugous.

callose. A carbohydrate polymer which is found in sieve plates, pollen tubes and on injured surfaces.

callus. 1: A lump of undifferentiated cells in a tissue culture.

2: A tissue produced on injured plant surfaces, Callus tissue contains callose.

Calvin cycle. The reductive pentose pathway of photosynthesis named after one of its discoverers. The cycle was worked out in the 1940s and 1950s. It is also known as the C_3 pathway.

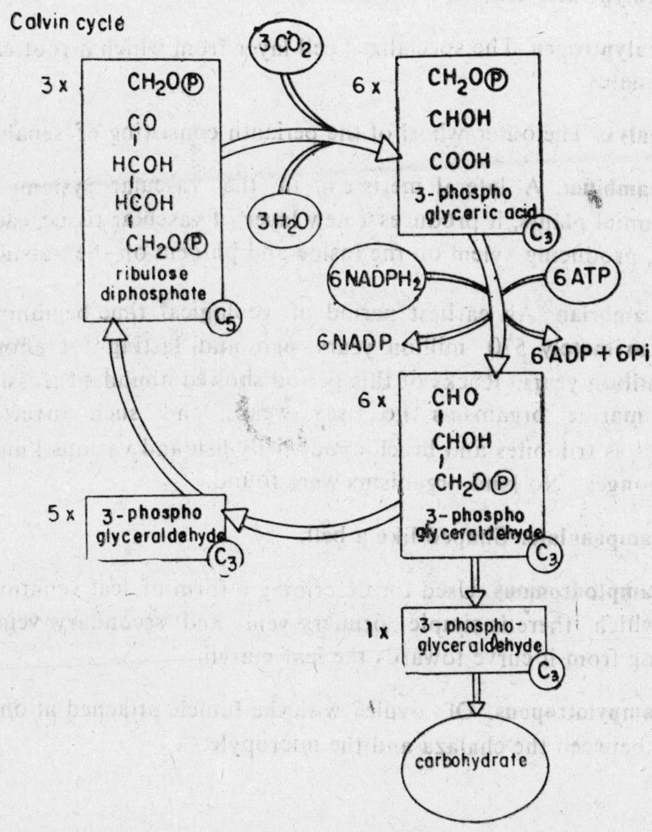

calycanthemy. Abnormal development of calyx structures into petals or petaloid structures.

calyculate. Having bracts that imitate a second, external calyx.

calyptra. 1: A membranous cap or hood-like covering, especially the remains of the archegonium over the developing sporangium or entire sporophyte of a moss.

2: Tissue surrounding the archegonium of a liverwort.

3: Root cap.

calyptrate. Having a calyptra.

calyptrogen. The specialized cell layer from which a root cap originates.

calyx. The outer whorl of the perianth consisting of sepals.

cambium. A lateral meristem in the vascular system. In perennial plants, it produces a new layer of vascular tissue each year, producing xylem on the inside and phloem on the outside.

Cambrian. An earliest period of geological time beginning approximately 570 million years ago and lasting for about 70 million years. Rocks of this period showed abundant fossils of marine organisms i.e., sea weads and such invertebrates as trilobites and brachiopods, jelly fish and various kinds of sponges. No land organisms were found.

campanulate. Shaped like a bell.

camptodromous. Used for describing a form of leaf venation in which there is simple primary vein and secondary veins arising from it curve towards the leaf margin.

campylotropous. Of ovules with the funicle attached at one side, between the chalaza and the micropyle.

campylotropous

micropyle

funicle

placenta

canada balsam. A yellowish, viscid turpentine produced by the balsam fir (*Abies balsamea,* fam. Pinaceae), This is used for mounting microscopic slides and a cement for optical lenses.

cane. 1: A hollow, usually slender, jointed stem, such as in sugarcane or the bamboo grasses.

2: A stem growing directly from the base of the plant, as in most Rosaceae, such as blackberry and roses.

canescent. Hoary; having a grayish epidermal covering of short hairs.

canker. A necrotic symptom of disease in woody plant parts like on stem, branch or twig. The necrosis is restricted to a definite area that is surrounded by callus.

cannabis. A drug obtained from the herbaceous plant called *Cannabis sativa* (Indian hemp, fam., Cannabinaceae). It is hallucinogenic and accumulates in the body. Cannabis is known under various names the world over, like marijuana, grass, pot, weed, tea, boo, Mary Jane, bhang, ganja, kif, hashish and dagga.

canopy. The top layer of a forest consisting of the crowns of trees.

cantala. A fiber produced from agave (*Agave cantala,* fam. Agavaceae) leaves; used to make twine; Cebu maguey; maguey; Manila maguey.

cantharophilous. Pollinated by beetles.

cap. 1: The pileus of a mushroom.

2: The calyptra of a moss capsule.

capillary water. The water held in soil pore spaces by capillary forces. This water is readily available to plants and soil organisms.

capillitium. Sterile, thread-like structures present among the spores in the fruiting bodies of many Myxomycetes and Gasteromycetes.

capitate. Like a head, e.g., when many flowers are clustered together in an inflorescence.

capitulum. 1: An inflorescence like a head, consisting of many sessile flowers, e.g., in Compositae; capitula.

2: Cell found in globule of charales. It gives rise to antheridial filaments.

3: A head, describing the crowded apical mass of branches in *Sphagnum*.

capreolate. Having one or more tendrils.

caprification. A method of pollination employed in the cultivation of figs in which fruits of the wild fig (capri fig) containing the fig wasp (*Blastophaga*) are placed in trees of cultivated figs. The insects on emerging enter the fiowers of edible figs, thus bringing about cross-pollination.

capsid. The protein coat of viruses forming the closed shell or tube that contains the nucleic acid and consisting of protein subunits or capsomeres.

capsomere. Also called a protein subunit; a small protein molecule that is the structural and chemical unit of the protein coat (capsid) of a virus.

capsule. 1: A dehiscent dry fruit, with at least one carpel, often with many small seeds e.g., in the family Orchidaceae.

2: The spore producing organ of the sporophyte of a bryophyte borne at the top of the seta.

3. A relatively thick layer of mucopolysaccharides that surrounds some kinds of bacteria.

carbohydrate. An organic compound containing carbon, hydrogen and oxygen in a ratio of 1:2:1. Starch and all sugars, in which plants store the energy obtained from light during photosynthesis, are carbohydrates.

carbon cycle. The pathway of the element carbon through ecosystems. Carbon dioxide is fixed from the atmosphere by plants during photosynthesis and used for the synthesis of organic compounds. These are passed through the food web and metabobilized by animals and decomposers and carbon dioxide is released back to the atmosphere by respiration.

carbonic anhydrase. The enzyme catalysing the formation of bicarbonate ions from carbon dioxide and water.

$$CO_2+H_2O \longrightarrow H^+ + HCO_3^-$$

carboniferous. The geological period 345-280 million years ago, during the Palaeozoic. The world's forests were dominated by tree-like pteridophytes which were fossilized to form coal.

carnivorous plant. See insectivorous plant.

carotenase. An enzyme that effects the hydrolysis of carotenoid compounds, used in bleaching of flour.

carotene. $C_{40}H_{56}$, Any of several red, crystalline, carotenoid hydrocarbon pigments occurring widely in nature convertible in the animal body to vitamin A, and characterized by preferential solubility in petroleum ether; carotin.

carotenoid. A class of labile, easily oxidizable, non-protein, yellow, orange, red, or purple pigments that are widely distributed in plants and animals and are preferentially soluble in fats and fat solvents. Some of the carotenoids are accessory pigments in photosynthesis.

carotol. $C_{45}H_{25}OH$, A sesquiterpenoid alcohol in carrots.

carpel. The female reproductive unit of a flower, consisting of the ovary with ovules. The carpels are the sporophylls of angiosperms and are like highly modified leaves. Many angiosperms have several carpels, which are joined together at their margins to form the ovary.

carpocephalum. The specialized, essentially radially symmetrical receptacle that surmounts the perpendicular stalk; it bears the sporangia in many marchantialean hepatics.

carpogonium. Oogonium with a trichogyne, found in red algae.

carpology. The study of the morphology of fruits and seeds.

carpophore. 1: Stalk which supports sporocarp in some fungi.

2: Sometime (esp. in France)=basidiocarp.

3: In the Umbelliferae, a stalk which supports mature carpels.

carposporangium. Unicellular sporangium borne singly and terminally by gonimoblast filaments of red algae.

carpospore. Spore of red algae found in carposporangium.

carposporophyte. Product of the fertilized carpogonium of some red algae.

carpotrophic. The movement of the flower-stalk after fertilization to bring fruit into a favourable position for the ripening and/or dispersal of the seed.

carrier. 1: An organism carrying a disease or infectious agent without showing typical symptoms and which is capable of passing the infection to another individual; *cf.* vector.

2: An individual that passes on a character or trait to the next generation without necessarily exhibiting the trait itself. This individual is heterozygous for a recessive gene.

carrageenin. A sulfated galacton phycocolloid found in some Rhodophyceae.

caruncle. Describing a horn-like outgrowth near the hilum of a seed which is formed from integuments e.g., the fleshy outgrowth of castor seed, also occurs in genera like *Euphorbia* and *Jatropha*. It is a water absorbing structure.

caryopsis. A small, dry, indehiscent fruit having a single seed with such a thin, closely adherent pericarp that a single body, a grain, is formed.

casparian strip. A band of suberin around the cells of the endodermis of the root, which stops the movement of substances from the cortex to the vascular cylinder other than through the cytoplasm of the endodermal cells.

catabolism, The destructive phase of metabolism; used for describing the break-down of complex molecules to simpler ones releasing energy; katabolism; *cf.* anabolism.

catalysis. The process in which natural chemical reactions are made more rapid, e.g., by enzymes; catalyze; catalytic

catalyst. A substance that increases the rate of a chemical reaction without itself being changed in the process, e.g., enzymes.

cataphyll. A small, scale-like leaf, as a bud scale.

catalase, An enzyme that catalyzes the decomposition of hydrogen peroxide into molecular oxygen and water. It is present in peroxisomes

catapult mechanism. A method of seed dispersal which depends on the jerking of a long stalk swaying in the wind.

catch crop. A quick-growing crop which is cultivated between the rows of a main crop, between the main crops of an ordinary crop rotation or in place of a main crop which has failed.

catenulate. Having a chain-like form.

catkin. A pendant, scaly spike of small, either male or female flowers, falling entire from the plant, e.g., in the willow family Salicaceae.

catothecium. An inverted perithecium having the asci han ging from its base. This name is sometimes given to the thyriothecium of the Trichothyriaceae; catathecium.

caudex. 1: Describing the persistent swollen stem base of certain herbaceous perennials.
2: Used for the trunk of a palm or a tree fern.

caulid. A term used in preference to the "stem" in bryophytes; because it lacks lignified vascular system.

cauliflorous. Of plants with flowers or inflorescences on the stem or trunk cauliflory.

caulome. The stem structure or stem axis of plant as a whole.

caulonema. The portion of the protonema capable of producing buds that initiate leafy gametophores in mosses.

cDNA. Single-stranded DNA synthesized from RNA by reverse transcription.

cell. A unit of protoplasm surrounded by a membrane. Nearly all living organisms are made of one or more cells. Cells are either prokaryotic or eukaryotic. Plant cells differ from animal cells by having cell walls and plastids in the case of eukaryotic cells;cellular.

cell culture. A population of cells grown *in vitro.*

cell cycle. The cyclic sequence of events in dividing cells including the G_1,S,G_2, and M periods.

cell division. The process in which a cell divides to form two new cells, each containing a nucleus. Cell division is either mitotic or meiotic.

cell membrane. The membrane which encloses a cell.

cell plate. New cell wall material laid down in the center of a dividing plant cell.

cellulase. Any of a group of extracellular enzymes, produced by various fungi, bacteria, insects, and other lower animals, that hydrolyze cellulose.

cellulolytic. Having the ability to hydrolyze cellulose, applied to certain bacteria and protozoans.

cellulose. A carbohydrate polymer made of glucose molecules, which is the most important substance in plant cell walls. Only class Oomycetes of fungi contain cellulose rest of the fungi contain chitin.

cell wall. The rigid wall which surrounds a plant cell, lying outside the cell membrane. Cell walls are made mainly of carbohydrate polymers such as cellulose. All plants, fungi and bacteria have cell walls, but animals do not.

central dogma. The relationship between DNA, RNA, and protein: DNA serves as a template for both its own replication and the synthesis of RNA: RNA, in turn, serves in protein synthesis.

central placentation. Having the ovules located in the center of the ovary.

centriole. A small granule outside the nuclear membrane, which divides at mitosis forming the two ends of the spindle. Centrioles are found in all animal cells, but in plants they can only be seen in motile male gametes.

centromere. The point of attachment of a chromosome to the spindle during cell division; kinomere; primary constriction.

centrosphere. The differentiated layer of cytoplasm immediately surrounding the centriole.

centrosome. A spherical hyaline region of the cytoplasm surrounding the centriole in many cells; plays a dynamic part in mitosis as the focus of the spindle pole.

centrum. The totality of structures enclosed by the ascocarp wall.

cephalosporin. Any of a group of antibiotics produced by strains of the imperfect fungus *Cephalosporium.*

cespitose. 1: Tufted, growing in tufts, as grass.
2: Having short stems forming a dense turf.

chalaza. A tissue in the region where the funicle is attached to the ovule.

chalazogamy. A process of fertilization in which the pollen tube passes through the chalaza to reach the embryo sac.

chamaephyte. A woody or herbaceous plant less than 25 cm tall, with buds perennating above the ground.

chartaceous. Of leaves which are like thick paper.

charophytes. A group of specialized green algae of fresh water habitats in which branches are borne in whorls at nodes.

chasmogamy. Pollination occurring after the opening of a flower; chasmogamic, chasmogamous; *cf.* cleistogamy.

chasmophyte. A plant that grows in rock crevices.

chemiosmosis. A process in which energy from the hydrolysis of ATP or the oxidation of organic molecules can be used to make an electrical and chemical gradient of protons across a membrane. This gradient can be used to drive energy—requiring reactions such as the uptake of ions or the synthesis of ATP.

chemoautotrophic. Obtaining metabolic energy by the oxidation of inorganic substrates, such as sulphur, nitrogen or iron as exhibited by some microorganisms; chemotrophic; chemoautotroph; *cf*. photoautotrophic.

chemostat. A steady state laboratory culture apparatus.

chemosynthesis. The synthesis of organic compounds using chemical energy derived from the oxidation of simple inorganic substrates; chemosynthetic; *cf*. photosynthesis.

chemotaxis. The directed reaction of a motile organism towards (positive) or away from (negative) a chemical stimulus; chemiotaxis; chemotactic.

chemotaxonomy. The application of biochemical techniques and the use of biochemical characters in taxonomy.

chemotherapy. Control of a plant disease with chemicals (chemotherapeutants) that are absorbed and are translocated internally.

chemotropism. Curving growth of a plant organ in response to a chemical stimulus or gradient; chemotropic.

chiasmata. (sing. chiasma) The points in a bivalent where the two chromosomes appear to be joined and crossed over. Chiasmata can be observed during diplotene.

chimaera. A plant whose tissues are of more than one genetic kind. This can happen due to mutations in a cell of a very young plant, or can be caused by grafting.

chitinolytic. Capable of degrading chitin.

chitosome. These are specialized organelle found in fungi associated with the enzyme chitin synthatase zymogen. These are spheroidal and measure 40-70 nm in diameter, its function is to deliver the enzyme for chitin synthesis at the cell surface.

chlamydeous. 1: Pertaining to the floral envelope.
2: Having a perianth.

chlamydospore: A hyphal cell, enveloped by a thick cell wall, which eventually becomes separated from the parent hypha and behaves as a resting spore. These spores contain food reserves and can survive periods when hyphae are unable to survive.

chloranemia. The necrotic symptom of yellowing; a loss of chlorophyll.

chloranthy. A reverting of normally colored floral leaves or bracts to green foliage leaves.

chlorenchyma. Chlorophyll containing parenchymatous tissue in parts of higher plants, as in leaves.

chlorobium chlorophyll. $C_{51}H_{67}O_4N_4Mg$, Either of two spectral forms of chlorophyll occurring as esters of farnesol in certain (*Chlorobium*) photosynthetic bacteria; bacterioviridin.

chloronema. This is the first formed protonema formed on spore germination in bryophytes. It is irregularly branched has hyaline cell walls, cross walls are transverse and have numerous spherical chloroplasts per cell; this part of the protonema often cannot produce "buds" that initiate gametophores in mosses.

chlorophylls. Magnesium containing green pigments found in the chloroplasts of all plants which trap light energy, of blue and red wavelengths, for photosynthesis. Chlorophylls give plants their green colour. The two most important chlorophylls are chlorophyll-a ($C_{55}H_{72}O_5N_4Mg$) and chlorophyll-b ($C_{55}H_{70}O_6N_4$ Mg.)

chloroplast. A green plastid containing chlorophyll. Chloroplasts are the site of photosynthesis. They contain their own DNA and reproduce themselves. Chloroplasts are found in the cells of leaf tissues and those of green stems.

chloroplast envelope. The double membrane surrouuding the chloroplast.

chlorosis (pl. chloroses). Yellowing of normally green tissue due to chlorophyll destruction or failure of chlorophyll formation; chlorotic.

chromatic adaptation. The ability of blue-green algae to modify their pigmentation to complement the incident light intensity and so maximize absorption.

chromatid. One of the pair of strands which result from the duplication of a chromosome during prophase and metaphase.

chromatophore. Plastids having chlorophyll-a along with other pigments but not chlorophyll-b, Present in algae and in photosynthetic bacteria.

chromatogram. The recorded result of a chromatographic analysis.

chromatography. A technique used to separate compounds based upon their differential solubilities in the solvents used to develop the chromatogram.

chromophyll. Any plant pigment.

chromoplasm. The pigmented, peripheral protoplasm of blue-green algae cells; contains chlorophyll, carotenoids and phycobilins.

chromoplast. A plastid containing pigments other than green pigment—the chlorophyll. e.g., the coloured plastids in the cells of the tissues of petals and fruits.

chromoprotein. Any protein, such as hemoglobin, with a metal-containing pigment.

chromosomal hybrid sterility. Sterility caused by inability of homologous chromosomes to pair during meiosis due to a chromosome aberration.

chromosome. Thread-like bodies containing DNA, RNA and protein found in the nuclei of all cells. They are usually only visible during cell division, during which they become shorter and thicker. All the vegetative cells in a plant, and in a species have the same number of chromosomes. Two similar chromosomes which pair with each other during meiosis are homologous chromosomes; have identical sequences of loci. Members of a pair of homologous chromosomes have centromeres in the same position and arms of the same length as each other.

chronic symptoms. Symptoms that appear over a long period of time.

chylophyte. A terrestrial plant rooted on a physically dry and hard substratum.

chrysolaminarin. Reserve food material of Chrysophyta.

Chytridiomycetes. A group of aquatic and soil-dwelling fungi commonly unicellular, which produce zoospores.

cingulum. Region of diatom frustule that connects two distal valves.

circinate. Coiled or rolled up, like the young frond of a fern.

circulative viruses. Viruses that are acquired by their vectors through their mouthparts, accumulate internally, then are passed through their tissues and introduced into plants again via the mouthparts of the vectors.

cistron. The genetic unit (deoxyribonucleic acid fragment) that codes for a particular polypeptide; mutants do not complement each other within a cistron.

citric acid cycle. See Krebs cycle.

citriculture. The cultivation of citrus fruits.

cirrhus (pl. cirrhi). A ribbon-like cylinder of spores held together by mucus as it issues from an ostiole.

cladophyll. A branch arising from the axil of a true leaf and resembling a foliage leaf; cladode.

cladoptosis. The annual abscission of twigs or branches instead of leaves.

cladus. A branch of a ramose spicule.

clamp connection. A small looped branch of a hypha which grows at the time of cell divison and septum formation in the dikaryon of many Basidiomycetes.

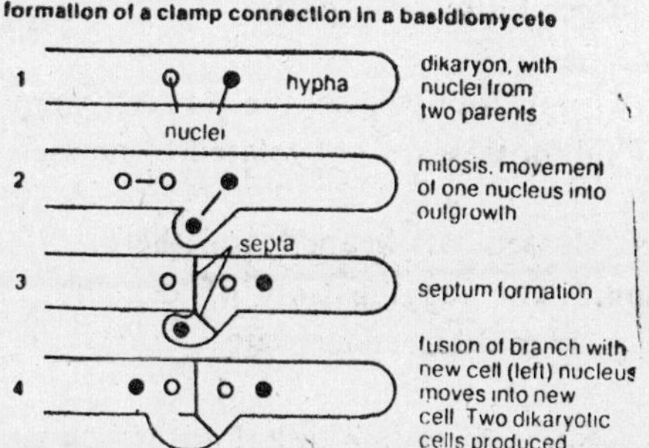

cleistogamy. The condition of having flowers, typically small and inconspicuous, which remain unopened and within which self-pollination takes place; kleistogamy; cleistogamic, cleistogamous; *cf.* chasmogamy.

cleistothecium (pl. cleistothecia). Completely enclosed ascigerous fruiting body of certain fungi belonging to Plectomycetes.

climacteric. A period of high CO_2 output, controlled by the hormone ethene at the beginning of fruit ripening.

climax. The last stage in a succession after which there are no further great changes in the structure or the species in a habitat.

cline. Continuous or gradual variation within a population, in some of the characters of a species. This variation is related to gradual changes in ecological conditions that occur, for intance, up the side of a mountain.

clone. A set of cells or individuals reproduced vegetatively from the same original cell or organism. All members of a clone have exactly the same genome or genetic material.

closed community. A plant community in which the niches are stable and 'full', not allowing the entry of extra species.

clove. 1: The unopened flower bud of a small, conical, symmetrical evergreen tree, *Eugenia caryophyllata*, of the myrtle family (Myrtaceae); the dried buds are used as a pungent, strongly aromatic spice.
2: A small bulb developed within a larger bulb, as in garlic.

clubmoss. A pteridophyte, of the order Lycopodiales, e.g., *Lycopodium.* Clubmosses are not related to mosses.

coacervate. Massed or heaped together.

coacervation. The aggregation of organic molecule to produce particular matter; coacervate.

cocaine. A white crystalline alkaloid which is found in the coca plant *Erythroxylon coca*, fam. Erythroxylaceae). Cocaine is a very potent local anaesthetic and is therefore of great value in minor surgical operations on eye, nose, ear etc.

coccoid. Of algae which are unicellular and non-motile.

codeine. An alkaloid which is a white crystalline solid. It has a similar effect to that of morphine, and is found together with morphine in opium.

coding. The process by which the sequence of nucleotides within a certain area of RNA determines the sequence of amino acids in the synthesis of the particular protein.

codon. A sequence of three nitrogen-containing bases in the triplet code on a molecule of messenger RNA which pairs with an anticodon on a molecule of transfer RNA during translation. Because each base only pairs with one other base, each codon has its own anticodon, e.g., a codon consisting of adenine, guanine and cytosine or AGC, will pair with an anticodon of uracil, cytosine and guanine or UCG.

coenobium. A colony with a regular shape, consisting of cells which do not divide vegetatively e.g., the alga *Volvox.*

coenocytic. Used for thalli, consisting of multinucleate masses of protoplasm. i e., they are aseptate as observed in classes Oomycetes and Zygomycetes in fungi and Xanthophyceae in algae.

coenzyme. A non-protein substance which some enzymes require to make them active. Different enzymes have different coenzymes, vitamins.

CO_2 fixation. The process in which CO_2 dissolved in the intercellular spaces is fixed into an organic molecule in the chloroplasts of plant cells. This is an important part of the dark reaction. Usually, CO_2 reacts with ribulose-diphosphate to produce two molecules of PGA.

colchicine. Alkaloid which inhibits the synthesis of the spindle in the metaphase stage of mitosis. It is used in many genetic experiments, as its effect is to produce tetraploid cells when the new nuclear membrane forms at the end of mitosis. It is extracted from the corm of the crocus (*Colchicum autumnale* fam. Liliaceae). It is also useful in the rheumatism and gout.

coleoptile. A sheath which protects the young shoot tip in grasses.

coleorhiza. A protective layer of cells which is present around the radicle of grass seedlings; coleorrhiza

collenchyma. Tissue of cells with thick cellulose cell walls especially at the angles of the cells, found in the stems of many herbs and in leaves. Its function is support.

colonization. The arrival and germination of a seed on a substrate or the spread of plants to places where they have not grown before. Successful colonization depends on growth to reproductive age.

colony. 1: Used loosely to describe any group of organisms living together or in close proximity to each other; more precisely it refers to an integrated society in which members may be specialized subunits, such as a continuous modular society.

2: A group of organisms that have recently become established in a new area, or that are occupying a particular site.

3; A group of cells of the sa ne kind forming a single organism, as in many algae.

4: In fungi, the term refers to many hyphae growing out from a single point and forming a round or globose thallus.

columella (*pl.* collumellae). 1: A central cylinder of sterile tissue in the capsule of a moss.

2: A sterile structure within a sporangium or other fructification in Myxomycetes often an extension of the stalk

commensalism. The kind of symbiosis in which the partners live in close association with each other and neither partner gains an obvious advantage.

community. The group of species of plants, animals, or both, living in the same habitat and interacting with each other.

comose. Having a tuft of soft hairs.

companion cell. A small, living cell next to the sieve elements in the phloem.

compatible. Of two plants which are able to breed with each other; compatibility.

competition. An interaction between two or more organisms in the same habitat when both, partly or wholly, share the same needs. Competition can be intraspecific or interspecific compete; competitor.

complanate. Flattened parallel to the substratum.

complete flower. A flower that has all four parts, pistil, stamen, petal and sepal; *cf*. incomplete flower.

composite. Of inflorescences where many small flowers are grouped together in a head, looking like a large single flower, e.g., in the daisy family Compositae.

compound. Of leaves which are divided into several or many leaflets without axillary buds.

compound oosphere. An oosphere with many functional gamete nuclei.

concentric. Forming one circle around another with a common centre.

cone. A group of sporophylls closely packed together around a central axis. Cones are the reproductive structures of all gymnosperms and many pteridophytes. In many cone-bearing plants, the male and female cones are separate.

conidiophore. A specialized hypha on which one or more conidia are poduced.

conidium (*pl*. conidia). Asexual fungal spore produced exogenously on a specialized hypha, conidiophore.

conifer A gymnosperm of the order Coniferales, which includes pines, yews, cedars, redwoods, etc. Most conifers are monoecious with separate male and female cones. They are mostly evergreen with narrow pointed leaves. coniferous.

conjugation. A process of sexual reproduction involving the fusion of gametes morphologically similar; conjugate.

conophorophilous. Thriving in coniferous forests; conophorophile, conophorophily.

consortism. The symbiotic relationship between the two partners (consortes) of a consortium.

consortium. A group of individuals of different species, typically of different phyla, living in close association.

continuous variation. The more or less continuous series of infinitely small steps exhibited by a character or trait having a continuous spectrum of values from one given extreme to another; *cf.* discontinuous variation.

control. Economic reduction of crop losses that are due to plant diseases.

coprophagous. Feeding on dung or faecal material; koprophagous; merdivorous; scatophagous.

coprophilous. Growing on dung.

coralloid. Resembling coral, or branching like certain coral as can be seen in coralloid roots of Cycas.

coremium (*pl.* coremia). A fungal fruiting structure consisting of a cluster of erect intertwined hyphae that bear conidia at their apices.

coriaceous. Of leaves which are thick and stiff, like leather.

cork. Tissue of dead cells with suberin cell walls forming part of the bark.

corm. A thickened stem base, usually underground, with buds in the axils of dead leaf bases. Corms are organs of vegetative reproduction and perennation.

cormatose. Having or producing a corm.

corolla. The inner whorl of the perianth of a flower composed of petals.

corollate. Having a corolla.

corolline. Relating to, resembling, or being borne on a corolla.

corona. 1: An appendage or series of fused appendages between the corolla and stamens of some flowers.
2. The region where stem and root of a seed plant merge; crown.

corpus. The inner layers of cells in the apical meristem of angiosperm shoots. Corpus cells divide anticlinally producing the inner tissues of the shoot.

cortex. The tissue between the vascular cylinder and the epidermis of a root or stem. The cortex usually has many layers of cells; cortical.

cortina. A curtain-like, cobwebby veil hanging from the margin of the cap of certain mushrooms.

corticicolous. See corticolous; corticicole.

corticolous. Inhabiting or growing on bark; corticicolous; corticole

corymb. A raceme whose lower stalks are longer than the upper ones, so the inflorescence has a fiat top; corymbose.

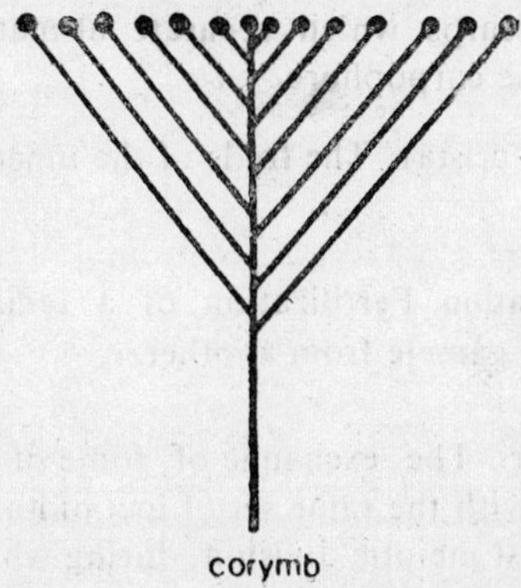

corymb

costa. The thickened midrib of a moss leaf or hepatic thallus.

cotyledon. Part of the embryo of a seed plant. The cotyledon sometimes becomes the first photosynthetic organ of the young seedling. Some plants, e.g., Leguminosae, have large cotyledons which store food. Angiosperms have either one or two cotyledons, gymnosperms have more. Angiosperms are classified into two classes, the monocotyledons which have one cotyledon and the dicotyledons which have two.

coulter counter. An electronic device for counting the number of cells in a liquid culture.

crassulacean acid metabolism. CAM. A kind of CO_2 fixation, found in many succulent plants, such as the family Crassulaceae. CO_2 is fixed by phosphoenolpyruvate, as in the C_4 pathway, to produce malate; this occurs in the night, when the stomata are open. During the day, when the stomata are closed, the CO_2 is released and fixed again by ribulose diphosphate. This reduces the water loss by transpiration on hot days.

cremocarp. A dry dehiscent fruit consisting of two indehiscent one-seeded mericarps which separate at maturity and remain pendant from the carpophore.

cristae (*sing.* crista). The folds of the inner membrane of a mitochondrion.

cross-fertilization. Fertilization of a female gamete of one plant by a male gamete from another.

crossing-over. The exchange of some of the corresponding segments each with the same set of loci of homologous chromosomes at the first meiotic division, during which chiasmata are formed. The result of crossing-over is recombination. Crossing over rarely occur during mitosis as in parasexual cycle in Ascomycetes, Deuteromycetes and Basidiomycetes.

cross-pollination. Pollination of one plant by pollen from another individual.

cross protection. The phenomenon in which plant tissues infected with one strain of a virus are protected from infection by other strains of the same virus.

crown. The top of a tree including the branches and leaves.

crozier formation. Process of ascus development from coiled tips of ascigerous hypahe.

cruciform. Intranuclear division in which the chromosomes are arranged in a ring around a dumbbell-shaped nucleolus as they divide. Found in the Plasmodiophoromycetes and some protozoa.

crustose. Of lichens whose thallus is closely pressed to the substrate or actually growing within it.

cryophyle. A plant growing in very cold conditions, e.g., on ice or snow. Cryophyles are usually algae or bryophytes.

Cryptogams. A general name for all plants except gymnosperms and angiosperms in an old classification of the plant kingdom. Cryptogams reproduce by spores.

cryptogram. In taxonomy, a shorthand summary of the properties of a virus taxon, comprising four pairs of symbols: (1) type of nucleic acid/strandedness; (2) molecular weight of nucleic acid/percentage of nucleic acid; (3) outline of particle/outline of nucleic acid and most closely surrounding protein, (4) host/vector.

cryptophyte. A plant with buds or shoot apices perennating underground or under water during unfavourable seasons.

cryptopolar. Describing a spore in which the germ tube emerges from either of the poles or at the equator.

cryptopore. Describing stomata that are immersed so that the pore is embedded within a chamber of overarching cells.

cryptostomata. Sterile conceptacles of some brown algae.

cucullate. Hooded or hood shaped, often used to describe the usual shape of the calyptra in mosses.

culm. The stem of a grass.

culmicolous. Growing on the stems of grasses; culmicole.

cultivar (CV). A variety of a plant produced and maintained by cultivation; also the corresponding taxon; cultivarietas, cultigen.

culture. To artificially grow micro-organisms on a prepared food material; a colony of microorganisms artificially maintained on such food material.

culture medium. The prepared food material on which micro-organisms are cultured.

cuprophyte. A plant adapted to, or tolerating high copper levels in the soils; frequently used as an indicator of this particuiar soil type.

cuticle. Layer of cutin on the surface of leaves and green stems, which prevents evaporation and protects the plant against attack from herbivores and pathogens.

cutin. A waxy substance comprising the inner layer of the cuticle.

cuticular transpiration. The loss of water vapour through the cuticle of leaves; *cf.* stomatal transpiration.

cutting. Piece of shoot cut from a plant which grows roots from its nodes when placed in soil.

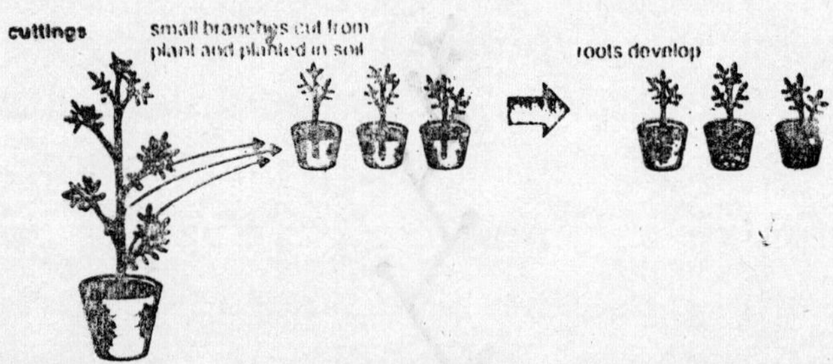

cyanophycin granules. Proteinaceous storage product of blue-green algae.

cycad. A gymnosperm of the order Cycadales. Cycads are dioecious with motile male gametes. They have palm-or fern-like leaves and are found mainly in tropical habitats. The cycads are a primitive group, and there are many fossils dating from the Mesozoic.

cyclic phosphorylation. A photosynthetic reaction cycle in which light energy is used to produce ATP from ADP and orthophosphate.

cyclosis. Massive rotational streaming of cytoplasm in certain vacuolated cells, such as the stonewort *Nitella* and *Paramecium* and in *Hydrilla* (a water plant).

cyme. A sympodial inflorescence growing by means of lateral branches, each with a flower at its apex; cymose.

cyme

cyst. 1: An encysted zoospore (fungi).
2: In nematodes, the carcass of dead adult females of the genus *Heterodera* which may contain eggs.

cystidium (*pl.* **cystidia**). A large sterile structure in the hymenium of a basidiomycete.

cystocarp. Fruiting body of red algae consisting of carposporophyte and pericarp or other surrounding envelope.

cystolith. A concretion of calcium carbonate arising from the cell walls of modified epidermal cells in some flowering plants. e.g., *Ficus benghalensis*.

cytochromes. A group of iron-containing proteins concerned with the electron transfer chain of photosynthesis and with the use of oxygen in aerobic respiration. The iron atom in a cytochrome molecule is at the centre of a porphyrin ring or haem.

cytochrome a_3. *See* cytochrome oxidase.

cytochrome oxidase. Any of a family of respiratory pigments that react directly with oxygen in the reduced state; cytochrome a_3.

cytology. The study of cells by microscopy.

cytokinesis. The division of a cell into two cells.

cytokinins. Group of plant growth regulating substances which control cell division.

cytophagous. Feeding on cells; cytophage; cytophagy

cytoplasmic androgamy. The fertilization of a male gamete by the cytoplasm of a female gamete.

cytoplasmic factor. Any genetic factor which is resident in the cytoplasm of a cell.

cytoplasmic gynogamy. The apparent fertilization of a female gamete by the cytoplasm of the male gamete.

cytoplasmic inheritance. Inheritance of characters through cytoplasmic factors i.e., in the mitochondria, chloroplast, or in other parts of cytoplasm rather than through the nuclear chromosomes.

cytoplasmic streaming. Intracellular movement involving irreversible deformation of the cytoplasm produced by endogenous forces.

cytosine. A pyrimidine base which pairs with guanine in DNA and RNA.

cytotaxonomy. Classification based on a study of cellstructure, with special reference to chromosome structure.

D

2,4-D. 2, 4-dichlorophenoxyacetic acid; a synthetic translocated hormone weed killer, used to control broad leaved herbaceous plants, and as a defoliant.

dacryoid. For spores having one end rounded and the other more or less pointed; pear or tear like in form.

dactyloid. Finger-like.

dagga. See cannabis.

damping off. Disease or necrotic symptom of disease in seedlings in which the seedling stem is decayed near the soil line and the seedling topples. Damping off pathogens may also prevent seed germination and kill the sprout before it emerges from the soil.

DAP. Diaminopimelic acid. A seven carbon diamino acid that occur as a component of cell wall peptidoglycan in some bacteria.

dark-field microscopy. A type of microscopic examination in which the microscopic field is dark and any objects, such as organisms, are brightly illuminatcd.

dark reaction. The part of photosynthesis controlled by enzymes rather than light, i.e., CO_2 fixation and the reductive pentose pathway.

dark respiration. Respiration in photosynthetic plants occurring during the night.

Darwin. Charles Darwin (1809-1882) proposed that evolution occurs by natural selection and the survival of the fittest. His book on the subject, *On the Origin of Species*, was published in 1858. Darwin's theory of evolution is often known as Darwinism.

day-neutral. A plant in which the flowering response is not dependent on photoperiodism.

DDT. Dichlorodiphenyltrichloroethane, a persistent organochlorine insecticide.

decalcification. Removal of calcium carbonate from the soil by leaching.

decay. The process of rotting and decomposition which takes place after the death of an organism. Decay involves the breakdown of organic compounds in the organism, by saprophytic bacteria and fungi. It is an important part of the cycle of nutrients and energy in ecosystems.

deciduous. Falling off, not persistent.

declinate. Bent or curved down or forwards.

declivate. Declivous; sloping.

decolourate. Colourless.

decomposer. An organism which breaks down organic matter, releasing carbon dioxide and inorganic compounds, e.g., nitrates, phosphates, ammonia. The main decomposers are bacteria and fungi; dccomposition; decompose.

decomposition. Metabolic degradation of organic matter into simple organic and inorganic compounds, with consequent liberation of energy; decomposer.

decoy. A crop of low value cultivated to attract pests away from a more valuable crop.

decumbent. Lying down on the ground but with an ascending tip, specifically referring to a stem.

decurrent. Running downward, especially of a leaf base extended past its insertion in the form of a winged expansion.

decussate. Of the arrangement of leaves, occurring in alternating pairs at right angles.

dedifferentiation. The process of regression to a less specialized structure or condition.

defective virus. A virus, such as adenoassociated satellite virus, that can grow and reproduce only in the presence of another virus.

deficiency. The lack of a nutrient required for growth and development. Deficiency can lead to poor growth and disease.

defoliant. A chemical, such as 2,4-D and 2,4,5-T, that causes leaves to fall from plants.

deforestation. The permanent removal of forest and undergrowth.

degenerate code. A genetic code in which more than one triplet sequence of nucleotides (codon) can specify the insertion of the same amino acid into a polypeptide chain.

dehisce. To split open spontaneously along a line. Many fruits, especially dry fruits, dehisce to release their seeds. Anthers dehisce to release their pollen.

dehiscence papilla. In Blastocladiaceae a family of lower fungi of the class Chytridiomycetes the zoosporangium or gametangium has a small rounded projection on its surface which later become a dehiscence pore.

deleterious. 1: Having an adverse effect.
2: Used of a trait which impairs survival, or of a mutation that reduces fitness.

deletion. The loss of a segment from a chromosome.

deletion mapping The use of overlapping deletions to map the positions of particular genes on a chromosome.

deliquescent. Of gills of agaric like *Coprinus comatus* becoming liquid after maturing.

deliquescence. The condition of repeated divisions ending in fine divisions, seen especially in venation and stem branching.

deltoid. Triangular or like a delta.

deme. Any population of organisms that is genetically isolated from other populations. Individuals in a deme breed amongst themselves, and there is no input of genetic material from other demes.

demicyclic. The life cycle of rusts in which uredospores are lacking as in *Gymnosporangium* in heteroecious rusts and *Xenodochus carbonarius* in autoecious rusts.

denaturation. Change in the native configuration of a macromolecule resulting, for example, from heat treatment, extreme pH changes, or chemical treatment Denaturation is usually accompanied by the loss of the molecule's biological activity.

dendrochronology. A method of dating using annual tree-rings, tree rings chronology.

dendroclimatology. The determination of past climatic conditions from the study of the annual growth rings of trees.

dendrocolcus. Living in, or growing on, trees; arboricolous; dendrocole.

dendrogram, A branching diagram in the form of a tree used to depict degrees of relationship or resemblance.

dendroid. Shaped like a tree; describing gametophores in which a leafy unbranched stem is terminated by an apical cluster of branches.

dendrology. The study of trees.

dendrophagous. Feeding on wood; hylophagous, lignivorous; xylophagous; dendrophage, dendrophagus, dendrophagy.

dendrophilous. Thriving in trees; living in orchards; dendrophile, dendrophily.

dendrophysis. A hyphal thread with irregular branching in certain fungi.

denitrification. The release of gaseous nitrogen, nitrous oxide and nitric oxide or the reduction of nitrates to nitrites and ammonia by the breakdown of nitrogenous compounds, typically by microorganisms when the oxygen concentration is low; on a global scale thought to occur primarily in oxygen deficient environments in the oceans.

denitrifying bacteria. Bacteria in the soil which reduce nitrate (NO_3^-) to nitrite (NO_2^-) and molecular nitogen (N_2).

de novo. Latin, meaning arising anew.

density-gradient centrifugation. A method of centrifugation in which different kinds of cells, particles or macromolecules are separated in layers according to their density.

dentate. With sharp spreading teeth.

denticulate. Minutely toothed.

denudation. 1: Erosion of surface material to expose underlying rock.
2: The removal of surface vegetation; denude.

deoperculate. Of mosses and liverworts, to shed the operculum.

deoxyribonuclease. An enzyme that catalyzes the hydrolysis of deoxyribonucleic acid to nucleotides.

deoxyribonucleic acid-directed RNA polymerase. An enzyme which transcribes a ribonucleic acid molecule complementary to DNA; required for initiation of DNA for replication as well as transcription of RNA.

deoxyribonucleic acid ligase. An enzyme which joins the ends of two DNA chains by catalyzing the synthesis of a phosphodiester bond between a 3′-hydroxyl group at the end of one chain and a 5′-phosphate at the end of the other.

deoxyribonucleic acid polymerase. An enzyme which catalyzes the addition of deoxyribonucleotide residues to the end of a DNA.

deoxyribonucleotide. A nucleotide consisting of a purine or pyrimidine base attached to the carbohydrate deoxyribose which is attached to a phosphate group.

deoxyribose $C_5H_{10}O_4$. A carbohydrate found in DNA. It is a colourless soluble crystalline solid.

deplasmolysis. A process in which a plasmolysed cell regain its turgidity when placed in a hypotonic solution.

dermatogen. The outer layer of primary meristem or the primordial epidermis in embryonic plants. protoderm.

dermatophyte. A fungus parasitizing keratinized tissue (hair, skin, nails) of man and animals and causing dermatophytosis. These fungi belong in their asexual state to Hyphomycetes a class of Fungi Imperfecti i.e., Deuteromycotina. The three main genera of dermatophytes are *Epidermophyton*, *Microsporum*, *Trichophyton*.

dermatophytid. A pustular allergic eruption of the skin at a distance from a primary infection by a dermatophyte.

dermatotrophic. Living and feeding on skin; dermatotroph, dermatotrophy.

desalinization. The removal of salts; as in the leaching of saline soils.

desert. A large, dry area of virtually uninhabitable, uncultivated land consisting almost entirely of sand and rock.

desertification. The development of desert conditions as a result of human activity or climatic changes; exsiccation.

determinate growth. Growth that is limited during the life span of an individual, so that the organism reaches a maximum size after which growth ceases.

deuterogamy. Secondary pairing of sexual cells or nuclei replacing direct copulation in many fungi, algae, and higher plants.

Deuteromycetes. A class of fungi that only reproduce asexually. They are common and some, e.g., *Penicillium*, are useful to man. Also known as Fungi Imperfecti.

development. The process of growth and differentiation of individual cells into tissues, organs and organisms.

devernalization. A process where stimulus to flowering received by treatment with low temperature (*i.e.* vernalization) can be reversed.

Devonian. A period of geologic time beginning approximately 395 million years ago and lasting for 50 million years; primitive lycopods, bryophytes and amphibians came into existence in this period; towards the close of this period seed ferns grew in the many low moist areas.

dextran. Any of the several polysaccharides, $(C_5H_{10}O_5)_n$ that yield glucose units on hydrolysis.

dextrin. A polymer of D-glucose which is intermediate in complexity between starch and maltose. Dextrins are used as adhesives.

dextrorse. Twining toward the right.

dextrose. $C_6H_{12}O_6.H_2O$. A dextrorotatroy monosaccharide obtained as a white, crystalline, odourless, sweet powder, which is soluble in about one part of water; an important intermediate in carbohydrate metabolism; used for nutritional purposes, for temporary increase of blood volume, and as a diuretic; corn sugar; grape sugar.

diadelphous. Having stamens in two groups, with the stamens in each group joined together by their filaments, e.g., pea.

diadromous. Having venation in the form of fanlike radiations.

diagcotropic. Transversely geotropic, said of roots or leaves which grow horizontally.

dialysis. The separtion of crystalloids from colloids by the use of selectively permeable membranes.

diakinesis. The last stage in the first meiotic prophase, when the chromosomes are shortest and thickest, the bivalents more apart and position themselves against the nuclear membrane which disappear in this stage only.

diandrous. Having two stamens in a flower.

diaheliotropism. Movement of plant leaves which follow the sun such that they remain perpendicular to the sun's rays throughout the day.

diarch. Having two xylem strands alternating with phloem strands in the vascular tissue of root.

diaspore. Any unit of dissemination e.g., a spore, fragment of mycelium, sclerotium. In lichens this particularly refers to vegetative propagules. See hormocyst, isidia, soredia.

diastase. An enzyme that catalyzes the hydrolysis of starch to maltose; vegetable diastase; amylase.

diatom. An alga of the division Bacillariophyta. Diatoms are mostly unicellular and their cell walls contain silicon. See also siliceous skeleton.

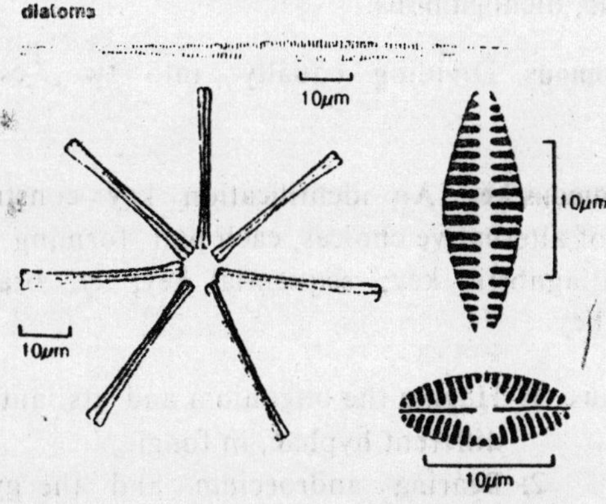

diatomaceous earth. Siliceous deposits of silica (SiO_2) walls of diatom frustules. This is used in sugar refining and brewing industry; car and silver polishing powders; in preparation of bleaching powder; in coal mines to reduce the danger of secondary explosion.

diatropism. Growth orientation of certain plant organs that is transverse to the line of action of a stimulus.

diauxic growth. Growth in two separate phases due to the preferential use of one carbon source over another; between the phases a temporary lag occurs.

diazotroph. An organism that carries out nitrogen fixation; examples are *Clostridium* and *Azotobacter*.

dicentric. Having two centromeres.

dichasium. A cyme producing two lateral shoots from the primary axis or shoot below the flower which terminates the apex, the process being repeated by each set of branches.

dichlamydeous. Having both whorls o the perianth i.e., calyx and corolla.

dichogamy. The maturation of male and female reproductive organs at different times in a flower or hermaphroditic organism, thereby preventing self-fertilization; heteracme; dichogamic, dichogamous.

dichotomous. Dividing equally into two, especially of branches.

dichotomous key. An identification key constructed as a sequence of alternative choices, each pair forming a character couplet; diagnostic key; sequential key; *cf.* bracketed key, indented key.

diclinous. 1: Having the oogonium and its antheridium on different hyphae, in fungi.
2: Bearing androecium and the gynoecium in separate flowers.

dicotyledon. An angiosperm of the class Dicotyledones. The seeds of dicotyledons have two cotyledons. Dicotyledons have secondary thickening in the stems. Most families and species of angiosperms are dicotyledons.

dictyoblastospore. A blastospore with both cross and longitudinal septa.

dictyochlamydospore. A non-deciduous multicelled chlamydospore composed of an outer wall which is separable from the walls of the component cells which are rather easily separated from each other.

dictyosome. A stack of cisternae that forms part of the Golgi apparatus. In plant cells, the term is often used for the entire Golgi apparatus.

Dictyosporae. A group of the imperfect fungi characterized by multicelled spores with cross and longitudinal septae.

dictyospore. A multicellular spore in certain fungi characterized by longitudinal walls and cross septa, a muriform spore as can be observed in the genus *Alternaria* of class Deuteromycetes,

Didymosporae. A group of the imperfect fungi characterized by two-celled spores (didymospore).

didynamous. An androecium having four stamens in two pairs, one of the pairs being shorter than the other.

dieback. Necrotic symptom of disease in which death of shoot tissues begins at the tip and progresses back toward the main stem.

dientomophilous. Used of a plant pollinated by two different insect species and having two kinds of flowers each adapted for one of the insect pollinators; dientomophily.

differential centrifugation. The separation of mixtures such as cellular particles in a medium at various centrifugal forces to separate particles of different density, size, and shape from each other.

differential hosts. The special species or varieties of host plants the reactions of which are used for determining physiologic races.

differential stain. A procedure using a series of dye solutions or staining reagents to bring out differences in microbial cells.

differentially permeable membrane. A membrane whereby the passage of diverse material is regulated into and out of the cell, organelles and vacuole at different rates depending on the relative affinity of the material for lipid and protein constituents of the membrane.

differentiated. Of cells which have developed a particular structure in relation to their function in a tissue or organ; differentiate; differentiation.

diffluent. Breaking up in water.

diffract. Surface of a pileus cracked into small areas; areolate.

diffuse. Widely or loosely spreading and having no distinct margin.

diffusion. The natural movement of molecules of a solute from regions of higher concentration to regions of lower concentration.

diffusion pressure deficit (DPD). The amount by which the diffusion pressure of the water in soil is less than that of pure water (under atmospheric pressure) at the same temperature.

digitate. Of leaves in which the lamina is divided like the fingers of a hand.

digitalis. *Digitalis purpurea*; (fam. Scrophulariaceae). The fresh and dried leaves of the foxglove—yield an active principle a glucoside—Digitoxin. This is used in the treatment of certain heart and circulatory disorders. It slows down the rate at which the heart beats; at the same time strengthening each heart beat.

digitipinnate. Having digitate leaves with pinnate leaflets.

dihybrid cross. A cross between two individuals heterozygous for two pairs of alleles; *cf.* monohybrid cross.

dihybrid inheritance. The inheritance of two pairs of genes.

dikaryon. A pair of haploid nuclei that occur in a cell; the nuclei undergo simultaneous division i.e., conjugate division upon formation of each new cell. Typified in the diploid phase (dikaryophase) of Basidiomycetes.

dikaryotization. The conversion of a homokaryon into a dikaryon typically by the fusion of two compatible homokaryons.

dimerous. Applied to flowers or whorls of a flower, where each whorl consists of two parts.

dimidiate. Appearing to lack one half, or having one half very much smaller than the other; a pileus without a stalk will be semicircular; gills, stretching only halfway to the stipe; and a perithecial wall, having the outer wall covering only the top part.

diminution. Increasing simplification of inflorescences on successive branches.

dimitic. Having two types of hyphae, generative and skeletal hyphae—thick walled, branched or unbranched, aseptate, straight or slightly curved, in the carpophore (sporophore) of Polyporaceae.

dimorphic. 1: Having two shapes, e.g., two different kinds of stamen in one flower.
2: Having two forms; yeast and mycelial forms as observed in *Histoplasma*, *Sporothrix* and other pathogens of man and animals.

dinoflagellate. A class of unicellular algae, usually yellow in colour, which have two flagella and thick cell walls arranged in a characteristic pattern of plates. Dinoflagellates are a major component of marine phytoplankton.

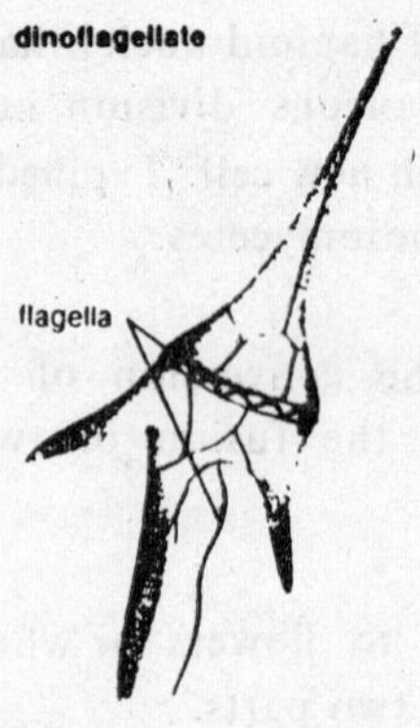

dioecious. Having male and female flowers or sex structures on different individuals of the same plant species. This is a way of avoiding self-fertilization; dioecy; dioecism.

dipicolinic acid. DPA A compound found in large amounts in endospores.

diplanetic. Refers to a species which produces two types of zoospores and in which two swarming periods occur with a resting stage in between as observed in zoospores of the class Oomycetes.

diplobiont. 1: A plant flowering twice in a single season.
2: An organism exhibiting a regular alternation of haploid and diploid generations during the life cycle; a diplohaplontic organism; *cf.* haplobiont.

diplohaplontic. A life-cycle where thereis a successful alter nation between haploid and diploid generations.

diploid. State of having homologous chromosomes in pairs in the nucleus so that twice the haploid number is present.

diplolepidous. Describing peristome teeth in which each segment of the outer face of the articulated tooth is formed of parts of the walls of two cells; thus each segment is composed of two "scales."

diplont. Of the diploid stage in a life cycle e.g., of the sporophyte.

diplospory. Polyploidy arising from diploid spores.

diplotene. The stage in the first meiotic prophase, when the centromeres of paired chromosomes move away from each other and crossing-over can be seen.

disaccharide. A sugar made of two monosaccharide units, e.g. sucrose.

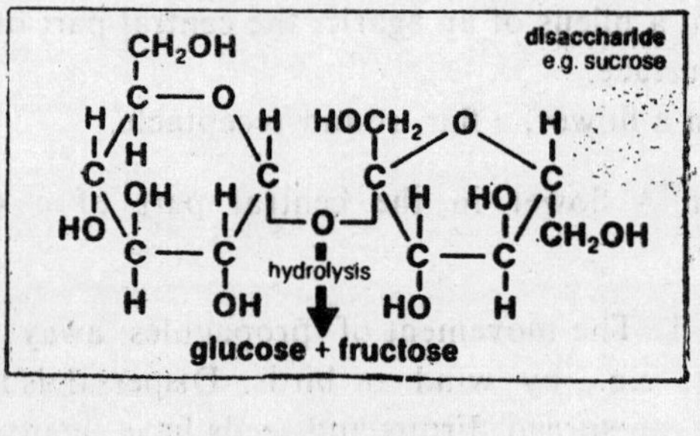

disaccharide e.g. sucrose

hydrolysis

glucose + fructose

discolourous. Of a different colour, as of two surfaces of a lichen thallus.

Discomycetes. An Ascomycete whose fruiting body is typically a sessile, open, saucer or cup-shaped apothecium (e.g., *Ascobolus*).

discontinuous variation. Variation in which individuals of a sample fall into two or more overlapping classes; *cf.* continuous variation.

disease. Harmful physiological processes caused by continuous irritation of a plant by a primary causal agent. It results in morbific cellular activity and is expressed in characteristic pathological responses called symptoms; continuous malfunctioning.

disinfest. To kill pathogens that have not yet initiated disease, but that occur in or on such inanimate objects as soil, tools, and so on, or that occur on the surface of such plant parts as seeds.

disjunction. The separation of the chromosomes during nuclear division.

disjunctor. A small cell or a projection between the chain of spores of certain fungi, which eventually breaks down and thus free the spores; a connective.

disk. 1: A flat, circular plate-like or curved spore producing part of a fruit body in Discomycetes.
2: In a pileus of an agaric, the central part of the top surface.
3: In a flower, a flat, round receptacle.

disk-floret. A flower in the central part of a composite inflorescence.

dispersal. 1. The movement of propagules away from the parent plant, e.g., by wind or birds. Dispersal is the way in which plants can spread. Fruits and seeds have many different adaptations for different kinds of dispersal.

dissemination. Scattering or spreading, as of infectious agents, seeds, spores or other propagules; distribution; "Effective dissemination" is synonymous with "inoculation".

disseminule. A disseminated propagule; a seed, fruit, spore or other structure modified for dispersal.

dissected. Of leaves which have many lobes.

dissociation. A mutation which increases the chromosome number.

dissophyte. A plant with xerophytic leaves and stems, and a mesophytic root system.

distribution. 1: Spread of a pathogen to areas outside of its previous geographical range.
2: The whole geographical range in which a taxon is found.

diurnal. 1: Active during daylight hours.
2: Lasting for one day.

diurnal rhythm. A biological rhythm having a periodicity of about 1 day length (24 h); circadian rhythm.

divaricate. Divergent at right angles.

diverticulum. A pocket-like side branch, as on mycelium of *Pythium* a genus belonging to Oomycetes.

division. A major taxon e.g., bryophytes, which is made up of classes. There are three divisions of land plants; these are Bryphyotes, Pteridophytes and Spermatophytes.

dixenous. Used of a parasite utilizing two host species during its life cycle; dixeny; *cf.* monoxenous.

DNA. Deoxyribonucleic acid. The main nucleic acid in the chromosomes of the nucleus of a cell. The DNA molecule consists of two chains of nucleotide polymer, arranged in a double helix. The sugar in the nucleotides of DNA is deoxyribose. DNA controls protein synthesis by the processes of transcription and translation. It is replicated by a self-copying process, and it is the hereditary material of all cellular organisms and some viruses.

DOC. Dissolved organic matter (carbon), expressed as grams carbon per litre.

dog lichen. *Peltigera canina* (L.) Willd.; used in folk lore for bites of a mad dog.

doliiform. Barrel-like in shape.

dolipore septum. A complex pore in the septum of a basidiomycete hypha which flares out in the middle portion forming a barrel-shaped structure with open ends as shown.

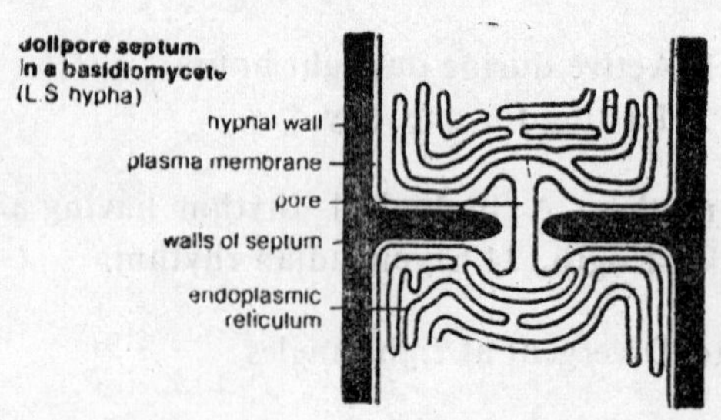

domestication. The adaptation of plants and animals for life in intimate association with man.

dominant. 1: Of alleles which have the same effects in the heterozygous and homozygous conditions; dominance.

2: Of the most common or the largest species in a community; dominate.

Domin scale. A scale for estimating cover and abundance of a plant species, comprising 11 categories: + (single individual), 1 (very few individuals), 2 (sparsely distributed, less than 1% cover), 3 (frequent but less than 4% cover), 4 (4--10% cover), 5 (11-25% cover), 6 (26-33% cover), 7 (34-50% cover), 8 (51-75% cover), 9 (76-90% cover) and 10 (91-100% cover).

dormancy. 1: A state of relative metabolic quiescence.

2: A state in which viable seeds, spores or buds fail to germinate under conditions favourable for germination and vegetative growth; dormant; *cf.* quiescence.

dormin. It is a growth inhibiting substance isolated from birch leaves under short-day conditions. This subtance when reapplied to leaves of birch seedlings, completely arrest apical growth. Later it was shown that abscissic acid and dormin were the same compound chemically.

dorsiferous. Of ferns, bearing sori on the back of the frond.

double fertilization. In most seed plants, fertilization involving fusion between the egg nucleus and one sperm nucleus, and fusion between the other sperm nucleus and the polar nuclei.

double-work. In plant propagation, to graft or bud a scion to an intermediate variety that is itself grafted on a stock of still another variety.

doubling time. The amount of time required for the number of cells in a culture to double during the exponential phase of the culture's growth curve.

drip tip. A long pointed tip to the leaf, which helps water to run off the leaf surface. Drip tips are common in wet tropical forests.

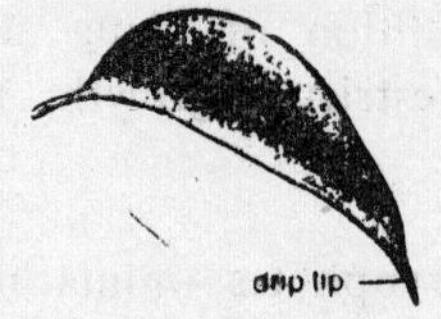

droplet nucleus. Airborne particles containing viable microbes.

drug resistance. A decreased reactivity of living organisms to the injurious actions of certain drugs and chemicals.

drupaceous. Pertaining to, or characteristic of a drupe.

drupe. A fruit with seeds which are covered by a hard, stony endocarp. Drupes usually have a fleshy mesocarp; and a thin or leathery exocarp; stone fruit.

drupelet. An individual drupe of an aggregate fruit; grain.

druse. A stellate cluster of large crystals.

dry rot. 1: A rapid decay of seasoned timber caused by *Serpula lacrymans* which cause the wood to be reduced to a dry, friable texture.
2: Any of various rot diseases of plants characterized by drying of affected tissues.

dry weight. The weight of an organism, part of an organism, or the organisms in a habitat or in an ecosystem, after drying. Because a large part of the biomass of most organisms is water, dry weight is usually small compared to biomass.

duriherbosa. Permanent tall grassland vegetation.

durilignosa. Broadleaved evergreen forest and scrub vegetation.

duvet. A soft, thick layer of hyphae formed in dermatophytes which appear like a brushed up cloth.

dwarfism. The condition of being stunted, much smaller than normal; having restricted growth; microsomia; nanism; dwarf.

dwarf male. 1: Of bryophytes, miniature antheridial gametophores that are attached to the normal-size archegonium bearing gametophores.
2: Of algae, small male filaments of nannandrous species of *Oedogonium* developed by the germination of androspore.

dynein. The protein arm of the microtubule doublet subfiber A of cilia and flagella that possesses AT Pase activity.

dysgonic. Term used for those dermatophytes which grow slowly in culture forming less aerial mycelium than a normal strain.

dystophic. Pertaining to an environment that does not supply adequate nutrition.

E

earthballs. Species. of *Scleroderma* belonging to Gastromycetes grows in acid woodland and heaths with *Pinus*, *Betula* and forms mycorrhiza with them.

earth stars. Species of *Geastrum* belonging to Gasteromycetes the young fruit body is onion-shaped but as it ripens the exoperidium splits open in a a stellate fashion and curve outwards.

ebracteolate. Lacking bracteoles.

ecad. A type of plant which is altered by its habitat and possesses nonheritable characteristics.

ecesis. Successful naturalization of a plant or animal popution in a new environment.

echard. That part of soil water not available for plant use.

echinomycin. $C_{50}H_{60}O_{12}N_{12}S_2$ A toxic polypeptide antibiotic produced by species of *Streptomyces.*

echinulate. Minute spines covering the surface of spores.

eclipse period. A phase in the proliferation of viral particles during which the virus cannot be detected in the host cell.

ecocline. A genetic gradient of adaptability to an environmental gradient; formed by the merger of ecotypes.

ecological efficiency. The efficiency of transfer of energy from one trophic level to the next.

ecological interaction The relation between species that live together in a community; specifically, the effect an individual of one species may exert on an individual of another species.

ecological niche. The concept of the space occupied by a species, which includes both the physical space as well as the functional role of the species; the biological space occupied by a species, and which is unique to the species; the dimensions of this space are the parameters of the niche; sometimes erroneously used as the equivalent of microhabitat; niche.

ecological pyramid. A pyramid-shaped diagram representing quantitatively the numbers of organisms, energy relationships, and biomass of an ecosystem; numbers are high for the lowest trophic levels (plants) and low for the highest trophic level (carnivores).

ecological succession. A gradual process incurred by the change in the number of individuals of each species of a community and by establishment of new species populations that may gradually replace the original inhabitants.

ecology. The study of organisms in relation to their environment; ecological.

ecophene. All the naturally occurring phenotypes produced within a given habitat by a single genotype; ecad; oecophene.

ecospecies. A group of populations or ecotypes having the capacity for free exchange of genetic material without loss of fertility or vigour, but having a lesser capacity for such exchange with members of other ecospecies groups; closely approximates to a biological species.

ecosphere. See biosphere.

ecosystem. An ecological system in which organisms interact with each other and with their non-living environment and in which there is a more or less closed cycle of nutrients.

ecosystem mapping. The drawing of maps that locate different ecosystems in a geographic area.

ecosystem respiration. (R_e) The total energy utilized during respiration by autotrophs (R_a) and heterotrophs (R_h) calculated as $R_e = R_a + R_h$

ecotone. The border between two habitats or types of vegetation. It is a fairly broad transition zone between adjacent biomes.

ecotype. A set of individuals or populations in a particular habitat that differ phenotypically from members of the same species in other habitats, the difference being genetically fixed.

ectal excipulum. Outer tissue or tissues of an apothecium charecteristic of the margins adjacent to the hypothecium and thecium.

ectogony. The influence of pollination and fertilization on structures outside the embryo and endosperm; effect may be on color, chemical composition, ripening, or abscission.

ectohydric. Mosses which lack well differentiated conducting strand are termed ectohydric. These are capable of absorbing water through any part of the external surface of the shoot or thallus. It include mosses such as *Rhacomitrium*, *Orthotrichum*, *Ulota*, *Cyphaea* and all the leafy liverworts.

ectoparasite. Parasite living on the outside of its host (for example, stubby-root and dagger nematodes). *cf.* endoparasite.

ectosymbiosis. Symbiosis in which one member (microsymbiote) develops on the outside of the other member. *cf.* endosymbiosis.

ectothrix. Living on the surface of hair.

ectotrophic. Of mycorrhizae whose hyphae do not grow into the cells of the host. Ectotrophic mycorrhizae form a sheath around the root of the host, and the mycelium also grows in the intercellular spaces of the root tissues.

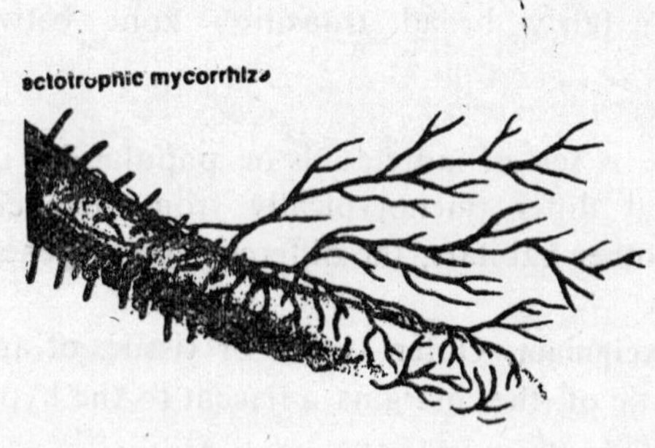

edaphic community. A plant community that results from or is influenced by soil factors such as salinity and drainage.

edaphic factors. The effects of the soil in an ecosystem. Different soils have different structural and chemical characteristics, and different plant species are adapted for growth on particular types of soil.

edaphon. The soil flora and fauna; those organisms living in the interstitial water and pore spaces of soil; edaphonekton.

edge effect. The influence of adjacent plant communities on the number of animal species present in the direct vicinity.

efflorescence. The period or process of flowering (blossoming).

effuse. Expanded; spread out in a definite form; as a film-like growth.

effuso-reflexed. Flat over the substratum and turned back at the margin to make a pileus in Hymenomycetes.

egg. 1: The female gamete.

2: In *Amanita* and phalloids the young basidiocrap before the volva is broken.

egg apparatus. A group of three cells, consisting of the egg and two synergid cells, in the micropylar end of the embryo sac in seed plants,

Egyptian cotton. Long-staple, high-quality cotton grown in Egypt.

elaioplast. An oil-secreting leucoplast.

elater. 1: One of a bunch of long, thin cells in the capsule of the sporophyte of a liverwort. Elaters have spiral thickening of the cell wall. They alter their position with changes in humidity, and help with the dispersal of spores from the capsule.

2: A free capillitium-thread, e.g., in myxomycetes.

elaterophore. A cylindrical mass of sterile cells to which elaters are attached in some liverworts sporangia for example in *Pellia*, *Aneura*.

elective culture. A type of microorganism grown selectively from a mixed culture by culturing in a medium and under conditions selective for only one type of organism.

electron microscope. A powerful instrument that uses electrons instead of light rays to magnify very small objects. The electron microscope can magnify more than 100,000 times, and can be used for observing very small details of cell structure.

electron transfer chain. 1: A set of redox reactions in the light reaction of photosynthesis involving plastocyanin, plasto quinone and cytochromes in which ATP is produced;

2: A set of redox reactions in aercbic respiration involving cytochromes and also producing ATP.

electrophoretic mobility. A characteristic of living cells in suspension and biological compounds (proteins) in solution to travel in an electric field to the positive or negative electrode, because of the charge on these substances.

electrophoretic variants. Phenotypically different proteins that are separable into distinct electrophoretic components due to differences in mobilities.

electrophysiology. The branch of physiology concerned with determining the basic machanisms by which electric currents are generated within living organisms.

elymoclavine. A clavine alkaloid in *Claviceps fusiformis* selerotia; an intermediate compound in the biosynthesis of ergot alkaloids.

emarginate. Gills having a cut at the point of attachment to the stipe.

emasculation. A technique of removal of anthers of a flower in order to avoid self polination.

Embden Myerhof Pranas Pathway (EMP pathway). See glycolysis.

EMB agar. A culture medium containing, sugar, eosin, and methylene blue, used in the confirming test for coliform bacteria.

embryo. The young plant contained in the seed. The embryo is the product of repeated mitotic divisions of the zygote. It consist of cotyledons, a plumule a hypocotyl and a radicle; embryonic

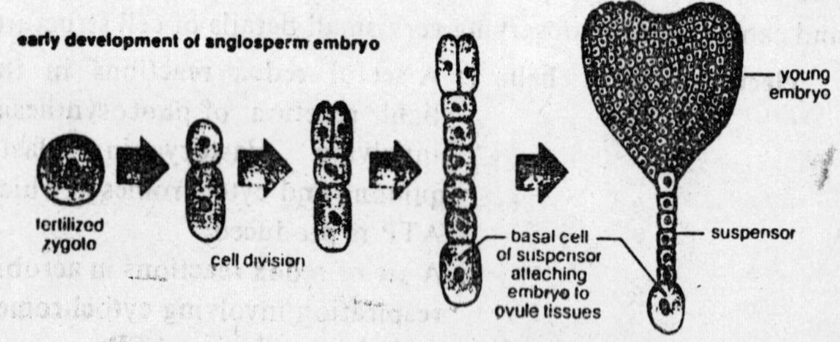

embryoids. Embryo formed in culture from cells other than a fertilized egg; are also known as accessory embryos, adventive embryos, somatic embryos and supernumerary embryos.

embryology. A branch of biology concerned with the study of the formation and development of embryo.

embryo sac. The female gametophyte in angiosperms, consisting of 8 haploid cells including the ovum, three antipodal cells, two synergids and two endosperm nuclei which fuse before fertilization. The embryo sac is contained inside the ovule.

emerged bog. A bog which grows vertically above the water table by drawing water up through the mass of plants.

Emerson effect. This is a kind of photosynthetic enhancement discovered by Emerson et al. where the efficiency of photosynthesis at wavelengths exceeding 680 nm can be restored by a simultaneous application of a shorter wavelength.

emigration. The movement of individuals or their disseminules out of a population or population area.

enation. Tissue malformation or overgrowth induced by certain virus infections.

encysted. Surrounded by a hard shell (cyst).

encystment. The process of forming or becoming enclosed in a cyst or capsule.

endarch. Formed outward from the center, referring to xylem or its development.

endemic. 1. Of taxa that are found only in one particular place or area; endemism.
2: Of disease restricted to a particular area.

endergonic. Of or pertaining to a biochemical reaction in which the final products possess more free energy than the starting materials; usually associated with anabolism.

endobiotic. Living within the host cell e.g., in *Diplophlyetis* a member of Chytridiomycetes the whole thallus, rhizoids and sporangium are formed within the host cell.

endocarp. The innermost layer of tissue in a fruit, surrounding the seeds.

endocarpinoid. Lichen perithecia when they are sunk into the tissue of the thallus, as in *Endocarpon*.

Endobiotic

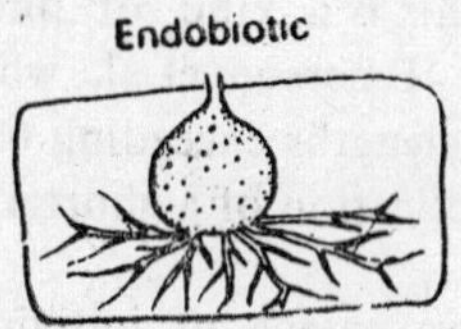

endocommensal. A commensal that lives within the body of its host.

endoectothrix. Making growth in and on a hair.

endodermis. The innermost tissue of the cortex of most plant roots and certain stems surrounding the vascular cylinder, consisting of a single layer of at least partly suberized or cutinized cells; functions to control the movement of water and other substances into and out of the stele.

endohydric. Mosses which has a well developed conducting strand, such as *Bryum, Mnium, Polytrichum*. These absorb water mainly by rhizoids at the bases.

endogamy. 1: Sexual reproduction between organisms which are closely related.
2: Pollination of a flower by another flower of the same plant.

endogenote. The genetic complement of the partial zygote formed as a result of gene transfer during the process of recombination in bacteria.

endogenous. Arising from within the generating structure e.g., endoconidia of *Thielaviopsis basicola. cf.* exogenous.

endogenous rhythm The repeated, regular, rhythmic changes of internal activity in an organism that are not due to external environmental factors.

endolithic. Living within or on rocks, as certain algae and coral.

endomitosis. Division of the chromosomes without dissolution of the nuclear membrane: results in polyploidy or polyteny.

endonuclease. Any of a group of enzymes which degrade deoxyribonucleic acid or ribonucleic acid molecules by attaching nucleotide linkages within the polynucleotide chain.

endoparasite. A parasite that lives inside its host.

endopeptidase An enzyme that acts upon the centrally located peptide bonds of a protein molecule.

endoperidium. The inner layer of peridium.

endophloeodic. Thallus of a crustaceous lichen almost entirely immersed in bark; endophoeodal; endophloeic.

endophllyous. Living within the leaves i.e., below the cuticle.

endophyte .A plant that lives within, but is not necessarily parasitic on, another plant.

endoplasmic reticulum. A system of membranes in the cytoplasm, where much of protein synthesis takes place. Endoplasmic reticulum may be rough (with ribosomes), or smooth (without ribosomes).

endosperm. Triploid tissue in the seed, resulting from double fertilization. The function of the endosperm is to store food for the seedling.

endospore. A resting cell formed within a cell and normally produced under unfavourable growth conditions. This is seen in the case of a few bacteria, the *Bacillus* and *Clostridium.* Usually endospores are oval or spherical structures and are highly resistant to heat, desiccation, disinfectants and X-rays. Germination of endospore takes place under favourable conditions producing a vegetative cell.

endosperm mother cell. The cell formed in the embryo sac by the fusion of the two haploid endosperm nuclei. The mother cell is diploid, and it is fertilized by a pollen nucleus, in angiosperms, to form the triploid endosperm.

endostome. 1: The opening in the inner integument of a bitegmic ovule.
2: The inner ring of peristome teeth in mosses.

endosymbiosis. Symbiosis in which one member (microsymbiote) lives within the other. without deleterious effect on the host; endosymbiont. *cf.* ectosymbiosis.

endothecium. 1: The middle of three layers that make up an immature anther; becomes the inner layer of a mature anther.
2: The inner tissue in a developing moss capsule from which the columella and archesporium develop.

endotoxin. A toxin that is produced within a microorganism and can be isolated only after the cell is disintegrated.

endotrophic. Of mycorrhizae which do not form a sheath around the root of the host. Endotrophic mycorrhizae usually grow into the cells of the host. e.g., in orchids.

energy pyramid. An ecological pyramid illustrating the energy flow within an ecosystem.

enhancer gene. Any modifier gene that acts to enhance the action of a nonallelic gene.

enolase. An enzyme that catalyzes the reversible dehydration of phosphoglyceric acid to phosphopyruvic acid.

enphytotic. A disease that occurs regularly among plants of a specific locality or geographical area from year to year.

enrichment culture. A medium of known composition and specific conditions of incubation which favors the growth of a particular type or species of bacterium.

ensiform. Sword-shaped, with sharp edges and tapering to a point.

entire. 1: Of leaves or bacterial colonies with a continuous edge or margin.
2: Complete; whole; without parts or divisions.

Entner-Doudoroff pathway. A sequence of reactions for glucose degradation, with liberation of energy; the distinguishing feature is the formation of 2-keto-3-deoxy-6-phosphogluconate from 6-phosphogluconate and the cleaving of this compound to yield pyruvate and glyceraldehyde-3-phosphate.

entomogenous. Growing on or in an insect body, as certain fungi belonging to Zygomycotina, Ascomycotina, Basidiomyco tina and Deuteromycotina.

entomophilic fungi. Species of fungi that are insect pathogens. e.g., *Entomophthora* sp. *Hypocrella, Aschersonia etc.*

entomophily. Pollination by insects. Flowers pollinated by insects are usually brightly coloured and scented. If they are pollinated by bees, they usually produce large amounts of pollen which the bees collect. If they are pollinated by butterflies or moths, they produce nectar; entomophilous.

environment. The living and non-living surroundings of an organism, and the events which take place in those surroundings.

enzyme. A protein which, in very small quantities, catalyses and controls the natural chemical reactions of metabolism. Enzymes are usually large complex molecules, and most are responsible for one or two particular reactions in the cell. Cells contain many thousands of different enzymes.

enzyme induction. The process by which a microbial cell synthesizes an enzyme in response to the presence of a substrate or of a substance closely related to a substrate in the medium.

enzyme inhibiton. Prevention of an enzymic process as a result of the interaction of some substance with the enzyme so as to decrease the rate of reaction.

enzyme repression. The process by which the rate of synthesis of an enzyme is reduced in the presence of metabolite, often the end product of a chain of reactions in which the enzyme in question operates near the beginning.

enzyme unit. The amount of an enzyme that will catalyze the transformation of 10^{-6} mole of substrate per minute or, when more than one bond of each substrate is attacked, $10^{-3}$of 1 gram equivalent of the group concerned, under specified conditions of temperature, substrate concentration, and pH number.

enzymology. A branch of science dealing with the chemical nature, biological activity, and biological significance of enzymes.

eosin. A red, flourescent, crystalline dyestuff used as a stain in microscopy.

Eocene. Which means 'dawn of recent' was that portion of Tertiary period that began about 54 million years ago and lasted about 16 million years. Most Eocene plants and invertebrates were like modern days plants. Most abundant plants were like Palm trees.

Ephedra. A genus of low, leafless, green-stemmed shrubs belonging to the order Ephedrales; source of the drug ephedrine.

Ephedrales. A monogeneric order of gymnosperms in the class chlamydospermopsida.

ephedrine. A white crystalline alkaloid obtained from *Ephedra* spp. It is used in thc treatments of colds, asthma, hay fever and as a cardiac stimulant.

ephemeral. Of plants which germinate, grow, reproduce and die in a very short moist season; e.g., desert plants.

ephydrogamous. Having water borne pollen grains transported at the water surface; or being pollinated by such pollen; *cf.* hyphydrogamous.

epibasidium. A lengthening of the upper part of each cell of the basidium of various Hymenomycetes e.g., *Tremella, Exidia.*

epibiotic. Living, usually parasitically, on the surface of plants or animals; used especially of fungi, *e.g.*, in *Rhizophydium* a member of Chytridiomycetes, the rhizoidal system only penetrates the host cell (often an alga or a pollen grain) and the sporangium is superficial; *cf.* endobiotic.

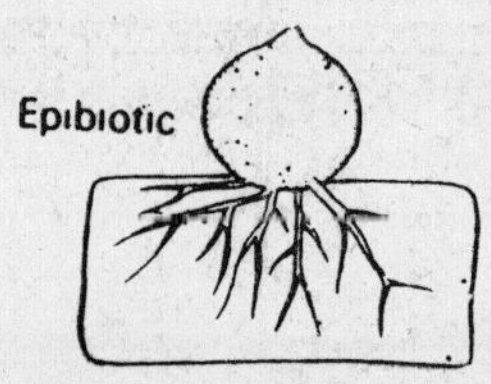

epiblem. A tissue that replaces the epidermis in most roots and in stems of submerged aquatic plants.

epicalyx. A ring of fused bracts below the calyx forming a structure that resembles the calyx.

epicarp. The outer layer of the pericarp; exocarp.

epicolous. Living attached to the surface of another organism but without benefit or detriment to the host; epibiontic; epicole.

epicotyl. Part of the embryo and seedling above the cotyledons. The first true leaves are produced on the epicotyl after germination.

epidemic. Used of a disease affecting a high proportion of the population over a wide area; *cf.* epiphytotic, epizootic.

epidemiology. The study of factors affecting the out break and spread of diseases in populations or communities.

epidermis. The outer layer of cells of leaves, green stems, young roots, etc. epidermal.

epigeal. 1: Used for Discomycetes where the apothecia are formed above ground; epigean, epigeic.

2: Of the kind of germination in which the cotyledons are borne above ground level, becoming the first photosynthetic organs of the seedling.

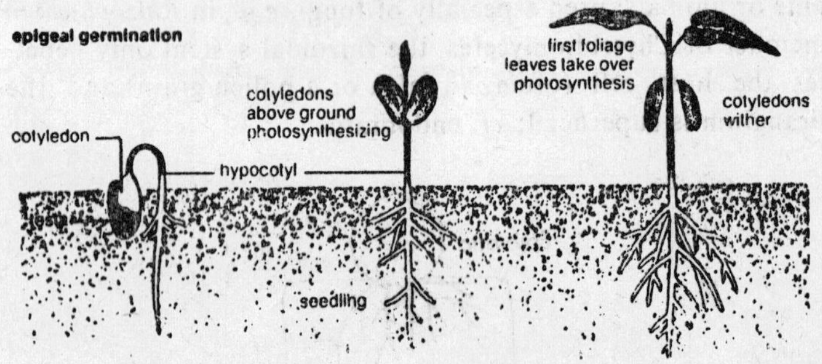

epigeic: Lichens which are not attached to any substrate but keep blowing about on the surface of the ground.

epigenetics. The study of those processes by which genetic information ultimately results in distinctive physical and behavioral characteristics.

epigynous. 1: Of flowers in which the ovary is within the receptacle, and the other floral parts attached above it; epigyny.
2: Of fungi, having the antheridium arising above the ogonium on one hypha.

epigenous. Developing or growing on a surface, especially of a plant or plant part.

epinasty. Downward curling of a leaf blade due to cell growth on the upper side of a petiole being more rapid than that on the lower side; often a hyperplastic symptom of plant disease e.g., the disease caused by potato virus Y in *Physalis floridana.*

epipetalous. Having stamens located on the corolla.

epiphragm. 1: A membrane-like expansion of the columella covering most of the mouth of the sporangium in Polytrichidae (hair-cap mosses).
2: Of fungi, a membrane over the young fruit body in the Nidulariales.

epiphyll. A plant that grows on the surface of leaves; epiphyllous.

epiphyte. 1: A plant growing on another plant (the phorophyte) for support or anchorage rather than for water supply or nutrients; aerophyte. e.g., various orchids, lichens, mosses etc.
2: Any organism living on the surface of a plant; epiphytically.

epiphytotic. 1: Any infectious plant disease that occurs sporadically in epidemic proportions.
2: Of or pertaining to an epidemic plant disease.

epiplankton. Plankton occurring in the sea from the surface to a depth of about 100 fathoms.

epiplasm. Pertaining to an ascus the cytoplasm not used up in the free cell formation of an ascospore.

epipodium. The apical portion of an embryonic phyllopodium.

episepalous. Having stamens growing on or adnate to the sepals.

episome. A circular genetic element in bacteria, presumably a deoxyribonucleic acid fragment, which is not necessary for survival of the organism and which can be integrated in the bacterial chromosome or remain free. Examples are F factors (plasmids) of *Escherichia coli* R-factor, col factor.

episperm. *See* testa.

epistasis. The suppression or the effect of one gene by another.

epithecium. (pl. epithecia). A layer of tissue on the surface of the hymenium of an apothecium, formed by the union of the tips of the paraphyses over the asci.

epixylous. Growing on wood; used especially of fungi; ligni colous.

epizoic. Living on the body of a animal.

equatorial plane. The plane in a cell undergoing mitosis that is midway between the centrosomes and perpendicular to the spindle fibers.

equitant. Of leaves, overlapping transversely at the base.

eradication. Principle of plant-disease prevention characteriz ed by destruction or removal of a pathogen already established in a given area.

erect stem. A stem that stands, having a vertical or upright habit.

eremacausis. The process of humus formation by the oxidation of plant matter.

eremad. A desert plant; eremophyte.

ergastic. Pertaining to the nonliving components of protoplasm.

ergastoplasm. A cytoplasm of component which shows an affinity for basic dyes; a form of the endoplasmic reticulum.

ergometrine. (D- lysergic acid propanol (amide) an ergot alkaloid from *Claviceps purpurea* sclerotia; used in medicine against migraine.

ergosterin. *See* ergosterol.

ergosterol. $C_{28}H_{44}O$. A crystalline, water-insoluble, unsaturated sterol found in ergot, yeast, lichens and other fungi, and which may be converted to vitamin D_2 on irradiation with ultraviolet light or activation with electrons, ergosterin.

ergot. Disease of certain grasses and cereals, especially rye, caused by *Claviceps purpurea*; also the spur-shaped sclerotium of C. *purpurea* that replaces the grain in a diseased inflorescence. The sclerotium contains toxic alkaloids, which have been used by physicians for medicinal purposes, including the induction of labor in women. Consumption of ergot-infested grain by animals, or human consumption of bread made from ergot-diseased grain, results in ergotism, a dreaded disease that is either paralytic or gangrenous.

ericophyte. A plant growing on heathland.

erineum. An abnormal growth of hairs induced on the epidermis of a leaf by certain mites.

eriophyllous. Having leaves covered by a cottony pubescence.

erosion. Wearing away; weathering; the removal of the land surface by water, ice, wind or other agencies; erode, erosive.

erumpent. Breaking through the substratum e.g., growth of sporodochia through the epidermis of susceptible plant organs.

escape. 1: Failure of inherently susceptible plants to become diseased, even though disease is prevalent.
2: The breaking away from, as spores or seeds from a capsule.

Escherichia coli (E. coli). A bacterium inhabiting the mammalian colon. It has been and still is widely used in genetic research, and as an indicator bacterium for faecal pollution or water of foodstuffs; colon bacillus

ethene. C_2H_4. A simple plant hormone which affects tropisms the inhibition of root growth, abscission, the ripening of fruit, and other growth processes, ethylene.

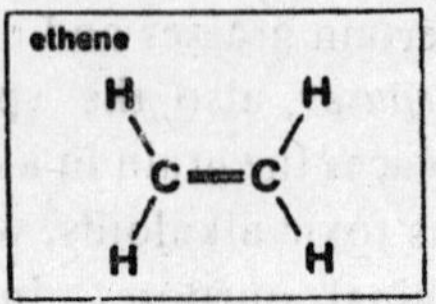

ethnobotany. Study of the use of plants by the races of man.

etioblast. An immature chloroplast, containing prolamellar bodies.

etiolation. The process of rapid growth, without the production of chlorophyll that occurs in shoots kept in the dark. Etiolated shoots are long. thin and pale, and their leaves are very small; etiolate.

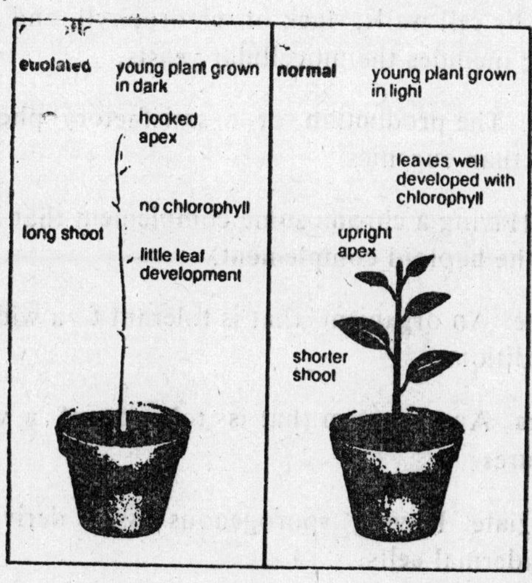

etiology. The study of cause; that phase of plant pathology which deals with the causal agent and its relations with the susceptible plant.

eucarpic. Forming reproductive structures on certain portions of the thallus, the thallus itself continuing to perform its somatic functions; *cf.* holocarpic.

euchromatin. The portion of the chromosomes that stains with low intensity, uncoils during interphase, and condenses during cell division.

eugenics. The use of practices that influence the hereditary qualities of future generations, with the aim of improving the genetic future of humanity.

euglenoid. An alga of the division Euglenophyta Euglenoids are unicellular flagellate and have paramylum instead of starch as their main storage product.

eukaryotic. Of cells with a nucleus surrounded by a nuclear membrane and distinct organelles; eukaryote; encaryote.

eumitosis. Typical mitosis.

eumycetes. The true fungi, a large group of microorganisms charaterized by cell walls, lack of chlorophyll, and mycelia in most species; includes the unicellular yeasts.

euphenics. The production of a satisfactory phenotype by means other than eugenics.

euploid. Having a chromosome complement that is an exact multiple of the haploid complement.

euryhaline. An organism that is tolerant to a wide range of osmotic conditions.

eurytherm. An organism that is tolerant of a wide range of temperatures.

eusporangiate. Having sporogenous tissue derived from a group of epidermal cells.

eustele. A modified siphonostele containing collateral or bicollateral vascular bundles, found in most gymnosperm and angiosperm stems.

eutrophic. Of habitats rich in nutrients.

eutrophication. A process which can occur in rivers and shallow lakes when the addition of extra nutrients e.g., from fertilizers, causes heavy growth of algae. When the algae die, their decay by bacteria reduces the concentration of oxygen in the water, so that aerobic organisms may not survive.

evapotranspiration. Evaporation from soil surfaces, lakes, and streams and by transpiration from plants into the atmosphere.

evergreen. Describes plants (trees and shrubs) which do not shed all their leaves at the same time and thus appear green all the year round.

evolution. The changes that occur in organisms over many generations and long periods of time. Evolution occurs by the natural selection of mutations; evolutionary; evolve.

exalate. Being without winglike appendages.

exalbuminous. Of seeds without albumen; as in beans and cucurbits; exendospermous.

exarch. A vascular bundle in which the primary xylem is centripetal.

excipulum (pl. excipula). The outer layer of the hypothecium.

excision. Recombination involving removal of a genetic element.

excision enzyme. A bacterial enzyme that removes damaged dimers from the deoxyribonucleic acid molecule of a bacterial cell following light or ultraviolet radiation or nitrogen mustard damage.

excitable. Describing a tissue or organism that exhibits irritability.

exclusion. The principle of plant-disease prevention in which the pathogen is prevented from entering a given region e.g., by seed disinfestation or quarantine.

exclusive species. A species which is completely or nearly limited to one community.

excretion. The process of removing waste products of metabolism from a cell or organism; excrete.

excurrent. 1: Having an undivided main stem or trunk.
2: Having the midrib extending beyond the apex.

exendospermous. Lacking endosperm; exalbuminous.

exendotrophic. Used of a flower pollinated by pollen from another flower of the same or different plant.

exergonic. Of or pertaining to a biochemical reaction in which the end products possess less free energy than the starting materials; usually associated with catabolism.

exine. The hard outer coat of a pollen grain. The patterns on the surface of the exine are often used as characters in the classification of seed plants.

exocarp. The outer layer of tissue of the fruit. The exocarp is often hard or skin-like.

exoconidium (pl. exoconidia). Any asexual spore formed on the surface of a hypha (exogenously).

exocytosis. The extrusion of material from a cell.

exodermis. Layer of cortical cells, with suberin in their cell walls. The exodermis is on the outer surface of the cortex, underneath the epidermis; exodermal.

exogamy. Union of gametes from organisms that are not closely related; outbreeding.

exogenote. The genetic fragment transferred from the donor to the recipient cell during the process of recombination in bacteria.

exogenous. 1: Due to an external cause, not arising within the organism.
2: Growing by addition to the outer surfaces.

exogynous. Having the style longer than and exserted beyond the corolla.

exon. That portion of deoxyribonucleic acid which codes for the final messenger ribonucleic acid.

exonuclease. Any of a group of enzymes which catalyze hydrolysis of single nucleotide residues from the end of a deoxyribonucleic acid chain.

exopathogen. An external, nonparasitic plant pathogen whose extracellular toxic metabolites cause disease in plants.

exopathogenesis. The external incitement of disease by a nonparasitic pathogen.

exoperidium. The outer layer of the peridium.

exosmosis. Passage of a liquid outward through a cell membrane.

exospore. The outer layer of a spore wall.

exosporic. A spore that undergoes cell division after the rupture of the spore coat.

exosporium. The outer of two layers forming the wall of spores such as pollen and bacterial spores, exine.

exostome. 1: The opening through the outer integument of a bitegmic ovule.
2: The outer row (or rows) of peristome teeth in the mosses.

exothecium. The outermost layer of cells in the sporangium.

exotic. Not endemic to an area.

exotoxin. A toxin that is excreted by a microorganism into the surrounding medium.

explant. The tissue removed in explantation.

explantation. The removal of living tissue from an organism and its cultivation in an artificial medium i.e., tissue culture.

exploitation. Removal of individuals or biomass from a population by predators or parasites.

explosive evolution. The splitting of a group or population into numerous lines of descent within a relatively short period of geological time; explosive radiation.

explosive fruit. One which bursts suddenly and violently, scattering seeds over a considerable area, as in the cucumber.

exponential growth. The period of bacterial growth during which cells divide at a constant rate; logarithmic growth.

exponential growth phase. The period of maximum population growth; log, growth phase *cf.* lag growth phase.

exserted. Protruding beyond the enclosing structure, such as stamens extending beyond the margin of the corolla.

exsiccatum. A dried specimen (pl. exsiccata; means a set of dried specimens).

extant. Of species which exist at present.

extinct. Of species which no longer exist.

extirpate To uproot, destroy, make extinct, or exterminate.

extracellular. Outside the cells; *cf.* intracellular.

extrachromosomal inheritance. *See* cytoplasmic inheritance.

extrafloral. Positioned away from the flower, e.g., an extrafloral nectary.

extramatrical. Living on or near the surface of the matrix or substratum.

extrorse. 1: Directed outward or away from the axis of growth.
2: Used of anthers facing outwards from the center of the flower.

extrorse dehiscence. Spontaneous opening of ripe fruit from the inside outwards; *cf.* introrse dehiscence.

exudate. The liquid exuded from pores and glands such as hydathodes.

exudativorous. Feeding on gum and other exudates from trees; exudativore, exudativory.

exude. To ooze out, or diffuse out liquid from pores e.g., in guttation or from a cut surface; exudation.

eyespot. 1: A small photosensitive pigment body in certain unicellular algae.

2: A dark area around the hilum of certain seeds, as some beans.

3: A fungus disease of sugarcane and certain other grasses which is caused by *Helminthosporium sacchari* and characterized by yellowish oval lesions on the stems and leaves.

F

F_1 generation. The first filial generation. The offspring of the parental generation at the begining of a genetic experiment.

F_2 generation. The second filial generation. The offspring resulting from sexual reproduction in the F_1 generation.

facilitated diffusion. Assisted transport of molecules through the plasma membrane along a concentration gradient.

facultative. Of organisms which can live under several kinds of conditions. For instance, a facultative epiphyte is a plant which can grow either on the ground or on other plants.

facultative gamete. A motile spore (zoospore) that can function as a gamete.

facultative aerobe. An anaerobic microorganism which can grow under aerobic conditions.

facultative anaerobe. A microorganism that grows equally well under aerobic and anaerobic conditions and for which oxygen is not toxic.

facultative halophytes. Plants capable of growing in salty as well as in non-salty soils, e.g., *Atriplex*, *Allenrolfea*, grow best where salt levels in the soil are high, as in deserts or in soils saturated with brackish waters on the sea coasts. They also grow in non-salty soils.

facultative parasite. An organism capable of infecting another living organism or of growing on dead organic matter, according to circumstances.

facultative parthenogenesis. The process by which some eggs develop parthenogenetically if not fertilized.

facultative photoheterotroph. Any bacterium that usually grows anaerobically in light but can also grow aerobically in the dark.

facultative saprophyte. An organism capable of growing on dead organic matter, or of infecting another living organism according to circumstances.

facultative thermophile. An organism requiring a temperature between 50 and 65°C for optimum growth but capable of growth at lower temperatures; *cf.* obligate thermophile.

FAD. flavin adenine dinucleotide. A hydrogen carrier in the Krebs cycle.

fairy ring. A ring of mushrooms on the ground representing the periphery of mycelial growth of a basidiomycete.

faithful species. A species occurring only in a particular community; exclusive species.

falcate. Shaped like a sickle.

falciform. Sickle-shaped.

false annual ring. A second ring of xylem formed in one season following abnormal defoliation, especially by insects.

false axis. A monochasium which looks like one axis, but is really a number of successive lateral branches running more or less in a line.

false branching. Branching in some blue-green algae formed by the breaking of a trichome from its sheath.

false fruit. A fruit whose structures include floral parts e.g., the receptacle or calyx, in addition to the gynoecium. Examples include the apples and pears etc.

false smut. 1: A fungus disease of palm caused by *Graphiola phoenicis* and characterized by small cylindrical protruding pustules, often surrounded by yellowish leaf tissue.

2. *See* green smut.

family. A taxon consisting of related genera used in the classification of living organisms. Similar families are grouped into the same order and orders into a class. The Latin name of families usually end with –aceae

farinaceous. 1: Having a mealy surface.

2: Of a mealy character.

3: Of an endosperm that is starchy.

farinose, farinosus. Covered with whitish, very short hairs, which are easily detached as a whitish dust.

farnesol $C_1H_{25}OH$. A colorless liquid extracted from oils of plants such as citronella, neroli, cyclamen, and tuberose; it has a delicate floral odor, and is an intermediate step in the biological synthesis of cholesterol from mevalonic acid in vertebrates; used in perfumery.

far red light. Solar radiation in the spectral range 700-800 nm.

fasciate. Massed or joined side by side.

fasciation. Hyperplastic symptom characterized by fusing (and flattening) of such plant organs as stems, for example: fasciation disease caused by *Corynebacterium fascians* occuring in various dicots like *Chrysanthemum*, garden peas and sweet peas.

fascicle. 1: A tuft of branches all arising from about the same place for example in *Sphagnum* the cluster of lateral branches are arranged in spirals on the main stem.

2. A tuft of leaves crowded on a short stem.

fasciculate. Arranged in tufts or fascicles.

fasciculation. Hyperplastic symptom characterized by a clustering of such plant organs as shoots into such structures as witches-brooms.

fastidious organism. An organism that is difficult to isolate or cultivate on ordinary culture media becaure of its need for special nutritional factors.

fastigate. Having many branches parallel to the main stem, and usually upright; fastigiate.

fathom. A secondary fps unit of length: used especially for measuring water depth; equal to 6 feet or 1.8288 m.

fatty acid. An organic acid with the general formula $C_nH_{2n}O_2$. A fatty acid molecule is a straight chain and the number of carbon atoms is usually even.

feedback. The process by which a product at or near the end of a metabolic pathway affects the reactions at the beginning of the same pathway. Feedback can be either positive or negative.

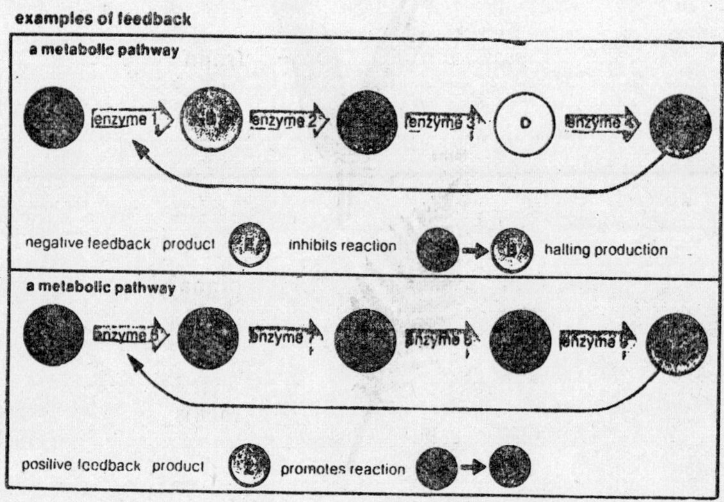

Fehling's solution. Fehling's A contain 34.66 gm. of $CuSO_4$ crystals, dissolved in 500 ml. of water, and Fehling's B, 173 gm potassium sodium tartarate, and 50 gm. of sodium hydroxide dissolved in 500 ml of water. They are used in equal parts and give a red-orange precipitate when boiled with a reducing sugar. This is due to the reduction of the cupric ions to cuprous.

female. Of individual tissues, organs etc. producing egg-cells.

fermentation. The breaking down of organic molecules, especially by yeast and bacteria under anaerobic conditions, to produce carbon dioxide and alcohol or lactic acid.

fern. A pteridophyte belonging to the order Filicales. Ferns have spirally arranged leaves, which are often pinnately compound. Ferns are homosporous with the sporangia borne in sori on the abaxial surface of the leaf.

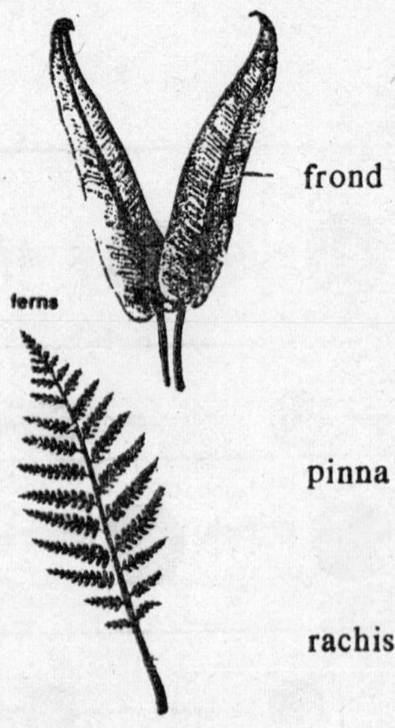

ferredoxin. A non-haem iron-containing protein in the chloroplast, which is involved in the photosynthetic light reaction.

ferrichrome. A cyclic hexapeptide that is a microbial hydroxamic acid and is involved in iron transport and metabolism in microoganisms.

fertile. Of organisms which produce offspring, or of reproductive organs which produce viable gametes; fertility.

fertility factor. A form of plasmid present in bacteria.

fertilization. The fusion of a male gamete with a female gamete to form a zygote; fertilize.

fertilization tube. A tube originating from the male gametangium and penetrating into the female, through which the male gametes (nuclei) are transferred.

Feulgen reaction. Staining reaction for DNA used in microscopy; based on the formation of a purple colored substance following treatment with Schiff's reagent (decolorized fuschsin) after acid hydrolysis.

fibre. A long, thick-walled cell in the sclerenchyma.

fibrovascular. Composed of conducting cells and their fibrous sheaths.

ficin. A proteolyic enzyme present in fresh fig latex, used as an anthelminthic against nematodes.

fig. A tree or shrub of the genus *Ficus* or the fruit it produces. Types indude Common figs, caprifigs, Smyrna figs and San Pedro figs.

filament. 1: The stalk of a stamen which supports the anther.

2: A chain of cells joined end to end, as in certain algae.

filamentous. Of algae consisting of long threads of cells, e.g., *Spirogyra*.

filamentous bacteria. Bacteria, especially in the order Actinomycetales, whose cells resemble filaments and are often branched.

filiform. Threadlike or filamentous.

filmy fern. A fern of the family Hymenophyllaceae. Filmy ferns have very delicate leaves, usually only one cell thick, and live in moist shady habitats.

fimbriae. Surface appendages of certain Gram negative bacteria, composed of protein subunits are shorter and thinner than flagella; pili.

fingerprinting. A technique used to separate peptides that combines paper chromatography with electrophoresis.

fir. The common name for any tree of the genus *Abies* in the pine family; needles are characteristically flat.

fission. A method of asexual reproduction among bacteria, algae, and protozoans by which the organism splits into two or more parts, each part becoming a complete organism.

fix. To kill, harden, or preserve a tissue, organ, or organism by immersion in dilute acids, alcohol, or solutions of coagulants.

flaccid. Deficient in turgor. Soft, flabby, or relaxed.

flagellate. Having one or more flagella.

flagelliform. 1: Like a whip-lash.
2: Pertaining to bryophytes a slender branch or stem, leafless or with leaves much smaller than those of the rest of the gametophore.

flagellin. The protein monomer of bacterial flagella.

flagellum. A long motile thread, consisting of a membrane enclosing a series of parallel microtubules. Flagella (*pl.*) are found in the cells of unicellular motile algae e.g., *Euglena*. *Chlamydomonas*, in zoospores of fungi and also in the male gametes of bryophytes, pteridophytes, and some gymnosperms. A flagellum can be whiplash or tinsel.

flagging. The loss of rigidity and drooping of leaves and tender shoots preceding the wilting of a plant.

flagilliflory. Of flowers, hanging down freely from ropelike twigs.

flavescence. Yellowing or blanching of green plant parts due to diminution of chlorophyll accompanying certain virus disease.

flavin. 1: A yellow dye obtained from the bark of *Quercus* trees.

2: Any of several water-soluble yellow pigments occurring as coenzymes of flavoproteins.

flavone. 1: Any of a number of ketones composing a class of flavonoid compounds.

2: $C_{15}H_{10}O_2$ A colorless crystalline compound occurring as dust on the surface of many primrose plants.

flavonoid. Any of a series of widely distributed plant constituents related to the aromatic heterocyclic skeleton of flavan.

flavonol. 1: Any of a class of flavonoid compounds that are hydroxy derivatives of flavone.

2: $C_{16}H_{10}O_2$ A colorless, crystalline compound from which many yellow plant pigments are derived.

flavoprotein. The group name for yellow coloured protein bonded to riboflavin involved in electron transfer reaction, in both plants and animals.

flax. *Linum usitatissimum.* An erect annual plant with linear leaves and blue flowers; cultivated as a source of flaxseed and fiber.

fleck. A minutespot.

flexuous. Said of a stem which is zig-zag, usually changing direction at the nodes.

flexuous hypha. Of the Uredinales, a branched, or unbranched hypha growing from a pycnidium, which may be diplodized by a pycnidium of the opposite 'sex'.

flimmer. Lateral fibrils of pleuronematic or tinsel flagellum.

floccose. 1: Cottony; covered with tufts of wool-llke hairs, that is easily detached from the plant.

flocculate. Having small tufts of hairs.

floccule. An adherent aggregate of microorganisms or other materials floating in or on a liquid.

floccus. A tuft of woolly hairs.

flora. 1: The total of plant species in a particular region, country, continent. etc.

2: In microbiology, the microorganisms present in a given situation e.g., intestinal flora, the normal flora of soil.

floral diagram. A diagram showing the position and number of all the parts of a flower in transverse section.

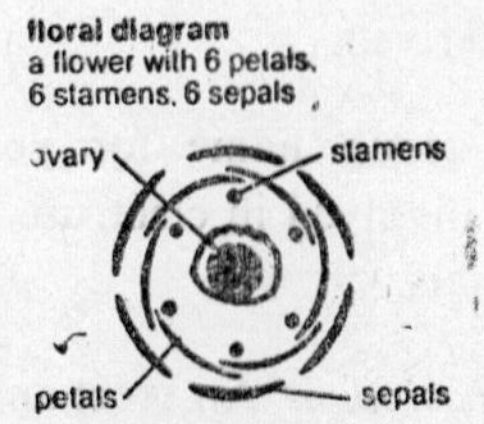

floral formula. A simple method of expressing in shorthand the information given in a floral diagram. K=calyx, C= corolla, A=androecium (stamens) G=gynoecium. The number following reprsents the number of parts.

( )means that the parts are fused, e.g., $C_{(5)}$ means 5 fused petals.

⌒indicats 1 whorl fused to another e.g., $\overset{\frown}{C_{(5)}A_5}$ would mean a corolla of 5 fused parts to which are joined 5 free stamens.

A line below the figure for the gynoecium, means that it is superior, e.g., $\underline{G_5}$, beside it, that it is perigynous e.g. $G_{(5)}^{-}$, and above it inferior e.g.,$\overline{G(3)}$.

floret. A small flower, usually in a large or composite inflorescence.

florey unit. A unit for the standardization of penicillin.

floribundus. Latin meaning 'producing many flower's.

floridean starch. An amylopectin found in the form of storage product in red algae. It stains red, rather than blue-black with iodine solution.

floridus. Latin meaning 'showy'.

floriferous. Blooming freely, used principally of ornamental plants.

florigen. A plant hormone that stimulates buds to flower. This may be synthesized in the leaves during a photo period.

floristic composition. A complete list of plants forming a plant community.

florula. Plants which grow in a small, confined habitat, for example, a pond.

flosculous. 1: Composed of florets.

2: Of a floret, tubular in form.

flosculus. A floret.

flow bog. A peat bog with a surface level that fluctuates in accordance with rain and tides.

flower. The reproductive shoot of an angiosperm consisting usually of four sets of modified leaves arranged in whorls. These are the sepals, petals, stamens and carpels. The function of a flower is to produce male gametes in pollen and female gametes, in ovules. After fertilization, the ovules develop into seeds. The reproductive shoots of conifers are also sometimes called flowers; floral.

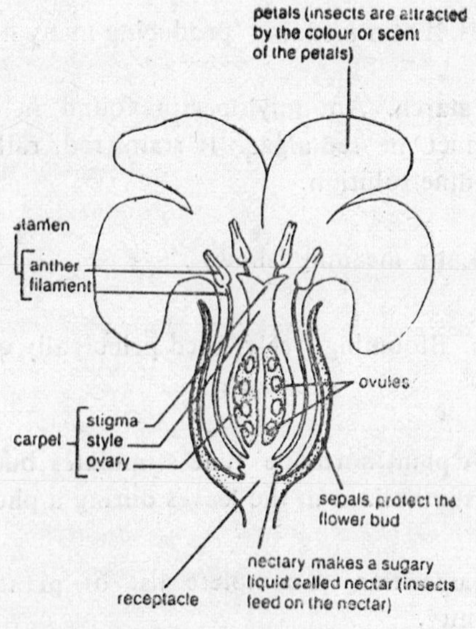

Parts of a flower

flowering plant. A plant which bears flowers. This term is usually used only for angiosperms but is sometimes used for some gymnosperms as well.

fluid-mosaic membrane. Model of the plasma membrane in which proteins are distributed in a phospholipid bilayer.

fluorescein. An organic, cyclic (aromatic), yellow compound having a brilliant greenish fluorescence. It is used to detect leakages in water systems and for colouring liquids in various instruments.

fluorescence. An instantaneous kind of luminescence whereby light is emitted from certain substances when these are irradiated by light or certain other radiations. The absorbed light is usually emitted at a greater wavelength than the incident light. Ultraviolet light striking a fluorescent substance gives rise to emission of visible light and chlorophyll appear red after it absorbs blue light. Several substances, such as fluorescein and uranyl (VI) sulphate, exhibit fluorescence; phosphorescence.

fluorescence microscopy. Microscopy in which cells or their components are stained with a fluorescent dye and thus appear as glowing objects against a dark background.

fluorescent staining. The use of fluorescent dyes to mark specific cell structures, such as chromosomes.

fluorochromasia. The immediate appearance of fluorescence inside viable cells on exposure to a fluorogenic substrate.

flush. An evergreen herbaceous or non-flowering vegetation growing in habitats where seepage water causes the surface to be constantly wet but rarely flooded.

fodder. Plants which are grown as food for domesticated animals. The plants may be dried e.g., hay.

fog climax. A community that deviates from a climatic climax because of the persistent occurrence of a controlling fog blanket.

fog forest. The dense, rich forest growth which is found at high or medium-high altitudes on tropical mountains; occurs when the tropical rain forest penetrates altitudes of cloud formation, and the climate is excessively moist and not too cold to prevent plant growth.

foliage. The leaves of a plant.

foliar. Of, pertaining to, or consisting of leaves.

foliation. 1: The process of developing into a leaf.
2; The state of being in leaf.

folic acid. $C_{19}H_{19}N_7O_6$. A yellow, crystalline vitamin of the B complex; it is slightly soluble in water, usually occurs in conjugates containing glutamic acid residues, and is found especially in plant leaves and vertebrate livers; pteroylglutamic acid (PGA).

folicaulicolous. Living attached to leaves and stem; folicaulicole.

foliferous. Producing leaves.

foliose. Of lichens with a leafy thallus. Foliose lichens have distinct upper and lower surfaces.

foliaceous. Consisting of or having the form or texture of a foliage leaf.

follicle. 1: A many seeded dry fruit, derived from a single carpel, and splitting longitudinally down one side at dehiscence.

2: A small bladder on the leaves of some mosses.

foliolate. Having leaflets.

fomites. Inanimate objects that carry viable pathogenic organisms.

food chain. The flow of energy and nutrients from one group of organisms to another in an ecosystem, e.g., from producers to consumers to decomposers.

food vacuole. A membrane-bound organelle in which digestion occurs in cells capable of phagocytosis; phagocytic vacuole; heterophagic vacuole.

food web. A set of interacting food chains e.g., an animal may feed on several plant species, and the animal may be fed on by several animal species which in turn may be fed on by other animal species.

foot. 1: The base of the sporophyte of a bryophyte and pteridophyte which is the part that attaches it to the gametophyte.

2: A small thick-walled segment of the hyphae, from which a conidiophore *of Aspergillus* arises.

3: The basal cell of the microspore *in Fusarium*.

forb. A weed or broadleaf herb.

forecasting. Based on the quantitative aspects of the ecology of disease; such as incidence as a function of time, distance from sources and number of sources and vectors: telling in advance to the farmers how epidemics are likely to develop, of when economic thresholds are likely to be exceeded and thus whether control is required and when.

forest. A habitat or type of vegetation in which large trees shrubs and shade tolerant herbs are dominant.

forest conservation. Those measures concerned with the protection and preservation of forest lands and resources.

forest ecology. The science that deals with the relationship of forest trees to their environment, to one another, and to other plants and to animals in the forest.

forest-tundra. A temperate and cold savanna which occurs at high altitudes and consists of scattered or clumped trees and a shrub layer of varying coverage.

forestry. The management of forest lands for wood, forages, water, wildlife, and recreation.

forficulate. Shaped like a scissors.

forma specialis. (*pl.* formae speciales; *abbr.* f.sp.): Special form; a biotype of a species of pathogen that differs from others in the ability to infect selected genera or species of susceptible plants; *cf.* physiologic race.

form-genus. A group of species which have similar morpho logical characters, but are not known certainly to be related by descent. The species are form species. It is for imperfect states i.e., Fungi Imperfecti.

formyl methionine. Formylated methionine, initiates peptide chain synthesis in bacteria.

fornicate. 1: Arched and hood like.
2: Of (*Geastrum*), having the fibrous and fleshy layers of the fruit body becoming arched over the cup-like mycelial layer.

fossil. The remains or marks left by dead organisms, converted to stone over geological time. Fossils provide important clues to the history and evolution of living organisms; fossilize.

fossil fungi Fossilized fruit-bodies, on or with fossilized stems or leaves. Most of the generic names are current names with the suffix—ites e.g., *Sphaerites*, *Rhytismites*, *Xylomites*; about 350 fungi are known.

fossil lichens. Very few fossilized lichens reported from Palaeozoic and Mesozoic but the first reliable report is from the Cenozoic. Lichens like *Alectoria succini* has been found preserved in amber, some are found on wood in peat.

foveola. A small pit.

foveolate Delicately pitted; dimpled; like a spore surface ornamented with regular circular depressions, giving it the appearence of a golf-ball.

fox glove. See digitalis.

fragmentation. The segmention of the thallus into a number of fragments each of which is capable of growing into a new individual. A method of asexual reproduction.

free. Said of gills of agarics that reach the stipe, but are not joined to it.

free central. Of a kind of placentation in which the ovules are borne on a central growth from the bottom of the ovary.

free energy. A component of the total energy of a system that can do work under conditions of constant temperature and pressure.

free-living. Living independently of any host organism; non-symbiotic; eleutherozoic.

free floating. Used of aquatic plants that are floating and not anchored to the substratum; pleustonic.

freestone. A fruit stone to which the fruit does not cling, as in certain varieties of peach.

freeze-fracture. Procedure for preparing materials for electron microscopy by rapid freezing and fracturing of the tissues; the exposed fracture faces are used to create a replica that is observed and photographed in the electron microscope; the fracture faces may or may not be further sublimed before the replica is made.

frequency. 1: A measure of how often an event takes place.
2: The number of time a particular class or value of a variable is recorded or observed in a series of sample.

frequency curve. A graphical representation of a continuous frequency distribution; the variate being the abscissa and the frequency being the ordinate.

frequency distribution. An arrangement of data grouped into classes, each with its corresponding frequency of occurrence.

fresh water. Water having a salinity of less than 0.5 parts per thousand, or alternatively, less than 2 parts per thousand.

fresh-water ecosystem. The living organisms and nonliving materials of an inland aquatic environment.

frill. A thin sheet of hyphae forming a horizontal, circular flange around the stipe of an agaric.

frond. 1. The leaf of a fern. Most ferns have pinnate or bipinnate fronds.

2: The leaves of palms are also called fronds.

3: The blade-like thallus of a sea-weed.

4: A lichen thallus.

frost cracks. Cracks in wood that have split outward from ray shakes.

frost ring. A false annual growth ring in the trunk of a tree due to out-of-season defoliation by frost and subsequent regrowth of foliage.

fructoscence. The period of fruit maturation.

fructiafiction. 1: The process of producing fruit.

2: A fruit and its appendages.

3: A general term for the body which develops after fertilization, and contains spores or seeds.

4: Any spore producing structure, whether developed after fertilization or vegetatively, as in fungi an ascocarp, basidiocarp or an ascervulus.

fructivorous. fruit eating; frugivorous.

D-fructopyranose. The pyranose form of fructose. The form in which it exists in solution. See fructose.

fructose. A hexose, ketose, monosaccharide, with formula $C_6H_{12}O_6$. The commonest and sweetest of sugars, found in free state in fruit juices, honey, and nectar of plant glands. With glucose it forms the disaccharide, sucrose. It is alsc known as D-fructo-pyranose; like glucose, fructose is also a reducing sugar.

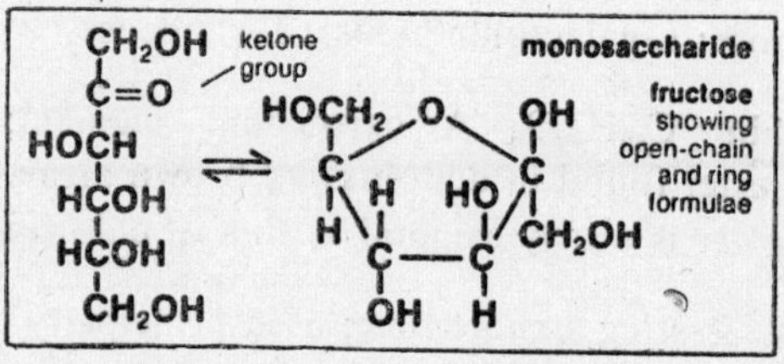

monosaccharide
fructose showing open-chain and ring formulae
ketone group

fruit. The organ of angiosperms containing the seeds. A true fruit is the product of the development of the ovary wall, and the seeds are fertilized ovules. The function of the fruit is to protect the seeds as they develop and to help in their dispersal. The term fruit or *fruiting body* can be used to describe any organ containing propagules in members of the plant kingdom.

fruit bud. A fertilized flower bud that matures into a fruit.

fruiting body. A complex structure that bears fungal spores, e.g., sporangia, coremia, sporodochia, ascervuli, pycnidia, apothecia, perithecia.

fruit sugar. See fructose.

frustule. A diatom shell; capsule of diatoms.

fruticose. 1: Bushy.

2: Said of lichen thallus which is attached by its base, and stands out from the substrate branching and having a bushy appearance; frutescent.

fucoidan. Sulphated water-soluble polysaccharides found in the cells of Phaeophyceae.

fucosan vesicles. Tannin and phenol-containing sac-like bodies founds in the cells of brown algae.

L-fucose. $C_6H_{12}O_5$ A methyl pentose present in brown algae and a number of gums and identified in the polysaccharides of blood groups and certain bacteria; 6-deoxy-L-galactose; L-fucopyranose; L-galactomethylose; L-rhodeose.

fucoxanthin. $C_{40}H_{60}O_6$ A carotenoid pigment; a partial xanthophyll ester found in diatoms and brown algae. The light absorbing power is nearly as good as that of chlorophyll.

fugacious. Lasting a short time; used principally to describe plant parts that fall soon after being formed.

fuliginous. Soot-coloured.

fumaginous. Smoky-coloured.

fumarase. An enzyme that catalyzes the hydration of fumaric acid to malic acid, and the reverse dehydration.

fumigant. Liquid or solid chemical that forms vapors that kill organisms; usually, they are used on soils or within closed structures to eradicate pathogens.

fumigatin. A benzoquinone antibiotic from *Aspergillus fumigatus*. It is active against bacteria, but not fungi.

fumigation. The application of a fumigant for disinfestation of an area.

fungi (sing. fungus). The large group of organisms, sometimes regarded as a separate kingdom which are set apart from other plants because they are heterotrophic, lack chlorophyll, and often have chitin in their cell walls. Most fungi consist of thread-like hyphae, which together form the mycelium although some, like yeast are unicellular. Fungi reproduce by spores. They are important as decomposers in ecosystems and many are parasitic.

fungicide. A chemical used to kill or control fungi.

fungicolous. Living in or on fungi; fungicole.

Fungi Imperfecti. A form class erected to contain those fungi whose sexual reproduction is unknown. Also known as Deuteromycotina. (a group)

fungi on fungi. A number of fungi are active parasites of other fungi. Some associations of fungi with fungi (not all parasitic) are:

'Host'	'Associated fungi'
Myxomycetes	*Nectriopsis violacea*
Mucorales	*Chaetocladium, Piptocephalis*
Erysiphaceae	*Cicinnobolus cesatii*

fungistasis. Inhibition of fungal growth.

fungivorous. Feeding on fungi; mecetophagous; mycophagous; fungivore, fungivory.

funicle. The stalk of the ovule, attaching it to the wall of the ovary. After the ovule is fertilized, the funicle becomes the stalk of the seed.

funicular. Cord-like.

funiculus. A thin cord by means of which the peridioles of some Nidulariales are attached to the basidiocarp which bears them.

furfuraceons. Covered with bran-like particles; scurfy.

6-furfurylaminopurine. A derivative of DNA which causes the acceleration of cell-division in tissue culture (in conjunction with auxin) so was given the name kinetin; also assist in shoot initiation, and retards the senescence of leaves. It may assist in the mobilization of amino acids.

furrowing. A cell division mechanism that involves a pinching-in, or cleavge, to form two daughter cells from the parent cell.

fuscous. Dingy brown.

fusiform. Spindle-shaped; tapering toward the ends.

fusiform initial cell. A cell type of the vascular cambium that gives rise to all cells in the vertical system of secondary xylem and phloem.

fusion. The joining together of two gametes to form a zygote. Fusion can mean the joining of the cells, the joining of the nuclei, or both.

fusion nucleus. The triplod, or 3n, nucleus which results from double fertilization and which produces the endosperm in some seed plants.

fusogenic agent. Chemical agents which are used to induce fusion of protoplasts. e.g., calcium nitrate, sodium nitrate, polyethyleneglycol (PEG); and lectins.

fusoid. Rounded in section; and tapering at each end; not markedly elongate.

G

galactan. Any of a number of polysaccharides composed of galactose units. Hemicelluloses, mucilages, pectins and gums yield galactose on hydrolysis; galactosan.

galactoglucomannan. Any of a group of polysaccharides which are prominent components of coniferous woods; they are soluble in alkali and consist of D-glucopyranose and D-mannopyranose units.

galactomannan. Any of a group of polysaccharides which are composed of D-galactose and D-mannose units, are soluble in water, and form highly viscous solutions; they are plant mucilages existing as reserve carbohydrates in the endosperm of leguminous seeds.

L-galactomethylose. *See* L-fucose.

galactosamine $C_6H_{14}O_5N$. A crystalline amino acid derivative of galactose; found in bacterial cell walls.

galactosan. *See* galactan.

galactose $C_6H_{12}O_6$. A monosaccharide occurring in both levo and dextro forms as a constituent of plant and animal oligosaccharides (lactose and raffinose) and polysaccharides (agar and pectin); cerebrose.

galactosidase. An enzyme that hydrolyzes galactosides.

galactoside. A glycoside formed by the reaction of galactose with an alcohol; yields galactose on hydrolysis.

galacturonic acid. The monobasic acid resulting from oxidation of the primary alcohol group of D-galactose to carboxyl; it is widely distributed as a constituent of pectins and many plant gums and mucilages.

galbalus. A strobilus with fleshy cone-scales.

galea. A helmet-shaped structure forming a part of the calyx and corolla in certain flowers.

galeate. 1: Shaped like a helmet.
2: Having a galea; galeiform.

gall. A tumefaction or tumor; hyperplastic symptom of plant disease characterized by localized swelling or outgrowth of tissue composed of unorganized cells; due to invasion by a parasite, such as fungi or bacteria, following puncture by an insect; insect oviposit and larvae of insects are found in galls.

gallery forest. A narrow strip of forest along the margins of a river in an otherwise unwooded landscape.

gallic acid: $C_6H_2(OH)_3.COOH$. 3, 4, 5-tri hydroxy benzoic acid. It is found in nut galls, tea, etc., and is obtained from tannins by hydrolysis.

gallicolous. Producing or inhabiting galls; gallicole.

galliphagous. Feeding on galls; galliphage, galliphagy.

gallivorous. Feeding on galls; gallivore, gallivory.

gallnut. A gall resembling a nut.

gallotannins. These are polymers containing numerous gallic acid molecules connected in various ways to one another and to sugars. They greatly inhibit plant growth, are commercially used to tan leather; act as allelopathic agents, inhibiting growth of other species around those plants that form and release them; and also act as feeding deterrents against various herbivores because of their astringency.

galvanotropism. A tropism in response to an electrical stimulus.

gametangial contact. A method of sexual reproduction in which two gametangia come in contact but do not fuse. The male nucleus migrates through a pore or fertilization tube into the female gametangium.

gametangial copulation. A method of sexual reproduction in which two gametangia or their protoplasts fuse and give rise to a zygote which develops into a resting spore.

gametangium (*pl.* gametangia). A structure in which gametes are produced, especially of thallophyta.

gamete. Germ cell; a reproductive cell whose haploid nucleus is capable of fusion with that of a gamete of the opposite mating type, to form a diploid zygote. In plants, gametes are produced by the gametophyte.

gametic copulation. The fusion of pairs of differentiated, uninucleate sexual cells or gametes formed in specialized gametangia.

gametophore. A branch that bears gametangia or sex organs.

gametophyte. 1: A phase of the life history of a plant that arises from a haploid spore resulting from meiosis in a diploid sporophyte; plants have haploid nuclei during the gametophyte phase.

2: An individual plant of this generation.

gametropism. Orientation movements of plant structures immediately before or after fertilization; gametropic.

gamma taxonomy. The level of taxonomic study concerned with biological aspects of taxa, including intraspecific populations, speciation, and evolutionary rates and trends.

gamma plantlets. These are plantlets formed from the seeds irradiated with gamma rays from a Cobalt-60 source. These seedlings have giant cells as the gamma rays stop DNA synthesis, mitosis and cell division. The seedlings can survive upto three weeks only.

gamodeme. An isolated breeding community.

gamogastrous. Said of a syncarpous gynoecium in which the ovaries are fused, but the styles and stigmas are free.

gamopetalous. Of flowers in which the corolla is a tube.

gamophase. The haploid phase of a life cycle; haplophase; *cf.* zygophase.

gamophyllous. Having the perianth members fused.

gemosepalous. Of flowers in which the sepals are united at their margins.

gamotropic. Used of flowers that alternate between open and fully closed; *cf.* agamotropic, hemigamotropic.

gamotropism. An orientation response of gametes to one another; the mutual attraction of gametes; gamotropic.

gangliform. Having knots; knotted.

ganja. *See* cannabis.

gasoplankton. Planktonic organisms which make use of gas-filled vesicles or sacs for buoyancy.

gas vacuole. A membrane-bound, gas-filled cavity in the cells of Cyanophyta (Blue green algae) and protozoans. They may contain gas or a viscous substance thought to control buoyancy.

Gause's principle. A statement that two species cannot occupy the same niche simultaneously; competitive exclusion principle.

geitonogamy. Pollination and fertilization of one flower by another on the same plant.

geminate. 1: Growing in pairs or couples.

2: Of two branches from the same node on the same side of the stem.

geminiflorous. Having flowers in pairs.

gemma (*pl.* gemmae). 1: A bud or outgrowth capable of developing into an independent organism.

2: Small green, multicellular, asexual reproductive units produced in cup-shaped structures on the surface of some thalloid liverworts. Gemmae are dispersed by splashes of rain.

3: A thick-walled cell similar to a chlamydospore in fungi.

gemmation. Asexual reproduction in plants by budding off a group of cells (a gemma) which later separates partially or completely from the parent to form a new individual; usually referred to in animals as budding.

gemmiform. Resembling a gemma or bud.

gemmiparous. Producing a bud or reproducing by a bud.

gemmule theory. The theory of heredity, that somatic cells contain particles (gemmules) that are responsible for carrying hereditary traits to the gonads from where they are transmitted to the offspring; *cf.* pangenesis.

gender. The sex of an individual. Gender can be male, female, or neuter.

gene. A section of the DNA that encodes a single polypeptide chain or one or more molecules of tRNA, mRNA, or rRNA, can be regarded as a unit of inheritance.

gene action. The functioning of a gene in determining the phenotype of an individual.

gene cloning. A technique used to produce millions of copies of a particular gene.

genecology. The study of species and their genetic subdivisions, their place in nature, and the genetic and ecological factors controlling speciation.

gene conversion. A situation in which gametocytes of an individual that is heterozygous for a pair of alleles undergo meiosis, and the gametes produced are in a 3:1 ratio rather than the expected 2:2 ratio, implying that one allele was converted io the other.

gene flow. The passage and establishment of genes characteristic of a breeding population into the gene complex of another population through hybridization and backcrossing.

gene frequency. The number of occurrences of a specific gene within a population in relation to all its alleles at the same locus.

gene library. A random collection of cloned deoxyribonucleic acid fragments in a vector; includes all the genetic information of the species.

gene pool. All the different genes present in a specific population at a given time.

generation. A group of organisms having a common parent or parents and comprising a single level in line of descent.

generation time. The time interval required for a bacterial cell to divide or for the population to double.

gene redundancy. The presence of many copies of one gene within a cell.

generic. Pertaining to or having the rank of a biological genus.

gene suppression. The development of a normal phenotype in a mutant individual or cell due to a second mutation either in the same gene or in a different gene.

genetic code. The genetic information in the nucleotide sequences in deoxyribonucleic acid represented by a four-letter alphabet that makes up a vocabulary of 64 three-nucleotide sequences, or codons; a sequence of such codons (averaging about 100 codons) constructs a message for a polypeptide chain.

genetic drift. The random fluctuation of gene frequencies from generation to generation that occurs in small populations.

genetic engineering. The intentional production of new genes and alteration of genetic constitution of individual by the substitution or addition of new genetic material.

genetic fingerprinting. Identification of chemical entities in animal tissues as indicative of the presence of specific genes.

genetic homeostasis. The tendency of Mendelian populations to maintain a constant genetic composition.

genetic load. The abnormalities, deformities, and deaths produced in every generation by defective genetic material carried in the gene pool of the human race.

genetic map. A graphic presentation of the linear arrangement of genes on a chromosome; gene positions are determined by percentages of recombination in linkage experiments; chromosome map.

genetic material. The ultramicroscopic particles or genes, first defined by H.J. Muller, the influences of which permeate the cell and play a fundamental role in determining the nature of all cell substances, cell structures, and cell effects; the genes have properties of self-propagation and variation.

genetics. The science that is concerned with the study of biological inheritance, and the control of the characteristics of an organism by its genes.

genetic variation. A variation inherited by an organism and caused by a change in the genes.

genome. The genetic material on the sets of chromosomes in a cell. The smallest genome consists of all the genes on a haploid set of chromosomes. A diploid cell, with two sets of chromosomes, is said to have a diploid genome.

genotype. 1: The genetic constitution of an organism, usually in respect to one gene or few genes relevant in a particular context.

2: The type species of a genus.

gentamicin. A broad-spectrum antibiotic produced by a species of *Micromonospora*.

gentianose. A trisaccharide, consisting of gentiobiose, and fructose, or sucrose and glucose. It is found in gentian roots.

gentiobiose. A disaccharide reducing sugar with a 1:6 linkage. 6-D glucose-1 β-D glucopyranoside. Gentiobiose is found in the form of its glycosides; amygdalin of almonds and crocin of *Crocus*; and is also present in the trisaccharide gentianose.

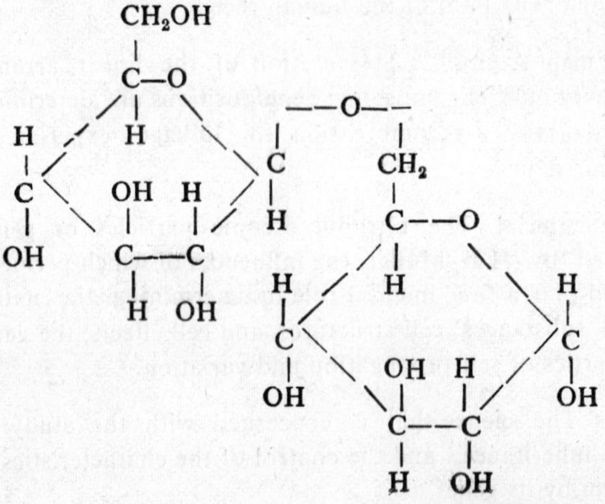

```
             CH2OH
               |
              C—O
             /|  \           ——O——
    H     /  H    \        |      |
    |   /           \   C       CH2
    C \      OH  H    / |        |
    |   \    |   | /    H       C——O
   OH       C———C               /|    \
            |   |         H   /  H      \   H
            H  OH         | /             \ |
                          C \              /C
                          |   \   OH  H  /  |
                         OH     \ |   | /   OH
                                 C———C
                                 |   |
                                 H   OH
```

genus. A taxonomic category that includes groups of closely related species; the principal subdivision of a family. The name of the genus is the first in a Latin binomial genera (pl.).

geobotany. Plant biogeography; the study of plants in relation to geography and ecology; phytogeography.

geocarpy. Ripening of fruits underground. The young fruits are pushed into the soil by a post-fertilization curvature of the stalk.

geographic speciation. Evolution of two or more species from a single species following geographic isolation.

geological epoch. A subdivision of a geological period.

geological era. A very long division of geological time, lasting tens of millions of years, whose beginning and end are recognized by major changes in layers of rocks and fossils in the earth. Geological eras are divided into geological periods and epochs.

geological period. A major subdivision of a geological era.

geological time. The sequence of geological eras in the earth's history, measured in millions of years. Geological time began about 4½ thousand million years ago, when the earth was formed.

geophilous. Living or growing in or on the ground.

geophyte. A perennial plant with buds or shoot apices perennating underground. Geophytes have rhizomes bulbs or corms, etc., geophytic; geocryptophyte.

geoplagiotropic. Growing in a direction at an angle to the ground surface.

geotaxis. Movement of a free-living organism in response to the stimulus of gravity.

geotropism. Curving growth of a plant organ due to gravity. Geotropism can be downwards (positive), e.g., in a taproot, or upwards (negative), e.g., in the shoot of a seedling; geotropic.

GERL. Golgi-associated endoplasmic reticulum involved in production of lysosomes.

germination. The first stage in the growth of a seed into a seedling, or a spore into a young plant. In seed plants germination begins with the imbibition of water and ends with the production of the first true leaves.

germ plasm theory. That there are two types of cells, germ plasm and somatic cells, and that germ plasm is potentially immortal through transmission from generation to generation; germ line theory.

germ tube. The hypha produced by a fungal spore when it begins to grow.

gerontology. The study of senescence, the processes and effects of ageing.

gibberellic acid $C_{18}H_{22}O_6$. A crystalline acid occurring in plants that is similar to the gibberellins in its growth-promoting effects.

gibberellins. Group of chemically complex plant hormones which have a gibbane skeleton, and were first isolated from the fungus *Gibberella fujikuroi*, important in the control of tropisms, the lengthening of cells during growth, germination and other processes.

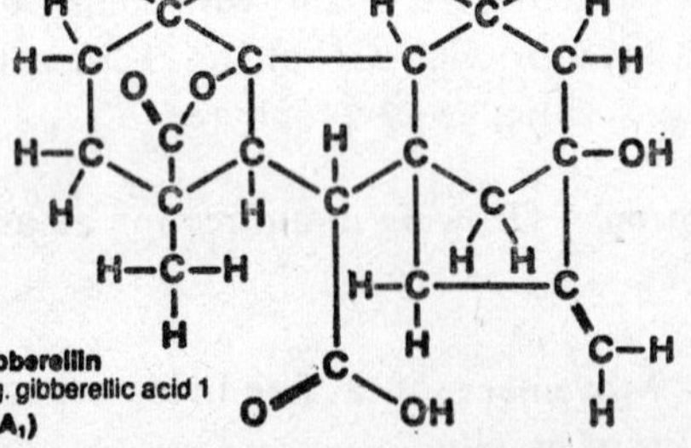

gibberellin
e.g. gibberellic acid 1
(GA_1)

Gibbs effect. The asymmetrical distribution of radiocarbon in glucose formed during photosynthesis is known as Gibbs effect. It suggests that the two halves of glucose are derived from different pools of triose and that fructose is not the precursor of glucose.

Gibbs-free energy. It is a measure of the maximum energy available for conversion to work at constant temperature and pressure.

gill. A flat, vertically-positioned structure on the underside of the cap of a mushroom or toadstool. The cap has many gills, radiating from the centre. The gills bear basidia on their surfaces.

gill fungi. Mushrooms; members of the family Agaricaceae of the class Basidiomycetes, which produce fruiting bodies consisting of a stipe that supports a pileus whose lower surface has radially arranged lamellae bearing the hymenium.

Ginkgoales. An order of gymnosperms with only one living species, *Ginkgo biloba* (maidenhair tree), which is found in China. They have motile male gametes, like cycads, but deciduous leaves.

girdle scars. A ring of scars on a twig where scale leaves of previous year's buds were attached.

glabrescent. 1: Very thinly covered with hairs.
2: Becoming nearly hairless as it matures.

glabrous. Having a smooth surface, specifically, having the epidermis devoid of hair.

gladiate. Sword-shaped.

gland. A group of cells on the surface of a plant, whose function is to secrete or excrete substances; glandular.

glareous. Growing in gravelly soil, refers specifically to plants.

glaucous. 1: Having a white or grayish powdery coating that gives a frosty appearance and rubs off easily.
2: Sea-green.

gleba. The inner, fertile portion of the fruiting body of the Gasteromycetes and Tuberales an order of Discomycetes.

gliding motility. A means of bacterial self-populsion by slow gliding or creeping movements on the surface of a substrate.

gliotoxin $C_{13}H_{14}O_4N_2S_2$. A heat-labile, antibacterial and antifungal antibiotic produced by species of *Trichoderma* and *Gliocladium* and by *Aspergillus fumigatus*; has been used as a seed dressing.

globular protein. Any protein that is readily soluble in aqueous solvents.

globule. Male reproductive organ of Charales.

globulin. A water insoluble protein which is soluble in dilute salt solution and coagulates when heated. Globulins are found in animal and plant tissue, e.g., fibrinogen and legumin.

glochidium. A barbed hair. These are produced by *Azolla* on the masses of microsporangia. They serve for attachment to the macrosporangium; glochid.

glomerule. 1: A cluster of branches of limited growth at a node in *Batrachospermum*.
2: A cluster of short-stalked flowers.

glomerulus. A cymose inflorescence in the form of a crowded head of small flowers.

glucan. A polysaccharide composed of the hexose sugar D-glucose. This yield only glucose on hydrolysis e.g., starch, dextrin, cellulose; glucosans.

glucokinase. An enzyme that catalyzes the phosphorylation of D-glucose to glucose-6-phosphate.

glucolipid. A glycolipid that yields glucose on hydrolysis.

glucomannan. A polysaccharide composed of D-glucose and D-mannose; a prominent component of coniferous trees.

glucosamine $C_6H_{13}O_5$. An amino sugar; the most abundant in nature, occurring in glycoproteins and chitin; may be obtained by hydrolysis of chitin. Glucosamine is one of the constituents of a molecule of streptomycin.

glucose. A hexose, aldose monosaccharide, with formula $C_6H_{12}O_6$; occurs free or combined and is the most common sugar. Glucose is the unit in polysaccharides such as starch and cellulose. It is a product of photosynthesis. With fructose, it forms the disaccharide sucrose; dextrose; grape sugar.

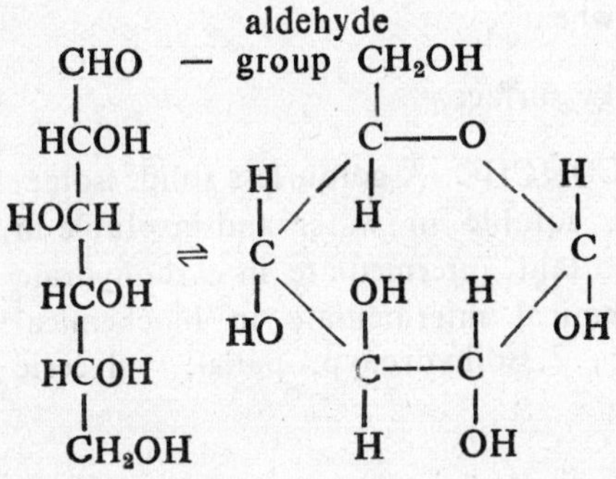

```
                 aldehyde
 CHO     —    group  CH2OH
  |                    |
 HCOH                  C——O                  glucose
  |          H       / |      \      H       showing
HOCH         |  /      H        \    |     open-chain
  |     ⇌    C                       C       and ring
 HCOH        |  \     OH    H    /   |       formulae
  |         HO    \   |     |  /    OH
 HCOH                 C———C
  |                   |     |
 CH2OH                H     OH
```

glucuronic acid $C_6H_{10}O_7$. An acid resulting from oxidation of the CH_2OH radical of D-glucose to COOH; a component of many polysaccharides and certain vegetable gums; glycuronic. acid.

glumaceous. Thin, brown and papery in texture.

glumes. The pair of outer bracts at the base of a grass spikelet.

glutaric acid $C_5H_5O_4$. A water-soluble, crystalline acid that occurs in green sugar-beets and in water extracts of crude wool.

glutathione $C_{10}H_{17}O_6N_3S$. A widely distributed tripeptide formed from the union of glutamic acid; cystine and glycine; that is important in plant and animal tissue oxidation reactions.

glutelin. A class of simple, heat-labile proteins occurring in seeds of cereals; soluble in dilute acids and alkalies; but insoluble in water, dilute salt solution, and 70 per cent ethanol.

gluten. 1: A mixture of proteins gliadin and glutenin found in the seeds of cereals; gives dough elasticity and cohesiveness.

2: The sticky coating on the pilei of some agarics.

glutenin. A glutelin of wheat.

glutinous. Having a sticky surface.

glyceraldehyde $CH_2OHCOHCHO$. A colourless solid, isomeric with dihydroxyacetone; soluble in water and insoluble in organic solvents; an important intermediate in carbohydrate metabolism; used as a chemical intermediate in biochemical research and nutrition; 2,3-dihydroxypropanal; glyceric aldehyde.

glyceride. An ester of glycerol and one or more organic acids (saturated and unsaturated). Fats are mainly composed of triglycerides of fatty acid, i.e., compounds in which the three hydroxyl groups in glycerol have reacted with three similar or different fatty acids.

glycerol. A trihydric alcohol with the formula $CH_2OHCHOHCH_2OH$. A compound which combines with fatty acids to form lipids.

glycogen. A nonreducing, white, amorphous polysaccharide built up of many units of glucose found as a reserve carbohydrate stored in muscles and liver cells of all higher animals, in cells of lower animals; and in fungi. It is known as animal starch.

glycogenesis. The metabolic formation of glycogen from glucose.

glycogenolysis. The metabolic breakdown of glycogen.

glycogen synthetase. An enzyme that catalyzes the synthesis of the amylose chain of glycogen.

glycolate pathway. A sequence of reactions that takes place in leaf peroxisomes in conjunction with the carbon cycle of chloroplasts. In these reactions glycolate is converted to 3-PGA via glyoxylate, glycine, serine, and other 3-carbon acids. The 3-PGA is then converted to sucrose and starch in chloroplasts.

glycolipid. Any of a class of complex lipids which contain carbohydrate residues. It is one of the principal lipid of most plant membranes, mainly chloroplast membranes.

glycolysis. The enzymatic breakdown of glucose or other carbohydrate, with the formation of lactic acid or pyruvic acid and the release of energy in the form of adenosinetriphosphate.

glycolytic pathway. The principal series of phosphorylative reactions involved in pyruvic acid production in phosphorylative fermentations; Embden-Meyerhof pathway; hexose diphosphate pathway.

glyconeogenesis. The metabolic process of glycogen formation from noncarbohydrate precursors.

glycopeptide. *See* glycoprotein.

glycophyte. A plant thriving in a soil of low salt concentration, typically less than 0.5 per cent sodium chloride *cf.* halophyte.

glycoprotein. Any of a class of conjugated proteins containing both carbohydrate and protein units; glycopeptide.

glycoside. An organic compound consisting of a sugar molecule bonded to another organic molecule by a glycosidic bond.

glycosidic bond. The chemical bond between the sugar monomers in a disaccharide or polysaccharide. The formation of the glycosidic bond is due to the reaction of the–OH group on the first carbon atom of one sugar molecule with any–OH group of another sugar molecule; H_2O is produced, and the sugars become linked by an oxygen atom.

glyoxylate cycle. A sequence of biochemical reactions related to respiration in germinating fatty seeds by which acetyl coenzyme A is converted to malic acid and succinic acid and then to hexose. These reactions take place in microbodies, termed glyoxysomes.

glyoxylic acid $CH(OH)_2COOH$. An aldehyde acid found in many plant and animal tissues, especially unripe fruit.

glyoxysome. A class of microbody found in plant endosperm that contains enzymes of the glyoxylate cycle in addition to catalase and oxidases. Two unique glyoxysome enzymes not known to be present in animals are; isocitrate lyase and malate synthetase.

Gnetales. A small order of gymnosperms made up of three genera, *Ephedra*, *Gnetum*, and *Welwitschia*. Gnetales are similar to angiosperms, because their wood contains vessels and their ovules do not have archegonia.

gnotobiology. That branch of biology dealing with known living forms; the study of higher organisms in the absence of all demonstrable, viable microorganisms except those known to be present.

gnotobiote. 1: An individual (host) living in intimate association with another known species (microorganism).
2: The known microorganism living on a host.

Golgi body. An organelle consisting of a group of membranes and vesicles. The Golgi body is often important in the synthesis of carbohydrates and the secretion of substances, especially glycoproteins, from the cell. In plants, it is usually called a dictyosome.

gonidium. 1: A non-motile reproductive cell of many Volvocales, e.g., *Volvox*.
2: An algal cell in a lichen thallus.

gonimoblast filaments. The filaments developed from the carpogonium in many red algae after fertilization. Each bears a terminal carposporangium.

gonophore. An elongation of the receptacle extending between the stamens and corolla.

gossypine. Cottony.

grossypol. A pigment found in cotton seeds.

graded topocline. A topocline having a wide range, or ranging into different kinds of environment, thus subjecting its members to differential selection so that divergence between local races may become sufficient to warrant creation of varietal, or even specific, names.

gradient. An increase or decrease of a measurable quantity over a given distance, e.g., a chemical gradient, the increase in concentration of a solute from one part of a plant to another, environmental gradient, the decrease in temperature with increasing height on a mountain.

graft. To join together artificially parts from two different plants, e.g., the shoot of one variety of a species onto the rootstock of another variety.

grain. 1: A rounded, granular prominence on the back of a sepal.
2: An individual drupe of an aggregate fruit.
3: Any member of the grass family which produces edible, starchy grains usable as food by man and his livestock.

graminoid. Of or resembling the grasses.

gram-negative. Of bacteria, decolorizing and staining with the counterstain when treated with Gram's stain.

gram-positive. Of bacteria holding the colour of the primary stain when treated with Gram's stain.

Gram's stain. A differential bacteriological stain; a fixed smear is stained with a slightly alkaline solution of basic dye e.g., crystal violet, treated with a solution of iodine in potassium iodide, and then with a neutral decolorizing agent, e.g., a mixture of propanone and ethanol and usually counterstained, e.g., safranin; bacteria stain either blue (gram-positive) or red gram-negative).

gram-variable. Pertaining to staining inconsistently with Gram's stain. Young cultures are Gram positive, but become negative as they age.

grana. *sing.* granum The stacks of flat vesicles or thylakoids in the chloroplast, where the pigments and enzymes of the light in reaction of photosynthesis are found.

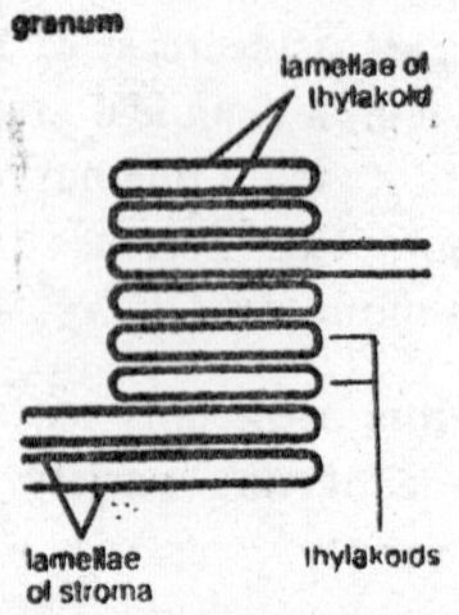

grape sugar. See glucose.

grassland. Vegetation in which grasses are dominant, e.g., prairies, savanna.

gravitational water. Water which drains by gravity through the soil and which is readily available to soil organisms and plants.

green algae. Algae of the division Chlorophyta. Green algae have chlorophyll b, produce starch, and have cellulose in their cell walls.

greenhouse. Glass-enclosed, climate-controlled structure in which young or out-of-season plants are cultivated and protected.

greenhouse effect. The effect experienced inside a green-house: infra-red radiation of short-wave-length from the Sun passes through glass and is absorbed by the contents which in turn radiate infra-red rays with a longer wavelength because of their relatively low temperature. These rays cannot pass through the glass so the temperature rises inside the glasshouse. The same effect is seen when short-wave infra-red rays from the Sun pass through a-layer-of CO_2 in the earth's atmosphere, whereas long wave infra-red rays radiated back from the Earth do not, i e., the solar energy is trapped by the Earth's atmosphere. It is possible that the increasing concentra-tion of CO_2 in the Earth's atmosphere may cause a permanant temperature rise. This could lead to catastrophes such as melting of polar ice caps, rise in ocean level, warming of Northern areas, changing of wind pattern, flooding of coastal areas, frequent occurrences of typhoones and cyclones.

green islands. These are green and starch rich areas formed on leaves around the necrotic areas when certain parasitic fungi (like rust fungi) infect leaves. These areas remain green even when the rest of the leaf has become yellow and senescent —thus known as green-islands. These are rich in cytokinins, probably synthesized by the fungus; cytokinins help maintain food reserves for the fungus and influence further progress of the disease symptoms.

green smut. A fungus disease of rice characterized by enlarged grains covered with a green powder consisting of conidia, and caused by *Ustilaginoidea virens*; false smut.

green sulfur bacteria. A physiologic group of green photosynthetic bacteria that are capable of using H_2S and other inorganic electron donors.

griseofulvin $C_{17}H_{17}O_6Cl$. A colourless, crystalline antifungal antibiotic produced by *Penicillium griseofulvum.*

gross primary production (GPP). The total assimilation of organic matter by an autotrophic individual, population or trophic unit, per unit time per unit area or volume; gross primary productivity; *cf.* nett primary production.

gross production rate. The speed of assimilation of organisms belonging to a specific trophic level.

ground cover. Prostrate or low plants that cover the ground instead of grass.

ground water. Water naturally contained in, and saturating the sub-soil.

growth. Increase in the quantity of metabolically active protoplasm, accompanied by an increase in cell number or cell size, or both.

growth curve. A graphic representation of the growth of a bacterial population in which the log of the number of bacteria or the actual number of bacteria is plotted against time.

growth factor. Any factor, genetic or extrinsic, which affects growth.

growth form. The habit of a plant determined by its appearance under a particular set of environmental conditions; life form.

growth hormone. Any hormone that regulates growth in plants; plant growth regulators.

growth rate. Increase in the number of bacteria in a population per unit time.

growth ring. A cylinder of secondary wood laid down in a single season in a stem or root. It is seen as a circle in transverse section and is not necessarily the same as an annual ring.

growth water. That portion of the total soil water available to plants; chresard.

growth regulator. A synthetic substance that produces the effect of a naturally occurring hormone in stimulating plant growth; an example is dichlorophenoxyacetic acid.

GTP. Guanosine triphosphate. A nucleotide similar to ATP, involved in the reactions of the Krebs cycle.

guanine $C_5H_5ON_5$. A purine base; occurs naturally as a fundamental component of nucleic acids and pairs with cytosine.

guanosine $C_{10}H_{13}O_5N_5$. Guanine riboside, a nucleoside composed of guanine and ribose; vernine.

guanosine monophophosphate. *See* guanylic acid.

guanosine phosphoric acid. *See* guanylic acid.

guanosine tetraphosphate. A nucleotide which participates in the regulation of gene transcription in bacteria by turning off the synthesis of ribosomal ribonucleic acid.

guanylic acid. GMP, A nucleotide composed of guanine, a pentose sugar, and phosphoric acid and formed during the hydrolysis of nucleic acid; guanosine monophosphate; guanosine phosphoric acid.

guard cell. Either of two specialized cells surrounding each stoma in the epidermis of plants; functions in regulating stoma size.

guayule. *Parthenium argentatum.* A subshrub of the family Compositae that is native to Mexico and the southwestern United States; it has been cultivated as a source of rubber.

guide cells. A row of enlarged cells that extends across a costa as seen in transverse section, continuous with the cells of the unistratose lamina of the leaf.

gullying. Erosion of soil by running water, when it is confined to narrow channels.

gum arabic. A water-soluble powder obtained from certain varieties of *Acacia.* It is used in pharmacy and in making adhesives.

gumivorous. Feeding on gum and other exudates from trees; gumivore, gumivory.

gummosis. Production of gummy exudates in diseased plants as a result of cell degeneration.

gums. These are colloidal plant products, which either dissolve or swell in water. On hydrolysis they give complex organic acids pentoses, and hexoses.

gum vein. Local accumulation of resin occurring as a wide streak in certain hardwoods.

gun-cotton. A highly flammable material made by treating cellulose with a mixture of conc. nitric acid and conc. sulphuric acid. It is used as an explosive and in lacquers. Sometimes gun-cotton which is cellulose trinitrate is inaccurately called as nitro-cellulose; however, it is not a nitro-compound but an ester.

gut flora. The microorganisms found in the gut of animals. The gut flora assists the animal in breaking down its food.

gutta-percha. A natural product obtained from the sap of certain tropical plants. Gutta percha has the same constitution as natural rubber but is a trans-isomer, whereas natural rubber is a cis-isomer. Gutta-percha is a tough, horn-like substance.

guttation. The process of exuding sap or water through hydathodes.

gymnocarpous. Having the hymenium uncovered on the surface of the thallus or fruiting body of lichens or fungi.

gymnogynous. Having a naked ovary.

gymnosperm. A spermatophyte of the subdivision Gymnospermae. Gymnosperms differ from angiosperms in their unprotected ovules, arrangement of reproductive organs in cones, archegonia and wood without vessels.

gynostemium. The column composed of the united gynoecia and androecium.

gypsophilous. Flourishing on a gypsum-rich substratum.

gynaecandrous. Having staminate and pistillate flowers on the same spike.

gynandrosporous. Species of *Oedogonium* in which oogonia and androsporangia develop on the same filament.

gynobase. A gynoecium-bearing elongation of the receptacle in certain plants

gynochoric. Pertaining to organisms dispersed by motile females; gynochore, gynochorous.

gynodioecious. Used of plants or plant species having female (pistillate) and hermaphrodite (perfect) flowers on separate plants in a population or species; gynodioecy.

gynoecium. The female part of a flower, consisting of one or more pistils.

gynomonoecious. Having female (pistillate) and hermaphrodite (perfect) flowers on the same plant; gynomonoecy.

gynomorphic. Having a morphological resemblance to females; gynomorphy.

gynopaedium. A family group in which the female parent remains with the offspring for some time.

gynophore. 1: A stalk that bears the gynoecium.

2: An elongation of the receptacle between pistil and stamens.

gynophyte. A female plant.

gynopleogamy. The condition of a plant species having three sexual forms, with either perfect flowers, pistillate flowers or staminate flowers.

gypsophilous. Thriving on chalk or gypsum-rich soils; gypsophile, gypsophily.

gypsophyte. A plant inhabiting chalk or gypsum-rich soils; gypsum plant.

gypsophytium. A plant community of limestone substrata.

gyrate. 1: Circinate.

2: Curved into a circle.

gyrogonite. A minute, ovoid body that is the residue of the calcareous encrustatior about the female sex organs of a fossil stonewort.

gyrose. Having a folded surface, marked with sinuous lines or ridges.

H

H-bodies. Pairs of sporidia of *Tilletia*, a member of Teliomycetes, fused in pairs while still attached to the promycelium.

habit. The external appearance, aspect or growth form of an organism; constitutional type; habitus.

habitat. The locality, site and particular type of local environment occupied by an organism, e.g., the habitat of an epiphyte is in the branches of trees, the habitat of algae is water; ece; local environment; oike; oikos.

habitat form. The characteristic growth form of an organism under a given set of environmental conditions; ecad.

hadromycosis. Any plant disease resulting from infestation of the xylem by a fungus.

haem. A porphyrin ring, with an atom of iron in its centre, e.g., in cytochromes.

Haldane's rule. The rule that if one sex in a first generation of hybrids is rare, absent, or sterile, then it is the heterogametic sex.

half-life. 1: The survival time of half the individual components of an unstable system.

2: A measure of radioactive decay; the time taken for the level of radioactivity to reduce by one-half.

3; The time taken for an individual or biological system to eliminate one-half of a given substance introduced into it.

hallucinogenic mushrooms. Mushrooms which contain a drug, so that when eaten they cause visions in colour and motion, e.g., *Amanita muscaria* and *Psilocybe mexicana*. Hallucinations seems to be due primarily because of muscimol, a cyclic acid together with muscazone and ibotenic acid two cyclic amino acids (from *A. muscaria*) and psilocybin, psilocin (from *Psilocybe* sp.)

halobiont. A marine organism or an organism living in a saline habitat; halobion, halobiontic.

halonate. 1: Of a leaf spot, having concentric rings; one of the 'frog-eye' type.
2: Pertaining to a spore, having a transparent coat around it.

halophillism. The phenomenon of demand for high salt concentrations for growth and maintenance.

halophilous. Thriving in saline habitats; halophile. halophily *cf.* halophobic.

halophobic. Intolerant of saline habitats; halophobe *cf.* halophilous.

halophreatophyte. A plant utilizing saline ground water.

halophyte. A plant living in saline conditions; a plant tolerating or thriving in an alkaline soil rich in sodium and calcium salts; a seashore plant; drymyphyte *cf.* glycophyte.

haloplankton. Marine or inland saltwater plankton; haliplankton.

halosere. An ecological succession commencing in a saline habitat; halarch succession.

hamate. Hook-shaped or hooked; uncinate.

hamus. A hook or a curved process.

handbook. A field guide or identification key not comprising taxonomic conclusions or nomenclatural data.

hanging-drop preparation. A techique used in microscopy in which a specimen is placed in a drop of a suitable fluid on a cover slip and the cover slip is inverted over a concavity on a slide.

hapaxanthic. Having a single flowering phase during the life cycle; hapanthous, hapaxanthous.

haplo. Prefix meaning simple, single.

haplobiont. 1: A plant flowering once per season.

2: An organism not exhibiting a regular alternation of haploid and diploid generations during the life cycle; *cf.* diplobiont.

haplocaulescent. Having a simple axis with the reproductive organs on the principal axis.

haplodioecious. Heterothallic.

haplo-heteroecious. Heterothallic

haploid. Of cells with a single set of chromosomes in their nuclei.

haplodiploidy. The genetic system found in some animals in which males develop from unfertilized eggs and are haploid, and females develop from fertilized eggs and are diploid; haplodiploid, haplodiplont.

haplolepidous. Describing jointed peristome teeth in which the outer face of each joint is composed of a single face of one cell.

haplomitosis. Type of primitive mitosis in which the nuclear granules form into threadlike masses rather than clearly differentiated chromosomes.

haplo-monoecious. Homothallic.

haplont. 1: The haploid phase of a life cycle; haplophase.

2: An organism having a life cycle in which meiosis occurs in the zygote to produce the haploid phase; only the zygote of haplonts is diploid; *cf.* diplont.

haplophase. The haploid stage of a life cycle; gamophase; haplont; *cf.* diplophase.

haplophyte. A haploid plant; gametophyte; *cf.* diplophyte.

haplosis. Reduction of the chromosome number to half during meiosis.

haplo-synoecious. Homothallic.

hapteron. 1: A disklike holdfast on the stem of certain algae, e.g., *Oedogonium*, *Fucus*.

2: An aerial organ of attachment of some fruticose lichens formed by a secondary branch which becomes attached to the substrate.

3: Attachment organ at the base of a funicular cord in Nidulariaceae.

haptotropism. Movement of sessile organisms in response to contact, especially in plants.

hardening. The process of acclimating plants to drought by being exposed to low water potentials, high light levels, high phosphorus and low-nitrogen fertilizers, so that they become drought tolerant or hardy compared to plants of the same species not treated in this way.

hard fiber. A heavily lignified leaf fiber used in making cordage, twine, and textiles.

hardwood forest. 1: An ecosystem having deciduous trees as the dominant form of vegetation.
2: An ecosystem consisting principally of trees that yield hardwood.

hartig net. The inter cellular hyphal network formed by a mycorrhizal fungus on the surface of a root. See also ectomycorrhiza.

harvest index. It is a ratio of the harvest yield to shoot yield.

hastate. Shaped like an arrowhead with divergent barbs.

Hatch-Slack pathway. A metabolic cycle involved in the non-light-requiring phase of photosynthesis in certain plants having specific metabolic and anatomical modifications in their mesophyll and bundle sheath cells which facilitate the temporary fixation of carbon dioxide (CO_2) into four-carbon organic acid; these acids are next broken down to three-carbon organic acids plus CO_2 in bundle sheath cells, where this freed CO_2 is then fixed into carbohydrates in a normal Calvin cycle pathway.

haustorium (*pl.* haustoria) 1: Absorbing organ of certain parasites. The haustoria of plant-parasitic fungi enter plant cells and invaginate the cytoplast as they enlarge into simple or branched structures.
2: The absoptive structure of a sporophyte that penetrates the gametophore and transfers nutrients and water from it to the sporophyte.

heartwood. The wood in the centre of a trunk or branch. Heartwood is usually densc and compact, and it helps to support the tree. It is often darker than the outer wood, and is unable to conduct sap.

heather rags. Common name for the lichen *Hypogymnia physodes.*

Helferich unit. A unit for the standardization of phosphatase.

helicoid cyme. A type of determinate inflorescence having a coiled cluster, with flowers on only one side of the axis.

helicosporae. A spore group of the Fungi Imperfecti characterized by spirally coiled, septate spores.

helicospore. A rolled up, cork-screw-shaped cylindrical spore, generally septate, and is found in members of Deuteromycotina, e.g., *Helicoon, Helicomyces.*

heliophyte. A plant that shows optimum growth under conditions of full sunlight; a species requiring full sunlight is sometimed called an obligate heliophyte and one that can tolerate full sunlight but grows best in shade is a facultative heliophyte; heliad; oread.

heliotaxis. Orientation movement of an organism in response to the stimulus of sunlight.

heliotrope. A plant whose flower or stem turn toward the sun.

heliotropism. Growth or orientation movement of a sessile organism or part, such as a plant, in response to the stimulus of sunlight.

helix. A thread or line coiled like a screw. Molecules of DNA have this shape with two helices coiled together; double helix.

helminthosporal. This is a naturally occurring growth promoting substance which possess gibberellin-like biological activity. It is isolated from the fungus *Helminthosporium sativum.*

helminthosporin $C_{15}H_{10}O_5$. A maroon, crystalline pigment formed by certain fungi growing on a sugar substrate.

helophyte. 1: A perennial plant with renewal buds, commonly on rhizomes, buried in soil or mud below water level; limnocryptophyte; *cf.* Raunkiaerian lifeforms.

2: Any marsh or bog plant; helad; helodad; limnophyte.

heloplankton. The floating vegetation of a marsh.

helotism. Symbiosis in which one organism is a slave to the other, as between certain species of ants; and in a fungus and algal association in a lichen.

helper virus. A virus that, by its infection of a cell, enables a defective virus to multiply by supplying one or more functions that the defective virus lacks.

hematochrome. A red pigment occurring in green algae, especially when plants are exposed to intense light on subaerial habitats.

hematoxylon. The heartwood of *Haematoxylon campechianum*; logwood.

hemerophyte. A cultivated plant.

hemi. Prefix meaning half.

hemiautoploid. A polyploid derived from intraspecific hybrids or by the differentiation of the chromosome sets of successful panautoploids.

hemicellulose. $(C_6H_{10}O_5)2n$. A type of structural polysaccharide found in plant cell walls in association with cellulose and lignin; it is soluble in and extractable by dilute alkaline solutions. The commonest sugars forming hemicelluloses are xylose, arabinose, galactose and mannose. These hemicelluloses are named from the monosaccharides which form them, like xylans and arabans (pentosans), and galactans and mannans (hexosans). Hemicelluloses vary greatly in constitution from source to source.

hemicleistogamic. Used of a plant having flowers that only open partially before pollination.

hemicryptophyte. A perennial plant with renewal buds at ground level or within the surface layer of soil; typically exhibiting degeneration of vegetative shoots to ground level at the onset of the unfavourable season; *cf.* Raunkiaerian life forms.

hemicydic. Of flowers, having the floral leaves arranged partly in whorls and partly in spirals.

hemiendophytic. Used of a fungus that occurs both external and internal to its host at different times.

hemiepiphyte. A plant that spends only part of its life cycle as an epiphyte, producing both aerial and subterranean roots at different times.

hemigamotropic. Used of flowers that alternate between open and partially closed; *cf.* agamotropic, gamotropic.

hemigamy. The activation of an ovum by a male nucleus without fusion; semigamy; hemigamous.

hemikaryotic. Used of a cell with the haploid number of chromosomes; hemikaryon.

hemiparasite. 1: A green plant whose roots grow into the tissues of another plant. Hemiparasites photosynthesize; but take some of their nutrients and water from the other plant, e.g., mistletoe; semiparasites.

2: A partial or facultative parasite that can survive in the absence of a host; meroparasite.

3: A parasitic plant that develops in the soil from free seeds; *cf.* holoparasite.

hemisaprophyte. A plant that may be autotrophic or saprotrophic at different times.

hemitropic Used of flowers adapted for pollination by particular insect species; also used of the insects that pollinate specifically adapted flowers.

henbane. *Hyoscyamus niger.* A poisonous herb containing the toxic alkaloids hyoscyamine.

henna. *Lawsonia inermis.* An Old World plant having small opposite leaves and axillary panicles of white flowers; a reddish-brown dye extracted from the leaves is used in hair dyes; Egyptian henna.

heptose. Any member of the group of monosaccharides containing seven carbon atoms.

heptulose. The generic term for a ketose formed from a seven-carbon monosaccharide.

herb. A plant having stems that are not secondarily thickened and lignified (non-woody) and which die down annually.

herbaceous. 1: Resembling or pertaining to a herb.

2: Pertaining to a stem with little or no woody tissue.

herbaceous dicotyledon. A type of dicotyledon in which the primary vascular cylinder forms an ectophloic siphonostele with widely separated vascular stands.

herbarium. 1: A collection of preserved (usually dried) plant specimens; *hortus siccus.*

2: The building in which such a collection is kept.

herbicide. A chemical used to kill weeds or herbage e.g., 2, 4 dichlorophenoxy acetic acid (2, 4-D), 2, 4, 5-trichlorophenoxy acetic acid (2, 4, 5-T) 2-methyl 4-chlorophenoxyacetic acid (MCPA). These are selective weed killers, used to kill broad leaf dicot weeds in cereal grains and lawns.

herbicolous. Living predominantly in herbaceous habitats; herbicole.

herbivorous. Feeding on plants; phytophagous; herbivore, herbivory.

herbosa. Communities of grasses and herbs; herbaceous vegetation.

hercogamous. Used of a flower having the stamens and stigma positioned in such a way as to prevent self-pollination hercogamy, herkogamous, herkogamy.

heredity. The passing of characters from one generation to the next; hereditary.

hermaphrodite. Having both male and female reproductive organs in the same individual (animal) or the same flower (plant); androgyne; bisexual; the maturation of the male organs before the female is protandrous hermaphroditism, the female before the male is protogynous hermaphroditism, and simultaneously is synchronous hermaphroditism.

hesperidium. A modified berry, with few seeds, a leathery rind, and membranous extensions of the endocarp dividing the pulp into chambers; an example is the orange.

heterandrous. Having stamens differing from each other in length or form.

heteroallele. An allele that carries mutational alterations at different sites.

heteroauxin. $C_{10}H_9O_2N$. A plant growth hormone with an indole skeleton.

heteroblasty. An adaptation of desert plants where they produce morphologically as well as physiologically different seeds from the same plants. Such seeds have different germination requirements, so only a few seeds in a given crop germinate at any given time.

heterocarpous. Producing two distinct types of fruit.

heterochromatin. Specialized chromosome material which remains tightly coiled even in the nondividing nucleus and stains darkly in interphase.

heterococcolith. A coccolith with crystals arranged into boat, trumpet, or basket shapes.

heterocyst. Clear, thick-walled cell occurring at intervals along the filament of certain bluegreen algae.

heteroduplex. 1: A double-stranded deoxyribonucleic molecul comprising strands from different individuals.

2: A doublestranded molecule of deoxyribonucleic acid in which the two strands show noncomplementary sections.

heteroecious. Pertaining to forms that pass through different stages of a life cycle in different hosts, *cf.* autoecious.

heterogamete. A gamete that differs in size, appearance, structure, or sex chromosome content from the gamete of the opposite sex; anisogamete.

heterogametic sex. That sex of some species in which the two sex chromosomes are different in gene content or size and which therefore produces two or more different kinds of gametes.

heterogamety. The production of different kinds of gametes by one sex of a species.

heterogamous. Of or pertaining to heterogamy.

heterogamy. 1: Alternation of a true sexual generation with a parthenogenetic generation.

2: Sexual reproduction by fusion of unlike gametes; anisogamy.

3: Condition of producing two kinds of flowers.

heterogeneity. The condition or state of being different in kind or nature.

heterogeneous ribonucleic acid. A large molecule of ribonucleic acid that is believed to be the precursor of messenger ribonucleic acid. Abbreviated H-RNA.

heterogenesis. Alternation of generations in a complete life cycle, especially the alternation of a dioecious generation with one or more parthenogenetic generations.

heteroegenetic induction. A process in which the new cells are different from the ones causing the changes, e.g., in some dicots, leaf hairs are located only over vascular bundles, suggesting that the presence of the bundle controls differentiation of the hair cells; *cf.* homeogenetic induction.

heterogenous. Not originating within the body of the organism.

heterogony. 1: Alternation of generations in a complete life cycle, especially of a dioecious and hermaphroditic generation.

2: The quantitative relation between a part and the whole or another part as the organism increases in size.

3: Having heteromorphic perfect flowers with respect to lengths of the stamens or styles.

heterokaryon. A bi-or multinucleate cell having genetically different kinds of nuclei.

heterokaryosis. The condition of a bi-or multinucleate cell having nuclei of genetically different kinds produced as a result of anastomosis.

heterokont. An individual, especially among certain algae, having unequal flagella.

heterolactic fermentation. A type of lactic acid fermentation by which small yields of lactic acid are produced and much of the sugar is converted to carbon dioxide and other products.

heteromerous. 1: Of a flower, having one or more whorls made up of a different number of members than the remaining whorls.

2: Pertaining to a lichen thallus, having the fungus and alga components in well marked layers, usually between the medulla and upper cortex.

3: Pertaining to trama in Russulaceae, having sphaerocyst nests among filamentous hyphae.

heteromixis. In fungi, sexual reproduction which involves the fusion of genetically different nuclei, each from a different thallus.

heteromorphic, heteromorphous. 1: Having synoptic or sex chromosomes that differ in size or form.

2: Having variation from normal stucture.

3: Having organs of different length.

4: Pertaining to agaric lamella edge, which is sterile due to the presence of cystidia.

heterophagic vacuole. *See* food vacuole.

heterophyllous. Of plants which have two different kinds of leaves, e.g., when the leaves of the young plant are different from the leaves of the old plant, as in many species of the ivy family, Araliaceae; heterophylly.

heterophyte. 1: A plant occurring in a wide range of habitats.

2: A dioecious plant.

3: A parasitic plant devoid of chlorophyll.

4: A plant that depends upon living or dead plants or their products for food materials.

heteroploidy. The condition of a chromosome complement in which one or more chromosomes, or parts of chromosomes, are present in number different from the numbers of the rest.

heteropolysaccharide. A polysaccharide which is a polymer consisting of two or more different monosaccharides.

heterosis. The increase in size, yield, and performance found in some hybrids, especially of inbred parents. hybrid; vigor.

heterosporous. Of plants which produce spores of two different sizes such as microspore and megaspores, as in some pteridophytes e.g., clubmosses and all spermatophytes. The large spore i e., megaspore develops into a female gametophyte and the small spore i.e, microspore develops into a male gametophyte; heterosporic; heterospory; *cf.* homosporous.

heterostemony. Presence of two or more different types of stamens in the same flower.

heterostyly. A polymorphism of flowers in which the length of the style and stamens varies between individuals of a species, helping to prevent self-pollination; *cf.* homostyly.

heterothallism. Of fungi and algae, a condition in which the fusion of two mycelia of different nuclear components (mating types) is required for sexual reproduction, even though each strain may produce male and female organs. Heterothallism has been used as an equivalent of haplodioecism, dioecism, self sterility or self incompatibility. Whitehouse (1949) has distinguished haplodioecism as 'morphological heterothallism' where male and female organs are produced on different individuals e.g., in *Dictyuchus monosporus*, and haploid incompatibility as 'physiological heterothallism' where, male and female organs are developed on one individual but they are self incompatible e.g., in *Ascobolus magnificus*; *cf.* homothallism.

heterotopia. An abnormal habitat.

heterotopic. Occurring in a wide variety of habitats.

heterotopic transplantation. A graft transplanted to an abnormal anatomical location on the host.

heterotopy. Hyperplastic symptom in which an organ develops in a position other than its normal one (e.g., the development of "ears" in the tassels of corn plants.)

heterotrichous. 1: Of diverse kinds of hairs; used to describe protonemata with branches that show differing form.

2: Of diverse kinds of filaments; as observed in some algae e.g., *Stigeoclonium*, *Ectocarpus*, *Fritschiella*, where the plant body consists of a prostrate system from which develops an erect system of filaments.

heterotrophic. Refers to organisms that are dependent upon outside sources for growth, being incapable of synthesizing required organic materials from inorganic sources using energy from light. Heterotrophs obtain their food from other organisms, living or dead. Fungi and animals are heterotrophic, as are many bacteria; *cf*. autotrophic.

heterotropic enzyme. A. type of allosteric enzyme in which a small molecule other than the substrate serves as the allosteric reflector.

heteroxenous. Requiring more than one host to complete a life cycle.

heterozygote. An individual that has different alleles at one or more loci and therefore produces gametes of two or more different kinds.

heterozygous. Having non-identical alleles at the same loci on two homologous chromosomes; heterozygosity.

hexapetalous. Having or being a perianth comprising six petaloid divisions.

hexenbesen. *See* witches' broom disease.

hexokinase. Any enzyme that catalyzes the phosphorylation of hexoses.

hexosamine. A primary amine derived from a hexose by replacing the hydroxyl with an amine group.

hexosan. *See* hemicellulose.

hexose. Any monosaccharide that contains six carbon atoms in the molecule, e.g., glucose, fructose.

hexose diphosphate pathway. *See* glycolytic pathway.

hexose monophosphate cycle. A pathway for carbohydrate metabolism in which one molecule of hexose monophosphate is completely oxidized.

hexose phosphate. Any one of the phosphoric acid esters of a hexose, notably glucose, formed during the metabolism of carbohydrates by living organisms.

hexulose. A ketose made from a six-carbon-chain monosaccharide.

Hfr. *See* high-frequency recombination.

hians. Latin meaning 'gaping'.

hibernaculum. 1: A winter shelter for plants or dormant animals.
2: A winter bud or other winter plant part.

hibernation. The slowing-down of metabolism which occurs in many organisms during winter; hibernate.

hicmatis. Latin meaning 'winter'.

high-frequency recombination. A bacterial cell type, especially *Escherichia coli*, having an integrated F factor and characterized by a high frequency of recombination. Abbreviated Hfr.

highmoor bog. A bog whose surface is covered by sphagnum mosses and is not dependent upon the water table.

Hill plot. A graphic representation of the Hill reaction.

Hill reaction. The name for the part of the light reaction of photosynthesis, after R. Hill who first observed it in 1937. It involves the release of molecular oxygen by isolated chloroplasts in the presence of light, water and a suitable electron receptor, such as ferricyanide and nicotinamide adenine dinucleotide phosphate (NADP).

$$2NADP + 2H_2O \xrightarrow{\text{light}} 2NADPH_2 + O_2$$

hilum. 1: The place on the seed marking the point where the funicle was attached to the ovule.

2: Pertaining to fungal spore, a mark or a scar at the point of attachment to a conidiophore or sterigma.

3: A small granule at the centre of a starch grain.

4: The lateral depression in which the flagella are inserted in reniform zoospores.

himantioid. Pertaining to a mycelium arranged in spreading fanlike cords.

hippocrepiform. Horseshoe-shaped.

hirsute. Shaggy; hairy.

his operon. A sequence of nine genes in the chromosomes of many bacteria which code for all the enzymes of histidine biosynthesis.

histogen. A clearly delimited region or primary tissue from which the specific parts of a plant organ are thought to be produced.

histogenous. 1: Produced from tissues.

2: Of spores, produced from hyphae or cells, without conidiophores.

histogram. A way of showing the frequency with which different values of a variable occur in a sample. The variable is divided into classes, and the frequency of each class is represented by the height of the bars on the chart.

histology. The study of individual tissues.

histone. Any of the strong, soluble basic proteins of cell nuclei that are precipitated by ammonium hydroxide.

histophyte. A parasitic plant.

hoary. 1: Having grayish or whitish color, referring to leaves.

2: Of a pileus or stipe, covered thickly with silk-like hairs; canescent.

holarctic. A distributional pattern in which a taxon is widely distributed in the Northern Hemisphere.

holdfast. 1: A suckerlike base which attaches the thallus of certain algae to the substratum

2: A disklike terminal structure on the tendrils of various plants used for attachment to a flat surface.

3: Of fungi, a process from the thallus for attachment, e.g., appressorium, hyphopodium, stigmatopodium, and stomatopodium; *cf.* rhizoid, hapteron, haustorium.

holendophyte. A parasitic plant which passes its entire life within its host.

holobasidium A basidium which is not septate.

holocarpic. Having the entire thallus developed into a fruiting body or sporangium.

holocellulose. The total polysaccharide fraction of wood, straw, and so on, that is composed of cellulose and all of the hemicelluloses and that is obtained when the extractives and the lignin are removed from the natural material.

holocoenosis. The nature of the action of the environment on living organisms.

holoenzyme. A complex, fully active enzyme, containing an apoenzyme and a coenzyme.

hologamy. 1: Condition of having gametes similar in size and form to somatic cells; gamontogamy; macrogamy; hologamete; *cf.* merogamy.

2: Condition of having the whole thallus develop into a gametangium.

holonecrosis. (*pl.* holonecroses). The condition of being dead *cf.* plesionecrosis.

holophyte. An organism that obtains food in the manner of a green plant, that is, by synthesis of organic substances from inorganic substances using the energy of light.

holosaprophyte. A true saprophyte.

holotype. A nomenclatural type for the single specimen designated as "the type" by the original author at the time of publication of the original description.

holozoic. Obtaining food in the manner of most animals, by ingesting complex organic matter.

homaxial. Having all axes equal,

homeogenetic induction. A process in which the new cells resemble the ones directing their differentiation. e.g., new cambial cells form from cortical cells adjacent to existing procambial cells and this might be caused by something released from the existing cells (redifferentiation); *cf.* heterogenetic induction.

homeomorph. An organism which exhibits a superficial resemblance to another organism even though they have different ancestors.

homeostasis. The maintenance of a relatively steady state or equilibrium in a biological system by intrinsic regulatory mechanisms; homeostatic.

homodichogamy. The condition of a species in which some individuals exhibit homogamy and others dichogamy; autoallogamy; homodichogamic.

homoecious. Having one host for all stages of the life cycle; autoecious; monoxenous; *cf.* heteroecious.

homogametic sex. The sex of a species in which the paired sex chromosomes are of equal size and which therefore produces homogametes.

homogamety. The production of homogametes by one sex of a species.

homogamous. Having male and female floral parts functioning at the same time; homogamy.

homogony. Condition of having one type of flower, with stamens and pistil of uniform length.

homoiochlamydeous. Having perianth leaves alike, not differentiated into sepals and petals.

homoiomerous. Having algae and fungi distributed uniformly throughout the thallus of a lichen as in *Collema cf.* heteromerous.

homokaryosis. The condition of a bi- or multinucleate cell having nuclei all of the same kind.

homologous. Pertaining to a structural relation between parts of different organisms due to evolutionary development from the same or a corresponding part.

homomorphism. Having perfect flowers, consisting of only one type.

homopetalous. Having all petals identical.

homoplastic grafting. Transplanting tissue (the homograft) from one individual to another of the same species.

homosporous. Having only one kind of spores, as in bryophytes and true ferns, isosporous; homosporic, homospory; *cf.* heterosporous.

homostyly. The condition in which all the flowers of a species have styles of similar length; homostylous; *cf.* heterostyly.

homothallism. A mode of reproduction in fungi and algae in which each thallus is self-fertile, not showing haploid incompatibility, and each produces both male and female sex cells; homothallic, homothallous; *cf.* heterothallism.

homotropic. Used of a flower fertilized by its own pollen.

homotropic enzyme. A type of allosteric enzyme in which the substrate serves as the allosteric reflector.

homotropous. Having the radicle directed toward the hilum.

homotype. A taxonomic type for a specimen which has been compared with the holotype by another than the author of the species and determined by him to be conspecific with it.

homozygote. An individual that has identical alleles at one or more loci and therefore produces gametes which are all identical.

homozyygous. Having identical alleles at the same loci on two homologous chromosomes, homozygosity.

honey guides. Coloured spots or lines on the petals of a flower, which may guide pollinating animals towards the sources of pollen and nectar.

horizon. A layer in a soil profile. Different soil profiles can be compared by examining the structure, thickness and chemical composition of their horizons.

horizontal resistance. A degree of resistance to all races of a pathogen; an epidemiological term.

hormocyst. A propagule or diaspore produced in a special hormocystangium comprising a few algal cells and fungal hyphae; produced in a few lichens.

hormogonium. Portion of a filament between heterocysts in blue-green algae, detaches as a reproductive body.

hormone. A substance which, in very small amounts, controls growth and development. Hormones are chemical messengers which are usually produced in one organ and transported to another part of the plant, where they have their effects. The five main groups of plant hormones are auxins, gibberellins, cytokinins, ethene and abscisic acid.

horsetail. A pteridophyte of the order Equisetales, which consists of abut 25 species of *Equisetum*.

hortal. Used of an ornamental plant that has escaped from cultivation and can be found growing wild.

horticulture. The art and science of growing plants.

hortinus. Pertaining to the plant growth of the current year; hornotinus, hornus.

hortus siccus: Herbarium.

hospitating. Offering refuge or acting as a host (hospitator) to another organism (the hospite).

hospitator. An organism offering refuge or acting as a host to another organism.

host. 1: An organism on or in which a parasite lives.

2: The dominant partner of a symbiotic or commensal pair.

host range. All the host species that may be infected by a given parasite; host list.

host specificity. The extent to which an adult parasite is restricted in the variety of host species utilized.

hot spot. A site in a gene at which there is an unusually high frequency of mutation.

hull. The outer, usually hard, covering of a fruit or seed.

hülle cells. Terminal or intercalary thick-walled cells which occur in large numbers in association with the perithecia of *Aspergillus nidulans.*

human community. That portion of a human ecosystem composed of human beings and associated plant and animal species.

humicolous. Of or pertaining to plant species inhabiting medium-dry ground.

humid transition life zone. A zone comprising the climate and biotic communities of the northwest moist coniferous forest of the northcentral United States.

humification. The transformation of dead plant material into humus.

humifuse. Spread over the ground surface.

humivore. An organism that feeds on humus.

humus. The layer of organic matter at the top of a soil profile. Humus is the habitat of most decomposers.

husbandry. The cultivation, production and management of plants and animals; used especially with reference to domesticated species; farming.

husk. The outer coat of certain seeds, particularly if it is a dry membranous structure.

hyaline. Transparent.

hybrid. A plant which results from the cross fertilization of two different species, subspecies, varieties, strains, etc., hybridize, hybridization.

hybrid enzyme. A form of polymeric enzyme occurring in heterozygous individuals that shows a hybrid molecular form made up of subunits differing in one or more amino acids.

hybridization. The act or process of producing hybrids.

hybrid vigour. *See* heterosis.

hydathode. A gland on the epidermis of higher plants specialized for exudation of water.

hydatoaerophyte. A plant growing in permanent water but with part of the assimilative organs floating with a dry upper surface, or emerging above the water.

hydatophytium. A submerged plant community.

hydrase. An enzyme that catalyzes the removal or addition of water to a substrate without hydrolyzing it.

hydroanemophilous. Having air borne spores that are discharged after wetting of the spore producing structure; hydroanemophily.

hydrobiology. The study of life in aquatic habitats.

hydrochory. Dispersal of disseminules by water.

hydrogenase. Enzyme that catalyzes the oxidation of hydrogen.

hydrogen bacteria. Bacteria capable of obtaining energy from the oxidation of molecular hydrogen.

hydroid. A specialized cell that conducts water internally in mosses.

hydrolase. Any of a class of enzymes which catalyze the hydrolysis of proteins, nucleic acids, starch, fats, phosphate esters, and other macromolecular substances.

hydrolytic enzyme. A catalyst that acts like a hydrolase.

hydrome. The system of water conducting cells (hydroids) that characterizes some mosses.

hydronasty. Orientation movement in plants in response to changes in atmospheric humidity; e.g., the flowers of dandelion open and close in response to such changes.

hydrophyte. 1: A plant that grows in a moist habitat.

2: A plant requiring large amounts of water for growth. A cryptophyte with buds or shoot apices perennating under water during unfavourable seasons; hygrophyte.

hydrophilous. 1: Thriving in wet or aquatic habitats; *cf.* hydrophobic.

2: Pollinated by water borne pollen; hydrophile, hydrophily.

hydrophobic. Intolerant of water or wet conditions; water repellent; hydrophobe, hydrophoby; *cf.* hydrophilous.

hydroponics. Growing of plants in a nutrient solution with the mechanical support of an inert medium such as sand; water culture.

hydrosere. An ecological succession commencing in a habitat with abundant water, typically on the submerged sediments of a standing water body; hydrarch succession.

hydrosis. (*pl.* hydroses) Necrotic symptom of disease characterized by water-soaking of tissues; hydrotic.

hydrotaxis. The directed response of a motile organism towards (positive) or away from (negative) a water or moisture stimulus; hydrotactic.

hydrotherophyte. An aquatic annual plant; an aquatic plant that completes its life cycle within a single limited vegetative period, overwintering as seeds.

hydrotropism. Orientation involving growth or movement of a sessile organism or part, especially plant roots, in response to the stimulus of moisture; hydrotropic.

hydroxylase. Any of several enzymes that catalyze certain hydroxylation reactions involving atomic oxygen.

hygrokinesis. A change in the rate of movement of an organism in response to a change in humidity; hygrokinetic.

hygrometer. An instrument to measure the humidity of air and other gases. Two common types are the wet and dry bulb hygrometer and the hair hygrometer.

hygrophyte. *See* hydrophyte.

hygroscopic. 1: Being sensitive to moisture, such as certain tissues, becoming soft in wet air, hard in dry.

2: Pertaining to a fruit body, opening and discharging spores in dry air.

hylodad. A plant of open woodland.

hylophyte. A wood or forest plant; hylad.

hylotomous. Used of wood-cutting insects.

hymenium (*pl.* hymenia). The outer, sporebearing layer i.e., asci or basidia in certain fungi or their fruiting bodies.

hymenophore. Portion of a sporophore that bears the hymenium.

hypanthium. A tube which results from growth at the edge of the receptacle, in some plants. The perianth i.e., sepals and petals and the stamens grow from the top of the hypanthium.

hyperosmotic. Pertaining to a solution that exerts a greater osmotic pressure than the solution being compared; *cf.* hypoosmotic, isosmotic.

hyperparasite. An organism parasitic upon another parasite; superparasite; hyperparasitic, hyperparasitism.

hyperplasia. Excessive growth due to increase in cell number, hyperplastic; *cf.* hypertrophy and hypoplasia.

hyperploid. Having one or more chrmosomes or parts of chromosomes in excess of the haploid number, or of a whole multiple of the haploid number; hyperploidy.

hypersaprophyte. A saprophyte only found on substrates invaded by other saprophytes.

hypersensitivity. Defensive reaction in which a plant restricts infection to necrotic lesions around points of entry of the pathogen. This word was used for the first time by Stakman, of reaction to rusts; intolerance.

hypertonic. Used of a cell or tissue having a greater osmotic pressure than the solution in which it is immersed; *cf.* tonicity.

hypertrophy. Excessive growth due to an abnormal increase in cell size; *cf.* hyperplasia.

hypha, (*pl. hyphae*) A filament of fungal mycelium (thallus) that is composed of one or more cylindrical cells and that increases in length by growth at its tip. New hyphae arise as lateral branches.

hyphidium. A sterile hymenial structure of hyphal origin.

hyphoid. Hyphalike.

hyphopodium. A short branch of one or two cells on epiphytic mycelium of Meliolales, etc., the end cell is pointed or rounded.

hyphydrogamous. Having water borne pollen transported below the water surface; hyphydrogamic; hypohydrogamic.

hypnoplasy. Arrested development resulting in failure to reach normal size.

hypnosis. A state of dormancy in seeds that retain the capacity for normal development; hypnotic.

hypobasal. Located posterior to the basal wall.

hypoblast. *See* endoderm.

hypychil. Lower portion of the lip in orchids; hypochillium.

hypochnoid. Having generally compacted hyphae.

hypocotyl. The part of the embryo and seedling below the cotyledons, bearing the radicle at its end.

hypocrateriform. Saucer-shaped.

hypodermis. 1: The outermost cell layer of the cortex of plants; exodermis.

2: The layer of cells that underlies and secretes the cuticle in arthropods and some other invertebrates.

hypogeal. Of the kind of germination in which the cotyledons remain below ground. Their stored food is used up in the early growth of the epicotyl and the hypocotyl.

hypogeous. Living or maturing below the surface of the ground.

hypogynous. 1: Having all flower parts attached to the receptacle below the pistil and free from it; hypogyny.

2: The condition of having the autheridium under the oogonium and on the same hypha.

hypomorph. An allele having an effect similar to the normal allele, but being less active.

hypomorphic allele. An allele that has reduced levels of gene activity.

hyponasty. A nastic movement involving inward and upward bending of a plant part, due to more rapid growth of the lower side of an organ than the upper side; *cf.* epinasty.

hypoplasia. Underdevelopment resulting from an abnormal paucity of cells; *cf.* hyperplasia.

hypoploid. Having one or more less chromosomes, or parts of chromosomes, than a whole multiple of the haploid number.

hypoploidy. The condition or state of being hypoploid.

hypostatic. Subject to being suppressed, as a gene that can be suppressed by a nonallelic gene.

hypotonic. Used of a cell or tissue having a lower osmotic pressure than the solution in which it is immersed; *cf.* tonicity.

hysterochroic. Having fruiting bodies which discolor progressively from base to apex with age.

I

I_1. First inbreeding generation; subsequent inbreeding generations denoted I_2, I_3...I_n.

IAA. See indoleacetic acid.

IAMS. International Association of Microbiological Societies.

IAPT. The International Association of Plant Taxonomy.

ICBN. International Code of Botanical Nomenclature.

iceland moss. The lichen *Cetraria islandica* habitually eaten by reindeer and caribou. It has been used as a substitute for flour during hard times in Iceland and Scandinavia as it is rich in digestible carbohydrates.

ICNB. International Committee on Nomenclature of Bacteria.

ICNV. International Committee on Nomenclature of Viruses.

icones. Figures; pictures; plates.

icosahedral virus. A virion in the form of an icosahedron.

icotype. A typical, accurately identified specimen of a species, but not the basis for a published description.

ICSB. International Committee on Systematic Bacteriology.

ICTV. International Committee on Taxonomy of Viruses.

ideotype. In taxonomy, a specimen examined by the author of a species but not collected from the type locality; idiotpye.

idioandrosporous. Species of *Oedogonium* which bear oogonia and androsporangia on separate filaments.

idiobiology. That aspect of biology concerned with study of individual organisms; autobiology.

idioblast. 1: A plant cell that differs markedly in shape or function from neighboring cells within the same tissue.

2: A supporting cell, found among chlorophyll tissue. It lacks chlorophyll, has thick walls, and is usually elongated.

idiochromosome. A sex chromosome.

idioecology. Autecology.

idiogamy. Self-fertilization; idiogamous.

idiogram. A diagrammatic representation of the chromosome complement of a cell.

idiotaxonomy. Taxonomic study of individuals, populations, species and higher taxa; traditional taxonomy; *cf.* syntaxonomy.

illegitimate pollination. Self pollination which takes place in spite of the flower appearing to be adapted for cross-pollination.

imbibition. The process in which water is taken up by a seed at the beginning of germination. It is a passive absorption or uptake of water; imbibe.

imbibitional water. That part of the soil water held within the lattice structure of colloidal matter.

imbricate. Having overlapping edges, such as scales, or the petals of a flower, or pilci, squamules etc.

immaculate. Not spotted.

immarginate. Having no well-defined edge.

immature. 1: Used of the developmental stages of an organism preceding the attainment of sexual maturity.

2: Used of any incompletely differentiated system.

immature soil. A recently formed soil in which profile development is incomplete for the prevailing climatic and biological conditions.

immersed Used of an aquatic plant growing entirely beneath the water surface; submersed.

immigrant. An organism that settles in a zone where it was previously unknown.

immigration. The one-way inward movement of individuals or their disseminules into a population or population area.

immission. Any pollutant discharged into the environment at a particular place and time.

immunity. In plants the ability to remain free from disease because of inherent structural or functional properties. *cf.* resistance.

immunization. In plants, complete resistance to disease. *cf.* resistance.

impaction. A process of vigorous shaking of seeds to dislodge the cork-like filling in a small opening in a seed coat. This treatment is generally given to dormant seeds of *Melilotus alba*, *Trigonella arabica and Crotallaria egyptica* in order to break their dormancy.

imperfect flower. A unisexual flower; a flower lacking either pistils or stamens.

imperfect stage. Of fungi, the period of the life cycle other than that in which spores are formed as a result of a sexual process. *cf.* perfect stage.

imperforate. Lacking a normal opening.

impermeable. Of membranes which do not allow the movement of substances from one side to the other.

impervious. Not permeable; usually with reference to the passage of water or air; impermeable.

impingement. The entrapment of aerosol particles on a solid surface in a sampling device.

implantation. The insertion of a graft into the host or stock.

impunctate. Lacking pores.

inaequi-hymeniferous. Pertaining to an agaric basidiocarp, where hymenium develops in an unequal manner, with basidia ripening in zones. In *Coprinus* a wave of gill maturation begins at the base and passes slowly upwards.

inaperturate. Lacking apertures.

inarching. A kind of repair grafting in which two plants growing on their own roots are grafted together and one plant is severed from its roots after the graft union is established.

inbreeding Breeding of closely related individuals; of species over many generations; self-fertilization, as in some plants, is the most extreme form.

incertae sedis. Of uncertain taxonomic position.

incipient. About to become or occur; nascent; used of the initial stage in the development of a structure or event.

incipient plasmolysis. This is the initial pulling away of the membrane from the cell wall. At this point the turgor pressure is zero.

incipient species. Populations that are in the process of diverging to the point of speciation but which still have the potential to interbreed although they are prevented from doing so by a specific barrier.

inclusion. A visible product of cellular metabolism within the protoplasm.

inclusion body. Any of the abnormal structures appearing within the cell nucleus or the cytoplasm during the course of virus multiplication.

incompatibility. 1: The inability to unite, fuse or form any homogeneous or viable association; incompatible.

2: Failure of either cross-fertilization or self-fertilization as a result of structural, physiological or genetic factors.

3: Failure of a graft to unite with the host or stock tissue.

incomplete dominance. The partial expression of both alleles at a given locus so that the phenotype of the heterozygote is intermediate between both homozygotes; codominance; semi-dominance; *cf.* complete dominance.

incomplete flower. A flower lacking one or more modified leaves, such as petals, sepals, pistils, or stamens.

incrusted. Having matter excreted on the walls of the hyphae.

incubation. Of a pathogen, its development, growth, and penetration of a plant prior to infection; or its activity within plant tissues subsequent to penetration and up to appearance of symptoms, signs, or both.

incubator. A laboratory cabinet with controlled temperature for the cultivation of bacteria, or for facilitating biologic tests.

incubous. Of a kind of growth in leafy liverworts, in which the front of each leaf lies on top of the leaf in front of it.

incumbent. Lying on or down.

indehiscent. 1: Remaining closed at maturity, as certain fruits.

2: Not splitting along regular lines.

indented key. A dichotomous key in which the first part of a contrasting couplet is followed by all subsequent couplets; each subordinate couplet being indented one step further to the right for clarity of presentation; yoked key; *cf.* bracketed key.

independent assortment. The random assortment of alleles at two or more loci on different chromosome pairs or far apart on the same chromosome pair which occurs at meiosis; first discovered by G. Mendel, and known as Mendel's second law which states that most of the characters of parents can appear in any combination in their offspring.

indeterminate. 1: Conidiophore continuing growth indefinitely.

2: Having the edge not well defined, especially of fruit bodies and leaf spots.

indeterminate growth. Growth of a plant in which the axis is not limited by development of a reproductive structure, and therefore growth continues indefinitely; *cf.* determinate growth.

Index of Atmospheric Purity (IAP). A numerical estimate of the purity of the air on the basis of the lichens present on trees.

index fossil. The ancient remains and traces of an organism that lived during a particular geologic time period and that geologically date the containing rocks.

indexing. 1: Process of determining the presence of disease in a plant by transferring inoculum from one plant to another in which diagnostic symptoms develop. The second plant is termed a test plant or an indicator plant.

2: A procedure to determine whether a given plant is infected by a virus. It involves the transfer of a bud, scion, sap, etc. from one plant to one or more kinds of (indicator) plants that are sensitive to the virus.

indicator. 1: An organism, species or community characteristic of a particular habitat, or indicative of a particular set of enviromental conditions.

2: A plant that reacts to certain viruses or environmental factors with production of specific symptoms and is used for detection and identification of these factors.

indicator medium. A usually solid culture medium containing substances capable of undergoing a colour change in the vicinity of a colony which has effected a particular chemical change, such as fermenting a certain sugar.

indicator plant. A plant used in geobotanical prospecting as an indicator of a certain geological phenomenon.

indicator species. A species, the presence or absence of which is indicative of a particular habitat, community or set of environmental conditions; characteristic species; guide species index species.

indole. Also known as 2,3 benzopyrrole, C_6H_4 (CHNH)CH. A decomposition product of tryptophan formed in the intestine during putrefaction and by certain cultures of bacteria.

indoleacetic acid (IAA). $C_{10}H_9O_2N$. A decomposition product of tryptophan produced by bacteria and occurring in urine and faeces, used as a hormone to promote plant growth. This is the most common auxin produced mainly in meristmatic tissues such as those of young developing leaves, flowers and fruits.

induced-fit hypothesis. This is a hypothesis put forward by Daniel E. Koshland for active site on an enzyme. Here the active site of the enzymes can be induced by close approach of the substrate (or product) to undergo a change in conformation that allows a better fit between the two.

indumentum. 1: A covering, such as one that is woolly.
2: A covering of hairs.

induplicate. Having the edges turned or rolled inward without twisting or overlapping; applied to the leaves of a bud.

indusium (*pl.* indusia). 1: The flap of tissue in a sorus, which protects the sporangia of ferns.
2: The annulus of fungi; a net like structure hanging from the top of the stipe under the pileus.

industrial microbiology. The study, utilization, and manipulation of those microorganisms, capable of economically producing desirable substances or changes in substances, and the control of undesirable microorganisms.

industrial microorganism. Any microorganism utilized for industrial microbiology.

industrial yeast. Any yeast used for the production of fermented foods and beverages, for baking, or for the production of vitamins, proteins, alcohol, glycerol, and enzymes.

inermis. Latin meaning 'unarmed, thornless.'

inermous. Having no spines or prickles.

infection. The process of invasion of a host by a symbiont, parasite or microorganism; the pathological condition resulting from the growth of infectious organisms within a host; infect.

infection court. The place on or in the susceptible plant where infection may be initiated.

infection thread. The specialized hypha of a pathogenic fungus that invades tissue of the susceptible plant.

infectious. 1: Refers to a pathogen that can be transmitted from one suscept to another by an external agent.
2: Refers to any disease whose pathogen can be so transmitted.

infective. Of an agent of inoculation, capable of transmitting inoculum.

inferior ovary. One which is beneath the point of attachment of the calyx, corolla and stamens of the flower.

infertile. Not fertile; non-reproductive; infertility.

infestation. Invasion by parasites or pests.

infested. Attacked by insects; sometimes used of fungi in soil or other substrata in the sense of 'contaminated'.

infiltration. The process by which water seeps into a soil, influenced by soil texture and vegetation cover.

inflated. 1: Distended, applied to a hollow structure.
2: Open and enlarged.

inflorescence. A shoot bearing flowers and no leaves. An inflorescence can have one to many flowers.

influent. An organism that disturbs the ecological balance of a community, by influencing another organism or group of organisms.

infrafoliar. Located below the leaves.

infraspecific. Of variation between individuals of a species.

infructescence, An inflorescence's fruiting stage.

infundibuliform. Tubular below, gradually opening upwards, i.e., funnel-shaped.

ingress. The act, by a plant pathogen, of gaining entrance into the tissues of a susceptible plant.

inhabited symbiont. The host symbiont in an endosymbiosis which contains the other symbiont within its body; *cf.* inhabiting symbiont.

inhabiting symbiont. The symbiont contained within the body of the host symbiont in an endosymbiosis; endosymbiont; *cf.* inhabited symbiont.

inheritance. 1: The acquisition of characteristics by transmission of germ plasm from ancestor to descendant; inherit.

2: The sum total of characteristics dependent upon the constitution of the fertilized ovum.

inhibition. The prevention or slowing down of a metabolic reaction, e.g., by an inhibitor or by temperatures that are too high or too low.

inhibitor. A substance which prevents or slows down a chemical reaction or process. Some inhibitors can slow down enzyme reactions by blocking the active site of the enzyme; inhibit.

initiation codon. A codon that signals the first amino acid in a protein sequence; usually *AUG*, but sometimes *GUG*; start codon.

initiation factors. Proteins required for the initiation of protein synthesis in prokaryotic and eukaryotic cells.

initiator codon. A codon that acts as a start signal for the synthesis of a polypeptide.

innovation. A new shoot that originates laterally after the sex organs are mature.

inoculation. 1: Introduction of a disease agent into an animal or plant to produce a mild form of disease and render the individual immune.

2: Introduction of microorganisms onto or into a culture medium; inoculate.

inoculum. 1: A small amount of substance containing bacteria or fungal spores or mycelium from a pure culture which is used to start a new culture or to infect an experimental animal.

2: The individual or group of individuals comprising the founders of a colony or newly established population; breeding stock.

inoculum potential. Regarding a fungus or other microorganism, the energy of growth available for colonization of a substratum at the surface of the substratum to be colonized.

inoperculate. Lacking an operculum.

inquinant. Stained; blackened; dirty.

insecticide. A chemical used to kill insects; used generally for any chemical agent used to destory invertebrate pests.

insectivorous. Feeding on a diet of insects.

insectivorous plants. Of plants which eat insects. Some plant species trap insects with sticky hairs (e.g. sundew), in bucket-shaped leaves (e.g. pitcher plants) or between hinged leaf-lobes (e.g. Venus fly-trap). Insectivorous plants obtain nutrients by secreting enzymes which break down the tissues and cells of the trapped insect. The insectivorous habit is an adaptation to habitats that are poor in nitrates.

integrated control. An approach that attempts to use all available methods of control of a disease or of all the diseases and pests of a crop plant for best control results but with the least cost and the least damage to the environment.

integration. Recombination involving insertion of a genetic element.

integuments. The outermost layers of the ovule, which become the coat of the seed after the ovule is fertilized.

interaction. The process in which two or more organisms, of the same or different species, act on each other, e.g., symbiosis; interact.

interaxillary. Located within or between the axils of leaves.

interbiotic. Living as a parasite on or near one or more living organisms, as certain rhizoidal chytrids.

intercalary. Referring to growth occurring between the apex and leaf.

intercalary meristem. A meristem that is forming between regions of permanent or mature meristem i.e., at the base of leaves and stems in monocotyledons.

intercellular. Between the cells.

intercellular space. The spaces between cells. In some tissues e.g., the spongy mesophyll of leaves, the intercellular spaces are large and filled with air.

interference phenomenon. Inhibition by a virus of the simultaneous infection of host cells by some other virus.

interferon. A protein produced by intact animal cells when infected with viruses; acts to inhibit viral reproduction and to induce resistance in host cells.

interferonogen. A preparation made of inactivated virus particles used as an inoculant to stimulate formation of interferon.

interfoliaceous. Between a pair of leaves, such as between those which are opposite or verticillate.

intergenic crossing-over. Recombination between distinct genes or cistrons.

intergenic suppression. The restoration of a suppressed function or character by a second mutation that is located in a different gene than the original or first mutation.

interkinesis. See interphase.

intermediate-day plants. These are plants that flower only on intermediate day lengths and remain vegetative when days are too short or too long, e.g., *Chenopodium album*, *Saccharum spontaneum* (sugarcane) and *Coleus hybrida*.

International Association of Microbiological Societies (IAMS). The body responsible for the organization of Microbiological Congresses and, through the ICSB, for the publication of the International Code of Nomenclature of Bacteria.

International Association of Plant Taxonomy (IAPT). The body responsible for the publication of the ICBN and for the organization of committees dealing with the nomenclature of plants, at Botanical Congresses.

International Bureau for Plant Taxonomy and Nomenclature. The publisher of the International Code of Botanical Nomenclature, Taxon and Regnum Vegetabile.

International Code of Botanical Nomenclature (ICBN). The internationally adopted set of rules governing botanical nomenclature.

International Code of Nomenclature of Bacteria. The internationally adopted set of rules governing the nomenclature of Bacteria; Bacteriological Code.

International Committee on Nomenclature of Bacteria (ICNB). The former name of the International Committee on Systematic Bacteriology.

International Committee on Nomenclature of Viruses (ICNV). The former name of the International Committee on Taxonomy of Viruses.

International Committee on Systematic Bacteriology (ICSB). The body responsible for the publication of the International Code of Nomenclature of Bacteria, and for the Judicial Commission and taxonomic subcommittees dealing with the nomenclature of particular groups of Bacteria.

International Committee on Taxonomy of Viruses (ICTV). The body responsible for the taxonomic subcommittees dealing with the nomenclature of viruses and for approval of newly proposed names of viruses.

International Journal of Systematic Bacteriology (IJSB). The official organ of the International Association of Microbiological Societies.

internode. The space on a stem between two nodes.

interphase. Also known as interkinesis.
1: The period between succeeding mitotic divisions.
2: The period between the first and second meiotic divisions in those organisms where nuclei are reconstituted at the end of the first division.

interspecies. A cross between two distinct species; interspecific hybrid.

interspecific. Between two or more distinct species; used of a hybrid between two distinct species.

interspersed. Scattered or spread irregularly amongst other objects or individuals; intersperse, interspersion.

intertidal. Organisms found between levels of low and high tides of sea-water.

intervening sequence. A region of a gene that is transcribed but is not included in the final transcript of a ribonucleic acid.

intracellular. Within a cell.

intracellar symbiosis. Existence of a self-duplicating unit within the cytoplasm of a cell, such as a kappa particle in *Paramecium,* which seems to be an infectious agent and may influence cell metabolism.

intracistron comlpementation. The process whereby two different mutant alleles, each of which determines in homozygotes an inactive enzyme, determine the formation of the active enzyme when present in the same nucleus.

intracytoplasmic. Being or occurring within the cytoplasm of a cell.

intrafascicular. Located or occurring within a vascular bundle.

intragenic. Within a gene, in referring to certain events.

intragenic recombination. Recombination occurring between the mutants of one cistron.

intragenic suppression. The restoration of a suppressed function or character as a consequence of a second mutation located within the same gene as the original or first mutation.

intramarginal. Within a margin.

intramatrical. Said of a fungus which lives in the substratum or in the matrix or inside the host cell.

intrapetiolar. 1: Enclosed by the base of the petiole.
2: Between the petiole and the stem.

intraspecific. Of competition within a species; between individuals or populations of the same species.

intravital. Occurring while the cell or organism is alive.

intravital stain. A nontoxic dye injected into the body to selectively stain certain cells or tissues.

intrinsic. Originating or occurring within an individual, group or system; *cf*. extrinsic.

intron. DNA sequences within a structural gene that are transcribed into RNA but do not give rise to amino acid sequences in the protein. Intron transcripts are excised during RNA processing.

introrse. Turned inward or toward the axis.

introrse dehiscence. Spontaneous opening of a ripe fruit from outside inwards; *cf.* extrorse dehiscence.

intumescence. Hyperplastic symptom characterized by blisterlike swelling due to excessive water on the surfaces of plant organs.

intussusception. The deposition of material between the microfibrils of a cell wall.

inulase. An enzyme produced by certain molds that catalyzes the conversion of inulin to levulose.

inulin. This is an unbranched nutrient polysaccharide found in the bulbs of such plants as antichokes, dahlias, and dandelions. It consists of repeating fructose units in β2→1 linkage.

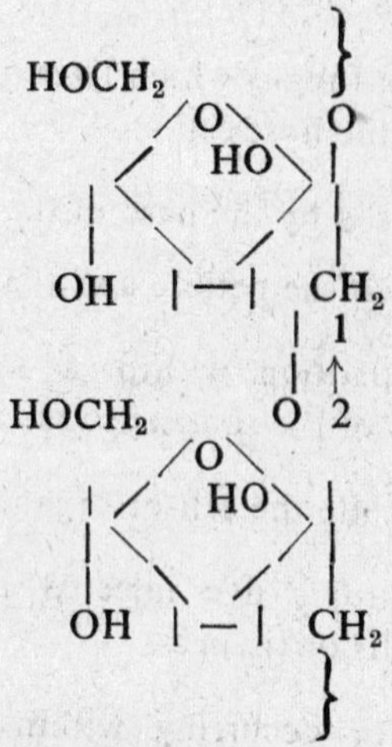

```
                        }
 HOCH2                 }
      /O\              O
     /    HO\          |
    |\        /        |
    | \      /         |
   OH  |  —  |         CH2
                      |  1
                      |  ↑
 HOCH2                O  2
      /O\
    |/    HO\  |
    |\        /|
    | \      / |
   OH  |  —  |  CH2
                 }
                 }
```

invagination. Retraction, under force of pressure, of an outer surface toward the inside.

invasion. Spread of a pathogen through tissues of a diseased plant.

inversion. 1: A type of chromosomal rearrangement in which two breaks take place in a chromosome and the fragment between breaks rotates 180° before rejoining.

2: The inside-out turning of the colony of the volvocales.

invertase. See saccharase.

invertin. See saccharase.

in vitro. Refers to biological processes when they are allowed to occur in isolation from the whole organism i.e., in an artificial apparatus; *cf. in vivo.*

in vivo. Refers to biological processes when they occur within the living organism; *cf. in vitro.*

involucrate. Having an involucre.

involucre. 1: A structure which protects or encloses another organ e.g., the bracts enclosing the developing inflorescence in Compositae.

2: The leaves joined together to protect the sex organs in leafy liverworts.

involute. Curving inward, as in leaves in which the margins curve inwards over the upper face of costa.

involution form. A cell with a bizarre configuration caused by abnormal culture conditions.

involvucel. A secondary involucre; involvucellate.

iodophors. Organic compounds of iodine; iodophors are mixtures of iodine with surface active agents which act as carriers and solubilizers for the iodine. Iodophors possess the germicidal characteristics of iodine and have the additional advantages of non-staining and low irritant properties. One of the agents used is polyvinylpyrrolidone (PVP), and the complex PVP-I.

ion-exchange chromatography. This is a technique to separate proteins and other macromolecules. A synthetic ion-exchange resin is packed in glass columns. As a solution of ions is passed through the column, the ions compete with each other for the charged sites on the resin. The differential rates of movement of ions through the column are the basis for separation.

ipecac. Any of several low, perennial, tropical South American shrubs or half shrubs in the genus *Cephaelis* of the family Rubiaceae; the dried rhizome and root, containing emetine, cephaeline, and other alkaloids, is used as an emetic and expectorant.

iron-porphyrin protein. Any protein containing iron and porphyrin, examples are hemoglobin, the cytochromes, and catalase.

irradiation. Subjection of a biological system to sound waves of sufficient intensity to modify their structure or function.

irregular. Lacking symmetry, as of a flower having petals unlike in size or shape.

irritability. Responsiveness to change in the physical or biological environment; a property of all living things.

isanthous. Having regular flowers.

IS element (insertion sequence). A short sequence of nucleotide pairs in DNA that facilitates transposition of itself and/or other genetic elements in the genome.

isidium (pl. isidia). A smooth outgrowth from the cortex of a lichen thallus. These are papilla, coral-petal or scale-like excrescences which are easily, broken away from thallus when touched, thus acting as a vegetative propagule.

isoaccepting tRNAs. Different tRNAs that activate the same amino acid during protein synthesis.

isoalleles. 1: Alleles which are so similar in their phenotypic effects that special techniques are required to distinguish them.
2: An allele that carries mutational alterations at the same site.

isoalloxazine mononucleotide. See riboflavin 5′-phosphate.

isobiochore. A boundary line on a map connecting world environments that have similar floral and faunal constituents.

isocarpic. Having the same numbe1 of carpels and perianth divisions.

isochromosome. An abnormal chromosome with a medial centromere and identical arms formed as a result of transverse, rather than longitudinal, splitting of the centromere.

isocitric acid. $HOOCCH_2CH(COOH)CH(OH)COOH$. An isomer of citric acid that is involved in the tricarboxylic acid cycle in bacteria and plants; 2-hydroxy-1, 2, 3,-propanetricarboxylic acid.

isodiametric. Having equal diameters or dimensions.

isoenzyme. Any of the electrophoretically distinct forms of an enzyme. representing different polymeric states but having the same function. Also known as isozyme. e.g., Lactic dehydrogenases, in which five different forms have been identified.

isoflavone. $C_{15}H_{10}O_2$. A colourless, crystalline ketone, occurring in many plants, generally in the form of a hydroxy derivative.

isogamete. A reproductive cell that is morphologically similar in both male and female and cannot be distinguished on form alone.

isogamous. Having gametes that are similar in size, shape and behaviour; having gametes not differentiated into male and female; *cf.* anisogamous.

isogamy. The fusion of gametes of similar size, shape and behaviour; isomerogamy; zygogamy. This is characteristic of some algae and fungi; *cf.* anisogamy.

isogeneic. Having the same origin and thus being genetically identical, as an inbred strain; isogenic; syngeneic.

isogony. Growth of parts at such a rate as to maintain relative size differences.

isograft. A tissue transplant from one organism to another organism which is genetically identical.

isolate. A single spore or culture and the subcultures derived from it. Also used to indicate collections of a pathogen made at different times.

isolation. The separation of one object from another, or the inability of two substances or organisms to mix with each other. Reproductive isolation is the inability of two or more populations to breed with each other, e g., because they live in different places or different habitats because they flower at different times of year, or because they have different genomes.

isolume. A line on a chart or map connecting points of equal light intensity.

isomerase. An enzyme that catalyzes isomerization reactions.

isomerism. The condition of having two or more comparable parts made up of identical numbers of similar segments.

isomerous. Characterized by isomerism.

isomers. Two or more molecules with the same numbers and kinds of atoms, but with different arrangements of atoms and sometimes different chemical properties; isomeric.

isomorphic. Used of a plant having an alternation between diploid and haploid generations which are morphologically similar in appearance; homomorphic; isomorphous; *cf.* heteromorphic.

isoosmotic. Having the same salt concentration or exerting the same osmotic pressure.

isophene. A line on a chart connecting those places within a given region where a particular biological phenomenon (as the flowering of a certain plant) occurs at the same time.

isophyllous. Having foliage leaves of similar form on a plant or stem.

isopycnic density gradient centrifugation. A centrifugal method used to separate particles in density gradients on the basis of differences in particle density.

isospore. A spore that does not display sexual dimorphism.

isostemonous. Having the number of stamens of a flower equal to the number of perianth divisions.

isotonic. 1: Having uniform tension, as the fibers of a contracted muscle.

2: Of a solution, having the same osmotic pressure as the fluid phase of a cell or tissue.

isosporous. Having asexually produced spores of only one kind; homosporous; isosporic, isospory; *cf.* heterosporous.

isosymbiosis. Symbiosis between two organisms of equal size (isosymbionts); *cf.* anisosymbiosis.

isotomic dichotomic branching. Branching in which both dichotomies from branches are of the same size so that the dichotomic pattern is visible even in older parts of the thalli, as in *Cladonia evansii.*

isotope farm. A carbon-14 growth chamber, or greenhouse, arranged as a closed system in which plants can be grown in a carbon-14 dioxide (CO_2^{14}) atmosphere and thus become labeled with C^{14}, isotope farms also can be used with other materials, such as heavy water (D_2O), phosphorus-35 (P^{35}), and so forth, to produce biochemically labeled compounds.

isotropic. 1: Having a tendency for equal growth in all directions.
2: An ovum lacking any predetermined axis.

isotypes. A series of antigens, for example, blood type, common to all members of a species but differentiating classes and subclasses within the species.

isozyme. See isoenzyme.

isthmospore. An asexual spore composed of four or more thick-walled cells separated by thin walled cells as in *Isthmospora*.

isthmus. 1: A passage or constricted part connecting two parts of an organ, as observed in two semicells of desmids.
2: The thickened medial perforated septum of a polarilocular ascospore.

jacket. The surface cells which covers an archegonium, capsule or sporogonium.

jacket cell. An outer cell, usually of a reproductive organ.

jack o' lantern. The basidiocarp of *Clitocybe illudens* which gives out light.

jamin's chain, jaminian chain. A series of short threads of water separated by bubbles of air, in the vessels of plants.

janthinellin. An antifungal antibiotic from *Penicillium ianthinellum*.

jasmonic acid. This acid and its methyl ester (methyl jasmonate) occur in several plants and also in the oil of jasmine. These compounds inhibit growth of certain plant parts and strongly promote senescence in detached oat leaves. Their functions in the plants that contain them remain to be discovered.

jarovization. See vernalization.

Java black rot. A fungus disease of stored sweet potatoes caused by *Diplodia tubericola*, the inside of the root becomes black and brittle.

Java cotton. See kapok.

jawaharene. An antitumour antibiotic from *Aspergillus niger*.

jelly fungus. The common name for many members of the Hymenomyceles, especially the orders Tremellales and Dacrymycetales, distinguished by a jellylike appearance or consistency.

Jerusalem antichoke. *Helianthus tuberosus*; Compositae. A hardy perennial; tubers appear like potatoes and are cooked, pickled, or eaten raw. It is a good source of a carbohydrate inulin, a good food for diabetics.

jet effect wind. A wind that is intensified as a result of funnelling through a narrow canyon or mountain gap.

Jew's ear. The basidiocarp of *Auricularia auricula* a member of Gasteromycetes, forms rubbery, ear-shaped fruiting bodies,

jimsonweed. *Datura stramonium* fam. Solanaceae. A tall, poisonous annual weed having large white or violet trumpet-shaped flowers and globose prickly fruits; apple of Peru.

jointed. Said of an elongated plant member which is constricted at intervals, and ultimately separates into a number of portions by breaking across the constrictions.

Jonathan freckle. A storage disease of apples characterized by small circular discolorations in the skin.

Jonathan spot. A disease of apples characterized by circular depressed necrotic areas around the lenticels.

joule (J). A derived SI unit of energy and work, defined as the work done to displace a point, by the application of a force of 1 newton, through a distance of 1 metre in the direction of the force; also the work done when a current of 1 ampere flows through a resistance of 1 ohm for 1 second. 1 joule=1 newton-meter=0.239 calories= 10^7 ergs.

juglone (5-hydroxynaphthoquinone). It is an allelopathic compound isolated from the roots and hulls of black walnut (*Juglans nigra*). It kills tomato and alfalfa plants (in the field, up to 25 m from a walnut tree), but bluegrass seems to be stimulated close to walnut trees. All sp. of walnuts do not produce juglone.

jugum. One pair of opposite leaflets of a planate leaf.

julaceous. 1: Cylindrical and smooth, i.e, worm-like.
2: Resembling a catkin.

junceous. Latin meaning 'rush-like'.

jungle. Dense seral vegetation especially characteristic of tropical regions with a high level of precipitation.

Jurassic. A geological period of the Mesozoic era which began about 190 million years ago and lasted about 59 million years. It was named after the Jura mountains on the borders of France and Switzerland. There were many distinctive plants in this period like cycads, conifers, ferns and Ginkgos, and the earliest flowering plants.

jute. Either of two Asiatic species of tall, slender, half-shrubby annual plants, *Corchorus capsularis* and *C. olitorius*, in the family Malvaceae, useful for their fiber.

juvenescence. The process of maturing at a stage of development normally immature.

juvenile leaf. The form of leaf found on a sporeling or seedling, when it differs markedly from the leaf of the adult plant; juvenility.

juvenile phase. A young plant that has leaves and other features characteristically different from those of the adult or mature phase of the same plant are said to be in the juvenile phase. This phase varies in pereninals (with respect to flowering) from one year to 40 years with values of 5 to 20 years common in trees.

juvenile stage. A special stage in the life history of some algae from which the ordinary plant develops as an outgrowth.

juvenilody. Condition in which tissues and organs remain immature.

juxtaposed. Side by side.

K

kaempferol. A flavonoid with a structure similar to that of quercetin but with only one hydroxyl in the B ring; acts as an enzyme cofactor and causes growth inhibition in plants.

kapok. Silky fibers that surround the seeds of the kapok or ceiba tree; ceiba; Java cotton; silk cotton.

kappa particle. A self-duplicating nucleoprotein particle found in various strains of *Paramecium* and thought to function as an infectious agent; classed as an intracellular symbiont, occupying a position between the viruses and the bacteria and organelles.

karyallogamy. Reproduction in which two morphologically identical gametes fuse; karyallogamous; karyallagy.

karyapsis. See karyogamy.

karyochorisis. A process observable in fungi where nuclear membrane does not break down during mitotic nuclear division but may constrict in the middle to separate into two sister nuclei.

karyogamete. The nucleus of a gamete.

karyogamy. The fusion of two haploid nuclei after plasmogamy. In some fungi, e.g., Basidiomycetes, karyogamy takes place many cell divisions after plasmogamy. Between plasmogamy and karyogamy, each cell is a dikaryon; caryogamy; karyapsis; karyogamic.

karyogenetic. Heritable; not subjected to direct environmental influences.

karyokinesis. Nuclear division characteristic of mitosis.

karyolymph. The clear material composing the ground substance of a cell nucleus.

karyolysis. Dissolution of a cell nucleus.

karyomite. See chromosome.

karyomitosis. See mitosis.

karyoplasm. *See* nucleoplasm.

karyoplasmic ratio. *See* nucleocytoplasmic ratio.

karyorrhexis. Fragmentation of a nucleus with scattering of the pieces in the cytoplasm.

karyotype. 1: The normal diploid or haploid complement of chromosomes, with respect to size, form, and number, characteristic of an individual, species, genus, or other grouping; caryotype.

2. A photograph or diagram of a complete complement of chromosomes from a cell or individual.

kasugamycin $C_{14}H_{28}ClN_3O_{10}$. A white, crystalline antibiotic used as a fungicide for rice crops; kasumin; kasugamycin hydrochloride.

kata. A Greek prefix meaning down, against; cata.

katabatic wind. A dense mass of cold air flowing down an incline in response to gravity.

katabolism. *See* catabolism; katabolic.

katatropism. An orientation response downwards; katatropic.

katharobic. Inhabiting pure water, or water of very low organic content; katharobe, katharobia.

Kbp. Abbreviation for kilobase pairs (i.e., 1000 base pairs of DNA).

keel. 1: A boat-shaped structure formed by two joined, anterior petals of the flowers of family leguminosae.

2: A prominent dorsal rib, as in some carpels or in glumes of grasses.

3: A longitudinal narrow outgrowth from the under-side of a leaf, or leaf-like structure.

kelp. The common name for brown sea-weed belonging to the Laminariales and Fucales.

kelvin (K). The standard SI base unit of temperature, defined as 1/273.16 of the thermodynamic temperature of the triple point of water.

kelvin scale (K). A temperature scale that takes absolute zero (−273.16°C) as its starting point zero K; each degree on the kelvin scale is equal to one degree Celsius; thus, 0°C= 273.16 K, 100°C=373.16 K, T=(273.16+t) K where t=°C.

kenapophyte. A plant colonizing cleared land.

keratin. A fibrous protein containing sulphur. It has great strength, is insoluble in water and forms the basis of claws, feathers, hairs and horns etc.

keratinolytic fungi. Fungi that have the ability to degrade keratinized substrata, like human hair, toe-nail clippings and hedge hog spines. Examples indude *Trichophyton*, *Microsporum* and *Epidermophyton*.

keratinophilic. Thriving on horny (keratin rich) substrates; kerotinophile, keratinophily.

kerion. An inflammatory form of ringworm of the scalp; tinea kerion.

kernel. 1: The inner portion of a seed.

2: A whole grain or seed of a cereal plant, such as corn or barley.

kerosene fungus. *Amorphotheca resinae* found on creosoted wood and corroding aircraft fuel tanks, creosote fungus.

ketoadipic acid $C_6H_8O_5$. An intermediate product in the metabolism of lysine to glutaric acid.

ketoglutaric acid $C_5H_6O_5$. A dibasic keto acid occurring as an intermediate product in carbohydrate and protein metabolism.

ketohexose. Any monosaccharide composed of a six-carbon chain and containing one ketone group.

ketolase. A type of enzyme that catalyzes cleavage of carbohydrates at the carbonyl carbon position.

ketose. Any monosaccharide in which one carbon atom is in a ketone (—CO—) group, e.g., fructose.

key. A tabulation of the distinguishing features of a taxonomic group to serve as a guide for establishing relationships and names of unidentified members of the group.

key character. Diagnostic character.

key classification. *See* artificial classification.

key fossil. *See* index fossil.

key gene. *See* oligogene.

key species. Any species that may be used to assess the degree of utilization, of the environmental balance, of a habitat or area.

keystone predator. The dominant predator, often the top predator in a given food web; a predator having a major influence upon community.

kieselguhr. *See* diatomaceous earth.

kif. *See* cannabis.

kinase. Any enzyme that catalyzes phosphorylation reactions.

kinetin $C_{10}H_9ON_5$. A cytokinin formed in many plants which has a stimulating effect on cell division.

kinetochore. Within the centromere, the granule upon which the spindle fibers attach.

kinetoplast. A genetically autonomous, membrane-bound organelle associated with the basal body at the base of flagella in certain flagellates, such as the trypanosomes; parabasal body.

kinetosome. *See* basal body.

kingdom. The highest category in the hierarchy of classification. In older systems of classification, there are only two kingdoms—plants and animals. In some newer systems, there are five—plants, animals, fungi, bacteria and viruses; earlier the fungi, bacteria and viruses were included with the plants. In the five kingdom classification of living organisms proposed by Whittaker and currently widely used is, Monera, Protista, Fungi, Plantae and Animalia.

kinoplasm. The substance of the protoplasm that is thought to form astral rays and spindle fibers.

kinopsis. Attraction within a group of social organisms resulting from visual perception of movement alone.

kipuka. An island of vegetation isolated by lava flows.

kleistogamy. *See* Cleistogamy; kleistogamous, kleistogamic.

klendusity. A special kind of disease escape, in which a susceptible plant avoids disease because of an intrinsic property of the plant itself that greatly reduces the chances of its being inoculated, even though there may be an abundance of inoculum in the area.

klinogeotropism. The drooping tendency displayed by the tip of a climbing plant during nutation; klinogeotropic.

klinokinesis. A change in rate of random movement of an organism (kinesis) in which the rate of change of direction (frequency of turning movements) varies with the intensity of the stimulus; *cf.* orthokinesis.

klinophyte A terrestrial plant that is dependent upon other plants for physical support.

klinostat. An instrument used for experimental investigation of geotropism and phototropism in seedlings and cut shoots. It consists of a cork disc attached to the spindle of a clockwork or an electric motor. The motor is adjustable to any angle from vertical to horizontal; clinostat.

klinotaxis. An orientation response of a motile organism (taxis) in which the direction of movement results from a comparison of the intensity of the stimulus on each side of the body by successive deviations from side to side; klinotactic; *cf.* tropotaxis, telotaxis.

klinotropism. An orientation response to a gradient of stimulation; clinotropism, klinotropic.

Koch's postulates. A set of laws elucidated by Robert Koch in vogue since about 1880 are still considered essential to reliable diagnosis of infectious diseases. They state that to establish a micro-organism as the cause of a disease:

1: it must be found in all cases of the disease
2: it must be isolated from the host and grown in pure culture
3. it must reproduce the original disease when re-introduced into a susceptible host
4. it must be found in the experimental host so infected.

koji. A Japanese food prepared by inoculating cooked rice with *Aspergillus oryzae*;; mould rice.

kojic acid. Characteristic metabolic product of *Aspergillus flavus-oryzae* and *tamarii* groups. It gives a blood-red colour with ferric chloride

kollaplankton. Planktonic organisms rendered buoyant by encasement in gelatinous envelopes; collaplankton.

koprophagous. See coprophagous; koprophage, koprophagy.

Kranz anatomy. Anatomically the leaves of C_4 plants (like sugarcane, maize, sorghum etc.) are characterized by a sheath of parenchyma cells that are radiallv arranged around each vascular bundle. The vascular bundle, in turn, is enclosed by loosely packed "spongy" mesophyll cells. This closed, vascular bundle sheath arrangement is known as Kranz anatomy.

Krebs cycle. A sequence of enzymatic reactions in aerobic respiration involving oxidation of a two-carbon acetyl unit to carbon dioxide and water to provide energy for storage in the form of high-energy phosphate bonds. The Krebs cycle takes place in mitochondria; citric acid cycle, tricarboxylic acid cycle.

kremnophyte. A plant which grows alongside a steep wall.

kurtosis. A measure of the departure of a frequency distribution from a normal distribution, in terms of its relative peakedness or flatness.

kybernetics. See cybernetics.

kymatology. The study of waves and wave motion.

L

labellate. Having a labellum.

labellum. The median membrane of the corolla of an orchid often differing in size and morphology from the other two petals.

labriform. Lip-shaped; frequently used for the terminal soralia of lichens having this shape.

laccal $C_{17}H_{31}C_6H_3(OH)_2$. A phenol compound which is found in the sap of lacquer trees, and which can be isolated in crystalline form.

laccase. Any of a class of plant oxidases which catalyze the oxidation of phenols.

laccate. Having a lacquered *i.e.*, polished or varnishd appearance.

lacerated. Having a deeply and irregularly incised margin or apex, as,if torn.

laciniate. 1: Having a fringed border.
2: Narrowly and deeply incised to form irregular lobes, which may be pointed.

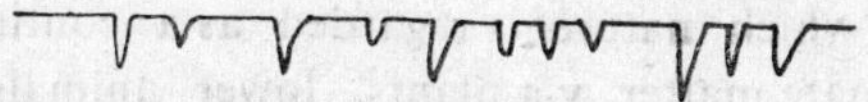

lac operon. A seqnence of three linked genes which code for the enzymes involved in lactose utilization in many bacteria,

lacrimiform. Like a tear drop.

lactase. An enzyme that catalyzes the hydrolysis of lactose to dextrose and galactose.

lactescent. Having a milky appearance.

lactic acid. $C_3H_6O_3$. A hygroscopic α-hydroxy acid, occurring in three optically isomeric forms; L. form, in blood and, muscle tissue as a product of glucose and glycogen metabolism; D form, obtained by fermentation of sucrose; and DL form, a racemic mixture present in foods prepared by bacterial fermentation, and also made synthetically; 2-hydroxypropanoic acid; α-hydroxypropanoic acid.

lactifer. A latex-carrying hypha.

lactin. See lactose.

Lactobacillus. Lactic acid bacteria, the single genus of the family Lactobacillaceae; found in dairy products, meat products, fruits, beer, wine, and other food products.

lactoflavin. See riboflavin.

lactose. $C_{12}H_{22}O_{11}$. A disaccharide composed of D-glucose and D-galactose which occurs in milk; lactin; milk sugar.

lacuna. A small space or depression.

lacquer tree. See varnish tree.

ladder of nature. The classification scheme devised by Aristotle in which nature is regarded as a continuum leading from inanimate matter via plants, lower animals, and higher animals, to mankind.

laevigate. Smooth.

lageniform. Flask-shaped; *i.e.*, swollen at the base, narrowed at the top.

lag phase. The period of physiological activity and diminished cell division following the addition of inoculum of bacteria to a new culture medium.

lake ball. A spherical mass of tangled, waterlogged fibers and other filamentous material of living or dead vegetation, produced mechanically along a lake bottom by wave action, and usually impregnated with sand and fine-grained mineral fragments; burr ball; hair ball.

Lamarck. Jean-Baptiste Lamarck (1744-1829) proposed that evolution occurred as a result of inheritance of changes which happen to an organism during its lifetime. This is known as the Theory of inheritance of Acquired Characters.

lamella (*pl.* lamellae) 1: In mushrooms, one of the thin, spore-bearing gills attached to the underside of the cap.

2: The membranes of the grana of chloroplast.

lamellar chloroplast. A type of chloroplast in which the layered structure extends more or less uniformly through the whole chloroplast body.

lamina. 1: Blade; used of the main part of the thallus in foliose lichens.

2: The parts of the leaf on either side of the midrib.

3: A thin layer; the smallest recognizable unit in a bedded sediment.

laminal placentation. Condition in which the ovules occur on the inner surface of the carpels.

laminar flow. The streamlined movement of water, in which water particles appear to move in smooh paths, and one layer of water slides over another; *cf.* turbulence.

lampbrush chromosome. An exceptionally large chromosome characterized by fine lateral projections which are associated with active ribonucleic acid and protein synthesis.

lanate. Like wool; covered with short hair-like processes; lanose; languinose.

lanatoside. Any of three natural glycosides from the leaves of *Digitalis lanata*; on hydrolysis with acid, it yields one molecule of D-glucose, three molecules of digitoxose, and one molecule of acetic acid, all three glyosides are cardioactive.

lanceolate. Shaped like the head of a lance.

lapachol. $C_{15}H_{14}O_3$. A yellow crystalline compound obtained from lapacho, a hard-wood in Argentina and Paraguay; 2-hydroxy-3-amylene-α-naphtho-quinone, lapachoic acid; targusic acid.

lapidicolous. Living under a stone.

latent bud. An axillary bud whose development is inhibited, sometimes for many years, due to the influence of apical and other buds; dormant bud.

latent period 1: The period between the introduction of a stimulus and the response to it.

2: The initial period of phage growth after infection during which time virus nucleic acid is manufactured by the host cell.

lateral bud. Any bud that develops on the side of a stem.

lateral meristem. Strips or cylinders of dividing cells located parallel to the long axis of the organ in which they occur; the lateral meristem functions to increase the diameter of the organ.

lateral root. A root branch arising from the main axis.

latex. A complex, coloured, sticky or milky liquid mainly containing hydrocarbons, carbohydrates and proteins, produced by specialized cells, which some plants exude when cut e.g., rubber. Natural rubber is a mixture of unsaturated hydrocarbons with a cis configuration.

laticifer. A latex duct found in the mid-cortex of cerain plants.

latiferous. Containing or secreting latex. latiferous.

lattice. Cross-barred; like a network.

laurel forest. See temperate rainforest.

laurisilva. See temperate rainforest.

laurophilus. Thriving in sewers and drains; laurophile, laurophily.

laurophyte. A sewer or drain plant; laurad; laurophyta.

law of minimum. The law that those essential elements for which the ratio of supply to demand (A/N) reaches a minimum will be the first to be removed from the environment by life processes; it was proposed by J. von Liebig, who recognized, phosphorus, nitrogen, and potassium as minimum in the soil; in the ocean the corresponding elements are phosphorus, nitrogen, and silicon; Liebig's law of minimum.

law of specificity of bacteria. See Koch's postulate.

layering 1: A propagation method by which root formation is induced on a branch or a shoot attached to the parent stem by covering the part with soil.

2: A stratum of plant forms in a community, such as mosses, shrubs, or trees in a bog area.

LD. (Lethal dose, LD_{50}). A concentration of a fungicide etc. which kills 50 per cent of the spores, cells or individuals of the test organism.

LDH. See lactic dehydrogenase.

leaf. A modified aerial appendage which develops from a plant stem at a node , usually contains chlorphyll, and is the principal organ in which photosynthesis and transpiration occur.

leaf fiber. A long, multiple-celled fiber extcarted from the leaves of many plants that is used for cordage, such as sisal for binder, and abaca for manila hemp.

leaf gap. A gap in the vascular cylinder of a stem, just above a node.

leaflet. 1: A small leaf-like division of a compound leaf.
2: A small or young foliage leaf.

leaf scar. The scar remaining on a stem at the point where a leaf has fallen from it.

leaf trace. The vascular tissue which branches off from the stem, at a node, into a leaf.

leafy liverwort. A liverwort in which the gametophyte has a simple stem, growing from the apex and bearing small leaves in rows along it. About 80 per cent of liverwort species are leafy.

succubous. Of a kind of growth in leafy liverworts, in which the front of each leaf lies underneath the leaf in front of it.

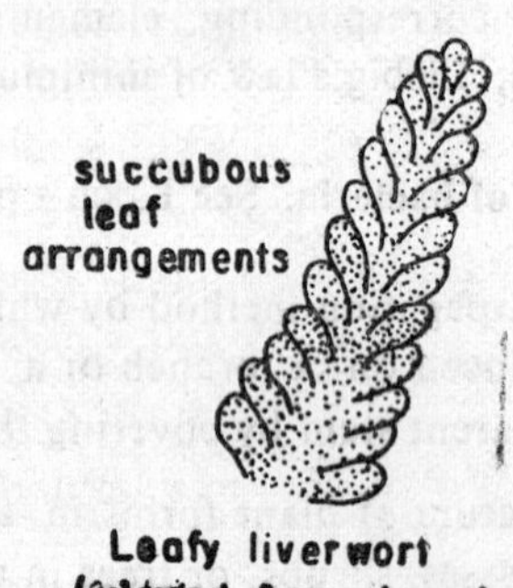

Leafy liverwort
(viewed from above)

leaky mutant gene. An allele with reduced activity relative to that of the normal allele.

lecithin. Any of a group of phospholipids having the general composition $CH_2OR_1.CHOR_2\,CH_2OPO_2OHR_3$, in which R_1 and R_2 are fatty acids and R_3 is choline, with emulsifying, wetting, and antioxidant properties, and is used in the manufacture of soaps. They are present in plasma membrane and play an active role in its function.

lecithinase. An enzyme that catalyzes the breakdown of a lecithin into its constituents.

lectin. Any of various proteins that agglutinate erythrocytes and other types of cells and also have other properties, including mitogenesis, agglutination of tumor cells, and toxicity toward animals; found widely in plants, predominantly in legumes, and also occurring in bacteria, fish, and invertebrates.

lectotype. A specimen selected as the type of a species or subspecies if the type was not designated by the author of the classification.

lecythiform. Spore, like a stoppered bottle; ninepin-shaped.

Lederberg technique. A method for rapid isolation of individual bacterial cells for demonstrating the spontaneous origin of bacterial mutants.

leghemoglobin. A hemoglobin-like oxygen binding red pigment in the root nodules of legumes which protects the nitrogenase enzyme complex from being destroyed by excess oxygen.

legume. 1: A dry, dehiscent fruit derived from a single simple pistil; e.g., *Medicago*, *Phaseolus*, *Arachis hypogoea* (peanuts).

2: A term used in agriculture for leguminous plants used as animal fodder.

legumin. A globulin found in the seeds of Leguminosae.

leiosporous. Having smooth spores.

lemma. (*pl.* lemmas) Either of the pair of inner bracts that are borne above the lumes and enclose the flower of a grass spikelet.

lentic. Of or pertaining to static, calm or slow moving waters such as lakes, reservoirs, ponds, and bogs; lenitic; *cf.* lotic.

lenticel. A loose structured openinig on the surface of the stems of some plants, allowing gas exchange between the stem and the atmosphere; lenticellate.

lenticular. Shaped like a biconvex lens.

lentigin. Freckled Spores i.e., having very small spots on surface; lentiginose; lentiginous.

lepidophyllous. Having scaly leaves.

lepidote. Covered with small scales.

leprose. Pertaining to lichens, having the surface or the whole thallus entirely dissolved into soredia e.g., *Lepraria*.

leptocaul. Having thin stems, without heavy primary thickening, as in most trees; leptocauly.

leptoids. These are special food conducting cells present in mosses and are collectively known as leptom; *cf.* hydroids, hydrom which only conduct water.

leptophyll. A Raunkerian leaf size class, for leaves having a surface area of 25 square millimeters of less; common in alpine and desert habitats.

leptotene. The first stage of meiotic prophase, when the chromosomes appear as thin threads having well-defined chromomeres.

leptotichous. Said of a tissue which is thin walled.

lesion. A localized area of diseased tissue. e.g., a spot, a scab.

lethal gene. A gene mutation that causes premature death in heterozygotes if dominant, and in homozygotes if recessive; lethal mutation.

leucoplast. A plastid containing no pigment. Leucoplasts may form pigments under certain conditions, e.g., leucoplasts in root cells form chlorophyll if exposed to light. There are different types of leucoplasts; amyloplasts, which store starch; aleuroplasts, which store proteins; and elaioplasts, which store fats.

leucosin. A simple protein of the albumin type found in wheat and other cereals.

leucosporous. Having white spores in the mass.

leukocidin. A toxic substance released by certain bacteria which destroys leukocytes.

levulose. Levorotatory D-fructose.

L-form bacteria. Bacteria that have, temporarily or permanently, lost thc ability to produce a cell well as a result of growth in the presence of antibiotics inhibiting cell wall synthesis.

liana. A woody or herbaceous climbing plant with roots in the ground, found in tropical forests.

lichen. A symbiosis between a green or blue-green alga and a fungus. Lichens are usually small plants, with a range of colour green, grey, bluish-green or yellow, which grow on rocks or as epiphytes, on tree trunks, old walls and roof. They are most universally distributed of all plants, being capable of growing in a variety of climates.

Lichenes Imperfecti. A class of the Lichenes containing species with no known method of sexual reproduction; Deuterolichenes.

lichenicolous. Growing on or in lichen thalli.

lichenology. The study of lichens.

lichenometry. A method for dating rocks based on a knowledge of the growth rates of encrusting lichens, used extensively by glaciologists; lichenometric.

lichenophilous. Thriving on, or having an affinity for, lichens or lichen-rich habitats; lichenophile, lichenophily.

licorice. *Glycyrrhiza glabra.* A perennial herb of the legume family (Leguminosae) cultivated for its roots, which when dried provide a product used as a flavoring in medicine, candy, and tobacco and in the manufacture of shoe polish.

Liebig's law of the minimum. *See* law of minimum.

life cycle. The complete succession of changes that occur from any stage in the life of an organism to the same stage in the life of its offspring. In bryophytes, pteridophytes and spermatophytes, the life cycle consists of an alternation of haploid and diploid generations; life history.

life form. The form characteristically taken by a plant at maturity.

life zone. A portion of the earths land area having a generally uniform climate and soil, and a biota showing a high degree of uniformity in species composition and adaptation.

ligase. An enzyme that catalyzes the union of two molecules, involving the participation of a nucleoside triphosphate which is converted to a nucleoside diphosphate or monophosphate; synthetase.

light microscope. An instrument that uses light rays passing through a system of lenses to magnify small objccts. The light microscope can be used for observing the arrangement of cells and tissues and the larger structures inside cells. It is not powerful enough for the observation of small details of cell structure; microscopy.

ligneous. Of, pertaining to, or resembling wood; lignose.

lignicolous. Growing on or in wood; lignicole.

lignify. To convert cell wall constituents into wood or woody tissue by chemical and physical changes.

lignin. A complex aromatic compound which is deposited in the cellulose cell walls of the xylem and sclerenchyma during the process of secondary thickening. Wood is made mostly of lignin, a colourless to brown substance removed from paper pulp sulphite liquor; lignify; lignified.

lignivorous. Feeding on wood; dendrophagous; hylophagous; zylophagous; lignivore, lignivory.

lignocellulose. Any of a group of substances in woody plant cells consisting of cellulose and lignin.

lignosa. Woody vegetation.

lignumvitae. *Guaiacum sanctum*. A medium-sized evergreen tree of the family zygophyllaceae that yields a resin or gum known as gum guaiac or resin of guaiac; hollywood lignumvitae.

ligulate. 1: Strap-shaped; flat and narrow; lorate; liguliform.
2: Having ligules.

ligule. 1: A small outgrowth in the axis of the leaves in Selaginellales.
2: A thin outgrowth of a foliage leaf or leaf sheath in many grasses.
3: The corolla of a ray-floret in a composite inflorescence.

lily. 1: Any of the perennial bulbous herbs with showy unscented flowers constituting the genus *Lilium*.
2: Any of various other plants having similar flowers.

limicolous. Living in mud.

limnetic. Of, pertaining to, or inhabiting the pelagic region of a body of fresh water.

limnonereid. A freshwater plant.

limnology. The science of the life and conditions for life in lakes, ponds, and streams.

limnoplankton. Plankton found in fresh water, especially in lakes.

limoniform. Lemon-shaped.

line. A unit of length, equal to 1/12 inch, or approximately 2.117 millimeters; it is most frequently used by botanists in describing the size of plants.

lineolate. Marked with fine lines.

lingulate. Tongue-or strap-shaped; linguiform.

linkage. The usual inheritance of two or more characters together, which happens when the genes controlling these characters are on the same chromosome. Linked genes can only be separated by crossing-over during meiosis.

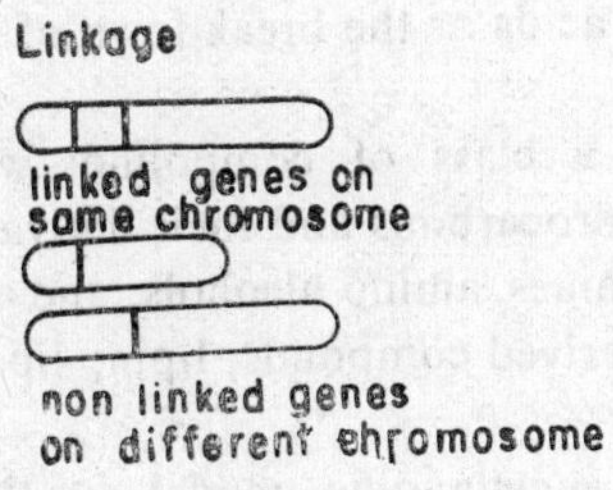

linkage group. The genes located on a single chromosome.

Linnaeus. Carl Linnaeus, (1707-1778) was responsible for the modern system of naming plants and animals. This is the binomial system. His most famous work was *Species Plantarum*, published in 1753, in which he described all the plants then known to man.

linoleic acid. $C_{17}H_{31}COOH$ A yellow unsaturated fatty acid, boiling at 229°C (14 mm Hg), occurring as a glyceride in drying oils; obtained from linseed, safflower, and tall oils; a principal fatty acid in plants, and considered essential in animal nutrition; used in medicine, feeds, paints, and margarine; linolic acid; 9,12-octadecadienoic acid.

linolenic acid. $C_{17}H_{29}COOH$. One of the principal unsaturated fatty acids in plants and an essential fatty acid in animal nutrition; a colourless liquid that boils at 230°C, soluble in many organic solvents; used in medicine and drying oils; 9,12,15-octadecatrienoic acid.

linseed oil. An oil extracted from the seeds of flax plants (Linum sp., family, Linaceae) *Linum* sp family Linaceae. It contains glycerides of unsaturated fatty acids and is therefore easily oxidised by oxygen in air attacking the double bonds. For this reason linseed oil is widely used in the varnish and paint industries.

lipase. An enzyme that catalyzes the hydrolysis of fats to glycerol and fatty acids or the breakdown of lipoproteins.

lipid. One of a class of compounds which contain long-chain aliphatic hydrocarbons and their derivatives, such as fatty acids, alcohols, amines, amino alcohols, and aldehydes; includes waxes, fats, and derived componds; lipin; lipoid.

lipid bilayer. Long-standing model for the structure of cell membranes based on the hydrophobic interactions between phospholipids. The polar heads face outward and the hydrophobic tails are clustered in the interior.

lipoid. 1: A fat-like substance.
2: *See* lipid.

lipopolysaccharide. Any of a class of conjugated polysaccharides consisting of a poly-saccharide combined with a lipid.

lipoprotein. Any of a class of conjugated proteins consisting of a protein combined with a lipid. Together with mucoproteins lipoproteins are important structural materials for cell membranes.

liposomes. Single or multilamellar membrane-bordered vesicles formed by vigorously dispersing lipids in water.

lipoxenous. Used of a parasite that leaves its host after feeding; lipoxeny.

lirella. A long, narrow apothecium with a medial longitudinal furrow, occurring in certain lichens, and in the order Hysteriales.

lithocyst. Epidermal plant cell in which cystoliths are formed.

lithodomous. Used of an organism that lives in holes in rock, or that bores into rock; lithotomous.

lithophagic. 1: Used of organisms that erode or bore into rock; further subdivided into epilithophagic, mesolithophagic and endolithophagic according to the degree of penetration; lithophagous, lithophage, lithophagy.

2: Eating small stones, as in some birds.

lithophyte. A plant that grows on rock; saxicolous.

lithosere. An succession of plant communities that originate on rock.

litmus. A amphoteric lichen dye obtained from *Ochrolechia tartarea*. *Roccella* sp.

litter. Dead plant and animal material on the surface of the ground, above the humus layer.

little-drop technique. A method for isolating single cells in which a drop of a cellular suspension containing a single cell, as determined by microscopic examination, is transfered with a capillary pipete to an appropriate culture medium.

littoral. Of habitats between the high and low tide marks on the sea shore.

littoral zone. Of or pertaining to the bio-geagrophic zone between the high-and low-water marks.

lituate. Having a forked member or part with the ends turned slightly outward, as in certain fungi.

liverwort. 1: One of the two groups of bryophytes. Liverworts differ from mosses in having less differentiated cells in the gametophyte, and in having elaters in the dehiscent capsule. The gametophyte is either thalloid or leafy; hepatic.

2: Sometimes used in common names for large foliose lichens e.g., ash coloured ground liverwort *Peltigera canina.*

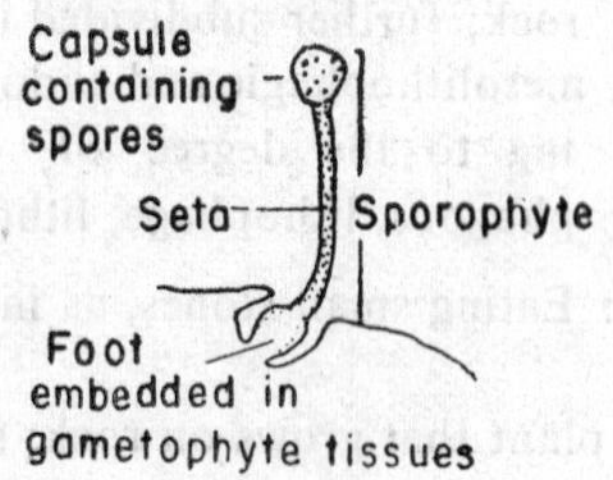

living fossil. A living species belonging to an ancient stock otherwise known only as fossils.

lobe. Flat, roundish piece of tissue, as at the margin of a digitate or dissected leaf. Also, petals are sometimes called corolla lobes.

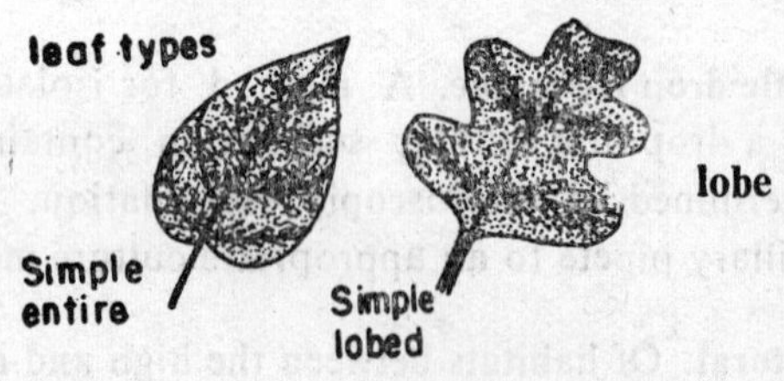

A lobed leaf.

lobule 1: A small lobe.

2: A division of a lobe.

local infection. An infection affecting a limited part of a plant.

locellus. 1: In some legumes, a secondary compartment of a unilocular ovary that is formed by a false partition.

2: One of the two cavities of a pollen sac.

lociation. One of the subunmits of a faciation, distinguished by the relative abundance of a dominant species.

loculicidal. Of the dehiscence of a capsule with several carpels, which split lengthways, exposing the seeds in each locule.

loculate. Having, or divided into loculi.

loculus. A small cavity or chamber.

locus. The fixed position of a gene in a chromosome.

logarithmic growth. *See* exponential growth.

logotype. The selection or designation of a genotype after the generic name was published.

logwood. *See* hematoxylon.

lomasome. Membranous vesicular material embedded in the wall external to the line of the plasmalemma. They are not peculiar to fungi but occur in other kinds of walled cells; *cf.* plasmalemmasomes.

loment. A dry, indehiscent single-celled fruit that is formed from a single superior ovary; splits transversely at maturity into several segments.

long-day plant. A plant which has a flowering period initiated and accelerated by exposure to a light regime with more than 12 h illumination daily; *cf.* short-day plant.

longicollous. Having a long beak or neck.

longitudinal section. A cut made through or along an organ or tissue, in the same direction as the main direction of growth L.S. (abbr.).

lophophilus. Thriving on hill tops; lophophile, lophophily.

lophophyte. A hill-top plant; lophad; lophophyta.

lophotrichous. Having a polar tuft of flagella.

lorate. Strap-shaped.

lotic. Of or pertaining to swiftly moving waters.

LSD (lysergic acid diethylamide). A very potent hallucinogenic drug which is active in minute doses. It is a derivative of lysergic acid and was originally obtained from ergot alkaloids i.e., sclerotia of *Claviceps paspali* found on *Paspalum* grass. It has been established that LSD can cause chromosomal damage in regular users. It has an unpredictable effect on the human mind. The phyrical effects include dilatation of the pupils, rise in blood pressure, and heart-rate, nausea, tremor, fever, and insomnia. LSD is not a drug of addiction and has been used in psychotherapy. It can now be fully synthesised in laboratory.

lucerne. *See* alfalfa.

lucicolous. Living in open habitats with ample light; lucicole; *cf*. umbraticolous.

luciferase. This is an enzyme present in luminescent fungi (or other organisms as well). This act on luciferin and light is emmitted. However, luciferase itself may be the light emitting molecule.

luciferin. A complex organic substance present in many luminous organisms that emits heatless light when combined with luciferase in the presence of oxygen. Luciferin causes bioluminescence.

luciferous. Light-producing; bioluminescent.

lucifugous. Intolerant of light; photophobic; lucifuged; *cf*. luciphilous.

luciphilous. Thriving in open, well-lit habitats; photophilic; lucipetal; luciphile; luciphily; *cf.* lucifugous.

luminescence. The emission of light by biochemical processes without the production of heat; *cf.* phosphorescence.

luminescent fungi. Mycelia and basidiocarp of some fungi emit light sometimes causing the attacked wood or leaves to become luminous, e.g., *Pleurotus*, *Clitocybe*, *Armillaria mellea*, *Xylaria hypoxylon*.

lunate. Crescent-shaped.

lungwort. Common name of a lichen *Lobaria pulmonaria*, because of its external resemblance to a human lung; lung lichen; lungs of oak. Previously used as a cure for pulmonary diseases.

lunule. A cresent-shaped organ, structure, or mark.

lutaceous. Pertaining to mud.

luticolous. Inhabiting mud; luticole.

lycopene. $C_{40}H_{50}$. A red, crystalline hydrocarbon that is the colouring matter of certain fruits, as tomatoes; it is isomeric with carotene.

lycopodium powder. A fine powder obtained from the spores of plants belonging to the genus *Lycopodium*. It is used in fireworks.

lygophilous. Thriving in dark or shaded habitats; lygophile, lygophily.

lyophilization. Freeze drying.

lyse. To undergo lysis.

lysigenous. Of or pertaining to the space formed following lysis of cells.

lysimeter. An apparatus for measuring the actual uptake of water by a plant.

lysis. Dissolution of a cell or tissue by the action of a lysin.

lysogenic bacterium. A bacterium that contains a prophage.

lysogenic viruses. Viruses that can become prophages.

lysogeny. The condition in which a bacteriophage is latent, integrated with the genetic apparatus of its host bacterium, yet capable of being activated by certain stimuli.

lysosome. An organelle surrounded by a single membrane, in which hydrolytic enzymes are stored. Lysosomes are common in animal cells, but probably not so common in plant cells. They have been observed in plants in onion seeds, corn seedlings and tobacco seedlings. The large vacuole of many plant cells is a modified lysosome.

lysozyme. This is an enzyme that is produced in both plant and animal tissue and was the first to have its complete 3-D structure revealed. The enzyme cleaves polysaccharide chains in the cell walls of certain bacteria by hydrolysis.

lytic infection. Penetration of a host cell by lytic phage, i.e., which lead to the lysis of the host cell.

lytic phage. Any phage that causes host cells to lyse; lytic viruses.

lytic reaction. A reaction that leads to lysis of a cell.

M

macrandrous. Having both antheridia and oogonia on the same plant; used especially for certain green algae, e.g., *Oedogonium*.

macroconsumer. A large consumer which ingests other organisms or particulate organic matter; biophage.

macrocyclic. Of a rust fungus, having binuclear spores as well as teliospores and sporidia, or having a life cycle that is long or complex, e.g., *Puccinia graminis.*

macroflora. 1: Plants which are visible to the naked eye.
2: Widely distributed flora.
3: Flora of a macrohabitat.

macrogamete. The larger of two anisogametes, usually female gamete produced by a heterogamous organism; megagamate; *cf.* microgamete.

macrohabitat An extensive habitat presenting considerable variation of the environmentt, containing a variety of ecological niches, and supporting a large number and variety of complex flora and faun.

macromolecule. A very large molecule. In most cells they constitute about 90 per cent of the total mass; vary in size from several hundred to several hundred million molecular weight units. Four major classes of macromolecules with their relative amounts in a typical cell are proteins (10-25 per cent), polysaccharides (1-5 per cent), lipids (2-10 per cent) and nucleic acids (0.5-5 per cent).

macronema (*pl.* macronemata). Rhizoids that originate from large cells that surround buds on a stem of mosses; have much branched uniseriate structure. Their function is protection of apical cell of a branch bud; *cf.* micronema.

macronutrient. An element, such as sulphur, phosphorus, magnesium, calcium, potassium and nitrogen, essential in large quantities for plant growth. Macronutrients are needed in concentrations of 1000 $\mu g\ g^{-1}$ of dry matter or more; *cf.* micronutrient.

macrophyte. A large macroscopic plant, used especially of aquatic forms such as kelp; macrophytic.

macrophyllous. Having large or long leaves.

macrophytophagous. Feeding on higher plant material only; macrophytophage, macrophytophagy.

macropodous. 1: Having a large or long hypocotyl.
2: Having a long stem or stalk.

macrosporangium. A spore case in which macrospores are produced; megasporangium; *cf.* microsporangium.

macrospore. The larger of two haploid spore types produced by heterosporous plants; the female gamete; megaspore; *cf.* microspore.

macrostylous. 1: Having long styles.
2: Having long styles and short stamens.

maculate. Marked with speckles or spots.

magnetotaxis. The reaction of certain aquatic bacteria in response to magnetic field lines. Such bacteria are influenced by the Earth's magnetic field. Magnetotactic bacteria (discovered in 1975) are able to synthesise tiny crystals of magnetite, Fe_3O_4. They are either north-seeking or south seeking. Magnetotactic bacteria are found in both freshwater and sea water. They include cocci, bacilli, and spirals.

male. Symbol: ♂ 1: A flower lacking pistils.
2: Of individuals, organs, tissues, etc. producing the gametes which fertilize egg cells produced by females.

malt. Grains such as barley, oats, and wheat, which has been germinated under controlled conditions and then dried. It is rich in proteins and is used in food manufacture and brewing.

maltase. An enzyme that catalyzes the conversion of maltose to dextrose. Maltase is found in intestinal juice and yeast.

Malthusianism. The theory that population increases more rapidly than the food supply unless held in check by epidemics, wars, or similar phenomena.

maltobiose. *See* maltose.

maltose. $C_{12}H_{22}O_{11}$. A crystalline disaccharide formed by the linkage of two D-glucose units in $\alpha(1\rightarrow4)$ glycosidic bond. It is a product of the enzymatic hydrolysis of starch, dextrin, and glycogen, does not appear to exist free in nature; maltobiose, malt sugar.

malt sugar. *See* maltose.

mangrove. A tropical tree or shrub of the genus *Rhizophora* characterized by an extensive, impenetrable system of prop roots which contribute to land building.

mangrove swamp. A tropical or subtropical marine swamp distinguished by the abundance of low to tall trees, especially mangrove trees.

manila hemp. *See* abaca.

mannan. Any of a group of polysaccharides composed chiefly or entirely of D-mannose units.

mannitol. $CH_2OH-(CHOH)_4-CH_2OH$. A white, crystalline polyhydric alcohol derived from mannose. It is found in many plants e.g., in brown algae as a storage carbohydrate and in the sap of various trees.

mannose. $C_6H_{12}O_6$. A fermentable monosaccharide obtained from manna.

manuscript name. In taxonomy, an unpublished scientific name.

marcescent. Withering without falling off.

marginal placentation. Arrangement of ovules near the margins of carpels.

marginal species. A plant species that occurs on the edge of a habitat or community.

marginate. Having a distinct margin or border.

maricolous. Living in the sea.

marihuana. *See* marijuana.

marijuana. The Spanish name for the dried leaves and flowering tops of the hemp plant (*Cannabis sativa*), which have narcotic ingredients and are smoked in cigarettes. Also spelled marihuana.

marine biocycle. A major division of the biosphere composed of all biochores of the sea.

marsh. A transitional land-water area, covered at least part of the time by estuarine or coastal waters, and characterized by aquatic and grasslike vegetation, especially without peatlike accumulation.

marsupium. A subterranean pouch of stem material that encloses the archegonia in some liverworts, termed a marsupidium when greatly elongated.

Mary Jane. *See* cannabis.

matric potential(ψm). It is measure of the tendency of a matrix to adsorb additional water molecules. This tendency is equal to the average tenacity with which the least tightly held layer of water molecules is adsorbed. Matric potential is expressed in units of water potential. A dry colloid or hydrophilic surface, such as filter paper, wood, soil, gelatin, or the stipe of a brown alga often has an extremely, negative matric potential, while the same colloid in a large volume of pure water at atmospheric pressure has a matric potential of zero.

matric pressure. Matric potential.

matrix. (*pl.* matrices) 1: A ground substance consisting of non-living intercellular material in which living cells are embedded.

2: The liquid inside a mitochondrion.

3: The substratum in or on which a fungus or lichen grows.

meadow. A vegetation zone which is a low grassland, dense and continuous, variously interspersed with broad leaved herbs but few if any shrubs.

mechanical inoculation. Of plant viruses, a method of experimentally transmitting the pathogen from plant to plant: juice from diseased plants is rubbed on test-plant leaves that usually have been dusted with Carborundum or some other abrasive material.

medium. A solid or liquid substrate containing all the materials necessary for growth, used by biologists for the cultivation of organisms such as bacteria, fungi and algae and, also for the growth of tissue cultures.

medulla. 1: The parenchyma or sclerenchyma inside the vascular cylinder of a stem or root. Its function is the storage of food.

2: The name given to the central part of the thallus of some algae and lichens.

3: In the sporocarp of fungi, the part composed mainly of longitudinal hyphae.

megaphyll. A leaf whose leaf trace makes a gap in the vascular system of the stem.

megaspore. A large spore produced in a megasporangium. in heterosporous plants. The megaspore develops into the female gametophyte. In angiosperms, the megaspore is the embryo sac; macrospore; *cf.* microspore.

megasporophyll. A sporophyll bearing megasporangia. In angiosperms, the carpels are the megasporophylls.

meiosis. Cell division that produces haploid sex cells from diploid cells. Meiosis involves two cell divisions: (1) the replicat-homologous chromosomes pair with each other on the spindle; ed crossing over happens at this stage. The chromosomes then separate to either end of the spindle. (2) The chromatids of each chromosome come apart at the centromere, and separate to each end of the second spindle. There is usually no interphase between the two divisions. Meiosis occurs in all organisms which reproduce sexually; meiotic.

meiospore. A haploid cell produced by meiosis that forms the gametophyte by mitotic division.

meiotaxis. The suppression of an entire whorl in phyllotaxis.

meiotherm. A plant thriving in cool-temperate habitats; meiothermous.

melanin. Any of a group of brown or black pigments occurring in plants and animals.

melanosome. An organelle which contains melanin and in which tyrosinase activity is not demonstrable.

melitriose. *See* raffinose. melitriose

membranaceous. Of leaves which are very thin.

membrane. A thin sheet of soft material which protects and encloses cells and organelles. Membranes control the movement of substances in and out of cells and organelles. Biological membranes are made of protein and phospholipid.

Mendel's Laws. The laws of inheritance worked out by an Austrian, Gregor Mendel, (1822-1884) in 1866. First law: that in sexual organisms the two members of an allele pair or pair of homologous chromosomes separate during gamete formation and that each gamete receives only one member of the pair; law of segregation. Second law: that the random distribution of alleles to the gametes results from the random orientation of the chromosomes during meiosis; law of independent assortment.

mentum. A projection formed by union of the sepals at the base of the column in some orchids.

merdicolous. Living on or in dung; stercoraceous; merdicole.

merdivorous. Feeding on dung and faecal matter; coprophagous; scatophagous; merdivore, merdivory.

mericarp. An individual, one-seeded carpel of a schizocarp.

meristem. Formatic plant tissue composed of undifferentiated cells capable of dividing and giving rise to other meristmatic cells as well as to specialized cell types; found in growth areas.

merosporangium. In Mucorales, a cylindrical out grouth from the swollen end of a sporangiophore in which a chain-like series of sporangiospores is generally produced.

mesarch. 1: Having metaxylem on both sides of the protoxylem in a siphonostele.

2: Originating in a mesic environment.

mesarch succession. An ecological succession beginning in a habitat with a moderate amount of water; mesosere; *cf.* hydrarch succession, xerarch succession.

mescal buttons. The dried tops from the cactus *Lophophora williamsii*; capable of producing inebriation (drunkenness) and hallucinations.

mesocarp. The layer of tissue in a fruit between the exocarp and the endocarp. The mesocarp is often fleshy or succulent.

mesophilic. 1: Thriving under intermediate or moderate environmental conditions; sometimes restricted to conditions of moderate moisture, or to moderate temperature; noterophilous; mesophilous; mesophilus; mesophil. mesophile, mesophily.

2: Used of microorganisms having an optimum for growth between 20-45°C.

mesophyll. 1: The parenchymatous tissue between the upper and lower epidermis of a leaf. In dicotyledons it is differentiated into palisade parenchyma and spongy mesophyll, but in most monocotyledons it is undifferentiated.

2: A Raunkiaerian leaf size class for leaves having a surface area between 2025-18 225 mm^2; sometimes subdivided into notophyll (2025-4500 mm^2) and mesophyll (4500-18 225 mm^2).

mesophyte. A plant requiring moderate amounts of moisture and teamperature, without major seasonal fluctuations for optimum growth; noterophyte!, mesad; mesophyta, mesophytic; *cf.* hydrophyte, hygrophyte, xerophyte.

mesopleustophyte. Any large plant floating freely between the surface and the floor of a lake.

mesosere. A sere originating in a mesic habitat and characterized by mesophytes.

mesosome. An extension of the cell membrane within a bacterial cell; possibly involved in cross-wall formation, cell division, and the attachment of daughter chromosomes following deoxyribonucleic acid replication.

mesotherm. A plant favouring intermediate temperature conditions, with a minimum of 22°C in the warmest month and a range between 6-18°C in the coldest month; mesothermophyte; mesothermic.

Mesozoic. The geological era 225·65 million years ago, following the Palaeozoic with gymnosperms as the dominant plants on land. Angiosperms first appeared during the Mesozoic.

messenger ribonucleic acid. (mRNA). A linear sequence of nucleotides which is transcribed from and complementary to a single strand of deoxyribonucleic acid and which carries the information for protein synthesis to the ribosomes.

metabiosis. An ecological association in which one organism precedes and prepares a suitable environment for a second organism.

metabolic pathway, A set of chemical reactions which follow each other in a sequence. Each reaction uses the product of the reaction before it.

metabolic poison. Any substance e.g., cyanide, that prevents the production of ATP in cells. Without ATP as a source of energy for metabolism, cells and organisms quickly die.

metabolism. The sum of the chemical reactions which occur in an organism or a cell. Metabolism involves the breakdown of organic compounds, releasing energy that is used in the synthesis of other compounds; metabolize.

metabolite. A substance produced by metabolism.

metacentric. Having the centromere near the middle of the chromosome.

metachromatic granules. Granules which assume a colour different from that of the dye used to stain them.

metallophyte. A plant confined to substrates with very high levels of heavy metals; *cf.* pseudometallophyte.

metaphase. 1: The second stage in cell division, in which the nuclear membrane breaks down, and the centromeres of the chromosomes position themselves at the centre of the spindle, forming the metaphase plate.

2: The phase of the first meiotic division when centromeric regions of homologous chromosomes come to lie equidistant on either side of the equator.

metaplasia. Changed condition of a structure or organ; hyperplastic class of symptoms characterized by overdevelopment other than that due to hypertrophy or hyperplasia, e.g., abnormal starch accumulation, virescence.

methanogen. An organism carrying out methanogenesis, requiring completely anaerobic conditions for growth; considered by some authorities to be distinct from bacteria.

methanogenesis. The biosynthesis of the hydrocarbon methane; common in certain bacteria; bacterial methanogenesis.

metoecious. Used of a parasite that is not host specific; heteroecious; metoecius,

metula. In *Penicillium* and *Aspergillus*, a sporophore branch having phialides.

micelle. 1: A submicroscopic structural unit of protoplasm built up from polymeric molecules.

2: The individual particles of the disperse phase of a colloid in which the dispersion medium is a liquid.

Michaelis constant. A constant K_m such that the initial rate of reaction V, produced by an enzyme when the substrate concentration is high enough to saturate the enzyme, is related to the rate of reaction v at a lower substrate concentration c by the formula $V = v\ (1 + K_m/c)$.

microaerophilic. Pertaining to those microoganisms requiring free oxygen but in very low concentration for optimum growth.

microbe. A microorganism, especially a bacterium of a pathogenic nature.

microbial insecticide. Species-specific bacteria which are pathogenic for and used against injurious insects.

microbiology. The science and study of microorganisms, including protozoans, algae, fungi, bacteria, viruses, and rickettsiae.

microbivorous. Feeding on microorganisms, in particular bacteria; microbivore, microbivory.

microbody. A membrane bounded cytoplasmic organelle with varied enzyme content and functions; e.g., peroxisomes and glyoxysomes.

microchemistry. The chemistry of individual cells and minute organisms.

microclimate. The climate of a small or limited surroundings or habitat e.g., the surface of the soil, or under the canopy of a forest; bioclimate.

microcomposer. *See* decomposer.

microconsumer. *See* decomposer.

microenvironment. The specific environmental factors in a microhabitat.

microfilaments. Narrow intracellular fibers that contain polymerized actin and that are thought to function in the maintenance of cell structure and movement.

microflora. 1: Microscopic plants.
2: The flora of a microhabitat.

microgamete. The smaller of two anisogametes or male gamete produced by heterogametic species; *cf*. macrogamete.

microgamy. Sexual reproduction by fusion of the small male and female gametes in certain protozoans and algae; merogamy.

micron (μ). One-thousandth of a millimeter. A millimicron ($m\mu$) is one thousandth of a micron and is equivalent to 10 Ångstroms (Å).

micronema. (*pl.* micronemata) Rhizoids that originate randomly on the moss stem; have fewer branches, and are usually smaller, important in capillary water conduction; *cf*. macronema.

micronutrients. Trace elements; nutrients required in minute quantities for optimal growth. These are needed in tissue concentrations equal to or less than 100 $\mu g\ g^{-1}$ of dry matter. Some of the micronutrients are: molybdenum, copper, zinc, iron, boron etc., *cf*. macronutrients.

microorganism. Microscopic or ultramicroscopic organism, including bacteria, protozoans, yeasts, fungi, mycoplasma, prions, viroids, viruses, and algae.

microparasite. A parasite of microscopic size.

microphagy. Feeding on minute organisms or particles.

microphyll. A leaf whose leaf trace does not make a gap in the vascular system of the stem,

microphyllous. 1: Having small leaves.

2: Having leaves with a single, unbranched vein.

microphyte. 1: A microscopic plant.

2: A dwarfed plant due to unfavorable environmental conditions.

micropyle. A hollow tube or pore in the integument at the tip of an ovule, through which the pollen tube enters. The micropyle can be seen in the testa of the mature seed between the hilum and the point of radicle. Water enters the micropyle at the beginning of germination.

microsomes. A membrane-rich fraction of a tissue homogenate produced during high-speed centrifugation of the mitochondrial supernatant. This include ribosomes and small fragments of intracellular membranes.

microsomia. Dwarfism; nanism.

microspore. A small spore, produced in a microsporangium, in heterosporous plants. The microspore develops into the male gametophyte. In angiosperms, the microspore is the pollen grain; *cf.* megaspore.

microsporophyll. A sporophyll bearing microsporagia.

microsurgery. Surgery on single cells by micromanipulation.

microtherm. A plant requiring a mean annual temperature range of 0-14°C for optimum growth.

microtome. An instrument for cutting thin sections of tissue to be examined under the microscope.

microtomy. Cutting of thin sections of specimens with a microtome.

microtubule. A very thin, hollow thread of protein in a cell. Microtubules have several different functions. They form the spindle in mitosis, control cell movements and form a structural part of the flagella, cilia and centrioles.

microzoophilous. Pollinated by small animals; microzoophily.

microzoophobous. Used of organisms that repel small animals.

middle lamella. The thin young cell wall formed between two new eukaryotic cells after cell division. The middle lamella is made of pectin(↓), and the thicker layers of cellulose are laid down on either side of it.

midrib. The large central vein of a leaf.

mildew. 1: A whitish powdery or downy growth on the surface of diseased plant or other substrates caused by a parasitic fungus.

2. Any fungus producing such growth.

milieu. The characteristic environs or surroundings of an organism or population.

millennial. Pertaining to periods of thousands of years..

millennium. A unit of time equal to 1000 years or 1 millicron.

millet. A common name applied to at least five related members of the grass family grown for their edible seeds.

milli (m). Prefix meaning thousand, thousandth; used to denote unit $\times 10^{-3}$.

mimicry. Assumption of colour, form, or behaviour patterns by one species of another species, for camouflage and protection,

minicell. A small daughter cell that arises from asymmetric septum formation during binary fiission and which lacks DNA.

minimal area. The smallest area in which a particular community develops the combination of species characteristic of that community; minimalraum; minimiared; minimum quadrat area.

minimum lethal dose (MLD). The minimum dose of an agent sufficient to cause 100 per cent mortality of the test population.

minimum quadrat area. Minimal area.

mire. An ecosystem in which the vegetation is rooted in wet peat; a collective term for various bogs and fens.

missense codon. A mutant codon that directs the incorporation of a different amino acid and results in the synthesis of a protein with a sequence in which one amino acid has been replaced by a different one; in so ne cases the mutant protein may be unstable or less active.

missense mutation A mutation that converts a codon coding for one amino acid to a codon coding for another amino acid.

mistletoe. 1: *Viscum album.* The true, Old World mistletoe having dichotomously branching stems, thick leathery leaves, and waxy-white berries.

2: Any of several species of green hemiparasitic plants of the family Loranthaceae.

mitochondrion. A round or rod-shaped organelle, found in the cytoplasm of all animal and plant cells except bacteria and blue green algae, and in which the reactions of the Krebs cycle and the electron transfer chain take place. Mitochondria have a smooth outer membrane and an inner membrane which is folded into cristae.

mitosis. Vegetative or somatic cell division, in which the chromosomes in the nucleus are duplicated into two chromatids. The nuclear membrane breaks down, the centromeres divide, and the chromatids move to either end of the cell on the spindle. Nuclear membranes re-form around each group of chromatids, and a new cell wall is laid down between them. In this way each new cell gains exactly the same chromosomes and genetic material. The four stages of mitosis are prophase, metaphase, anaphase and telophase; mitotic.

mitospore. A diploid spore produced by mitosis.

mitotic index. The number of cells undergoing mitosis per thousand cells.

mitotic inhibitor. A compound that inhibits mitosis.

mitrate. A term used to describe a calyptra shaped like a bishop's miter; i.e., conic and undivided or equally lobed at base.

mitriform. Shaped like a miter.

mixotrophic. Used of organisms that are both autotrophic and heterotrophic; mesotrophic; mixotroph, mixotrophy.

modifier gene. A gene that alters the phenotypic expression of a nonallelic gene.

molasses. The uncrystallised residue remaining after sugar crystallises out of the juice from sugarcane and sugarbeet. It is used in the manufacture of rum and treacle, and as a food for domestic animals.

mold Superficial fungal growth on various substrates.

molecular biology. That part of biology which attempts to interpret biological events in terms of the physicochemical properties of molecules in a cell.

molecular genetics. The approach which deals with the physics and chemistry of the processes of inheritance.

monadelphous. Having the filaments of the stamens united into a tube which surrounds the style, e.g , lupin.

monandrous. 1: Having one stamen.
2: Used of a female that mates with a single male; monandry; *cf.* polyandrous.

monaxenic. Used of a mixed culture of an organism with one prey species; *cf.* axenic, dixenic.

moniliform. Constructed with contractions and expansions at regular alternating intervals, giving the appearance of a string of beads.

monocarpic. Producing a single fruit, or having only one fruiting period, during the life cycle, e.g. most annual plants; monotokous; *cf.* polycarpic.

monocarpous. Having a single ovary.

monochlamydous. Referring to flowers having only one set of floral envelopes, that is, either a calyx or a corolla.

monoclimax. A climax community controlled primarily by one factor as climate.

monoclinic. Having both stamens and pistils in the same flower.

monoclinous. Used of a flower having both male and female organs; hermaphrodite; perfect; monocliny; *cf.* diclinous.

monocolpate pollen. Pollen grains having a single furrow.

monocotyledon. An angiosperm of theclass Monocotyledones. Their seeds have one colyledon. Monocotyledons do not have secondary thickening and most are small herbaceous plants with parallel venation and floral parts in whorls of three or multiples of three, e.g., grasses, sedges, orchids, palms,

monoecious. 1: Used of individuals having both male and female reproductive organs; hermaphrodite; bisexual.

2: Used of a bisexual plant species having separate male and female flowers (unisexual) on the same individual; autoicous; ambisexual; monecious; monoecism, monoecy; *cf.* dioecious.

monoecious polygamy. The possession of both single sexed and perfect flowers on the same individual.

monogenic. 1: Used of characters or traits controlled by a single gene; monofactorial, unifactorial; *cf.* digenic, oligogenic, polygenic, trigenic.

2: Producing only male or only female offspring; monogeny.

monograph. A comprehensive published work on a single group or subject; monographic.

monogynopaedium. An assemblage of one female parent with her immediate progeny; *cf.* polygynopaedium.

monogynous. Having only one pistil.

monogyny. 1: The mating of a male with only one female.

2: The presence of a single queen in a colony of social insects; monogynous; *cf.* oligogyny polygyny.

monohybrid. A hybrid individual heterozygous for one gene or a single character.

monohybrid cross. A cross between two individuals heterozygous for only one pair of alleles; *cf.* dihybrid cross.

monohybrid inheritance. The inheritance of one pair of genes.

monomorphic. Having or exhibiting only a single form.

mononuclear Having only one nucleus.

monophagous. Utilizing only one kind of food; feeding upon a single species or food plant; host specific; monotrophic; univorous monophage, monophagy; *cf.* oligophagous, polyphagous.

monophyletic Derived from the same ancestral taxon; used of a group sharing the same common ancestor; monophyly, monophyletism; *cf.* diphyletic, polyphyletic.

monoplanetic. Having a single motile stage during a life cycle; monoplanetism; *cf.* diplanetic, polyplanetic.

monoploid. 1: Having only one set of chromosomes.
2: Having the haploid number of chromosomes.

monopodial. A kind of growth in which the main axis of the plant is formed by continuous growth of the same shoot apex, with lateral branches arising from it.

monopodium. A primary axis that continues to grow while giving off successive lateral branches.

monosaccharide. A simple sugar, with between three and seven carbon atoms.

monosome. The monomeric form of a ribosome that may contain no mRNA, or nascent polypeptide chain.

monosomic. Pertaining to an aneuploid, having an abnormal chromosome complement in which one chromosome of one homologous pair is absent; monosomy, monosome.

monospermous. Having or producing one seed.

monosporangium. A sporangium producing monospores.

monospore. A simple or undivided nonmotile asexual spore; produced by the diploid generation of some algae.

monotokous. Having only one offspring per brood; fruiting only once during a life cycle; monocarpic; uniparous; monotocous; *cf.* ditokous; oligotokous, polytokous.

monotrichous. Of bacteria, having an individual flagellum at one pole.

monotype. A single type of organism that constitutes a species or genus.

monotypic. Used of a taxon comprising a single immediately subordinate principal taxon, as in a family genus or comprising a single genus or species respectively; monobasic; *cf.* bitypic, polytypic.

monoxenous. Used of a parasite utilizing a single host species during its life cycle; monoxeny.

montane. Of, pertaining to, or being the biogeographic zone composed of moist, cool slopes below the timberline and having evergreen trees as the dominant life-form.

montane forest. A forest on a mountain. Montane forests have smaller trees than lowland forests, and the trees get smaller towards the tree line.

monticolous. Living in mountainous habitats; monticole.

moor. An open elevated region of wet acidic peat typically dominated by heathers, sedges and some grasses; moorland.

mor. A very acid humus which hardly mixes with the inorganic soil underneath it.

moribund. Dying; close to death.

morphine. An organic, aromatic alkaloid obtained from opium, *Papaver sominiferum* fam. Papaveraceae. It is a narcotic and a powerful pain-killer leading to addiction. Morphine represses the cough reflex.

morphogenesis. 1: The totality of the process of embryological development and growth i.e., development of shape and structure of organs and tissues.

2: The evolution of morphological structures; morphogenetic, morphogeny; to ogenesis.

morphology. The study of the shape and arrangement of organs and tissues of an organism at any stage of its life history.

mosaic. 1: An organism comprising tissues of two or more genetic types; usually used with reference to plants; chimaera.

2: A symptom of certain viral diseases of plants: a patchy pattern of green and light green or yellow in leaves or fruit that are normally green.

moss. One of the two main groups of bryophytes. Mosses differ from liverworts in having more differentiated cells in the gametophyte. The gametophyte usually has a stem with leaves, and is often branched. The capsule of the sporophyte is also more differentiated in mosses, and the spores are released through a peristome.

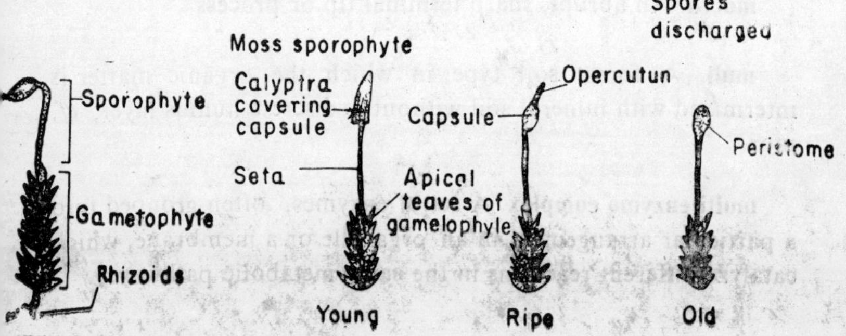

moss forest. *See* temperate rainforest.

moss land. An area which contains abundant moss but is not wet enough to be a bog.

motile. Able to move, as in cells with flagella.

mottle. An irregular pattern of indistinct light and dark areas.

mould. 1: A cavity or space in a sediment remaining after the organic material of an organism has been removed by the action of soil water; mold.

2: The general name for a fungal growth on a surface.

m-RNA. *See* messenger ribonucleic acid.

mucivorous. Feeding on plant juices or mucilage; mucivore, mucivory.

muck. Highly decomposed plant material typically darker and with higher mineral content than peat.

mucopolysaccharide. Any of a group of polysaccharides containing an amino sugar and uronic acid; a constituent of mucoproteins, glycoproteins, and blood-group substances.

mucoprotein. Any of a group of glycoproteins containing a sugar, usually chondroitinsulphuric or mucoitinsulphuric acid, combined with amino acids or polypeptides.

mucro. An abrupt, sharp terminal tip or process.

mull. A forest soil type in which the organic matter is intermixed with mineral soil without a discrete humus layer; *cf.* mor.

multi-enzyme complex. A set of enzymes, often grouped into a particular arrangement in an organelle or a membrane, which catalyze different reactions in the same metabolic pathway.

multiple epldermis. Epidermis that is several layers thick, occurring in many species of *Ficus*, *Begonia*, and *Peperomia*.

mummy. A dried, shriveled fruit.

mural. Pertaining to, or growing on, walls; rupestral; murarius.

muramic acid. An organic acid found in the mucopeptide (murein) in the cell walls of bacteria and blue-green algae.

murein. The peptidoglycan of bacterial cell walls.

muricate. Rough with short, hard outgrowths.

muriform. 1: Resembling the arrangement of in a brick wall, especially having both horizontal and vertical septa; as seen in conidia of *Alternaria*.

2: Pertaining to or resembing a rat or mouse.

muscarian Used of flowers that attract flies by a putrid odour.

muscicolous. Living on or in mosses or moss-rich communities, muscicole, muscicoline.

muscology. The study of mosses; bryology.

mushroom. A fleshy fruiting body of a fungus, especially of a Basidiomycete of the famiy Agaricaceae.

muskeg. A grassy bog habitat, with scattered, stunted conifers.

mutagen. Any agent that produces a mutation or enhances the rate of mutation above the spontaneous rate. e.g., X-rays, gamma rays or certain chemicals; mutagenic, mutagenesis.

mutant. An individual bearing an allele that has uudergone mutation and is expressed in the phenotype.

mutase. An enzyme able to catalyze a dismutation or a molecular rearrangement.

mutation. The general term for a change in the sequence of nucleotides in the DNA of a cell e.g., the replacement of a pair of nitrogen containing bases in the DNA chain by another pair, the turning round of a sequence of nucleotides in the chromosome, or the loss of a whole chromosome or piece of DNA. Mutations can be inherited if they are present in the gametes. They can be harmless, useful, or can cause death, depending on where they occur in a chromosome. They occur rarely, but this rate can be speeded up by mutagens. Mutations result in variation between individuals, and the natural selection of this variation leads to evolution.

mutilous. Harmless; without defensive structures.

muton. The mutational site within a gene; the smallest part of DNA that will produce a mutation when transformed.

mutualism. A kind of symbiosis is between two species that are beneficial to both species.

mycelium. (*pl.* mycelia): The aggregate of the hyphae that constitute the thallus or vegetative body of a fungus.

-mycetes. The ending of a name of a class of fungi.

mycetidae. The ending of a name of a subclass of fungi.

mycetogenic. Produced by fungi.

mycetology. The study of fungi; mycology.

mycetophagous. Feeding on fungi; fungivorous; mycophagous mycetophage, mycetophagy.

mychogamy. Self-fertilization.

mycobiont. The fungal partner of a lichen (algal/fungal symbiosis); *cf.* phycobiont.

mycology. The branch of botany that deals with the study of fungi.

mycophagous. Feeding on fungi.

mycophthorous. Used of organisms that destroy fungi.

mycoplasmas. Smallest free living pleomorphic microorganisms that, like the bacteria, lack an organized and bounded nucleus but, unlike the bacteria, also lack a true cell wall and the ability to synthesize the substances to form a cell wall. Mycoplasmas cause many diseases in animal and in plants.

mycoplasmalike organisms. Micro-organisms found in the phloem and phloem parenchyma of diseased plants and assumed to be the causes of the disease; they resemble mycoplasmas in all respects except that they cannot yet be grown on artificial nutrient media.

mycosis (*pl.* mycoses): Disease caused by a fungus.

-mycota. The ending of a name of division of fungi.

mycotic. Produced by the agency of fungi.

-mycotina. The ending of a name of a subdivision of fungi.

mycotrophic. Used of plants living in a symbiosis with a fungus, and nutritionally dependent upon the fungus, as in those plants with mycorrhizal associations; mycotrophy.

mycovirus. A virus that infects fungi.

myiophilous. Pollinated by flies; myiophily.

myosin. A muscle protein, comprising up to 50 per cent of the total muscle proteins; combines with actin to form actomyocin. The same protein is known to cause rhythmic reversible movements of granules in the plasmodium of Myxomycetes.

myrmecobromous. Myrmecotrophic.

myrmecochorous. Dispersed by the agency of ants; myrmecochore, myrmecochory.

myrmecoclepty. A symbiosis between ant species in which the guest species steals food directly from the host species.

myrmecodomatia. Plant structures inhabited by ants or termites.

myrmecodomus. Used of a plant affording shelter to ants.

myrmecology. The study of ants.

myrmecolous. Living in ant or termite nests; myrmecocole.

myrmecophagous. Feeding on ants or termites; myrmecophage, myrmecophagy.

myremcophilous. 1: Thriving in association with ants; used of organisms that spend part of their life cycle in ant or termite nests; myrmecophile, myrmecophily; *cf.* myrmecophobous.

2: Pollinated by ants or termites.

myrmecophobous. Used of organisms that repel ants or termites; myrmecophobe' myrmecophobic, myrmecophoby; *cf.* myrmecophilous.

myrmecophyte. A plant having specialized structures for sheltering ants or termites, or having a mutual interdependence with ants or termites; myrmecoxenous plant.

myrmecosymbiosis. A symbiosis between an ant and its host plant; myrmecosymbiotic.

myrmecotrophic. Pertaining to plants and animals that provide food for ants; myrmecobromous,

myrmecoxenous. Used of plants that provide both food and shelter for ants and termites; myrmecoxeny.

myxamoeba. A naked cell capable of amoeboid movement; characteristic of the vegetative phase of myxomycetes and such Plasmodiophoromycetes as *Plasmodiophora brassicae.*

myxohydric mosses. This is a group of mosses which has features of both ectohydric and endohydric types. e.g., *Funaria hygrometrica.* In these mosses both external and internal conduction takes place. These are found prodominantly in the bryopytic vegetation of forests and related shady habitats.

Myxomycetes. Plasmodial (acellular or true) slime molds, a class of microorganisms; that are on the borderline of the plant and animal kingdoms and have a non-cellular, multinucleate, jellylike, creeping, assimilative stage (the plasmodium) which alternates with a myxamoeba stage. They produce brightly coloured sporangia with capillitial threads.

myxospore. A desiccation-resistant resting cell of myxobacteria; microcyst.

myxotrophic. Used of an organism obtaining nutrients through the ingestion of particles; myxotrophy.

N

NAD. Nicotinamide adenine dinucleotide. A derivative of nicotinic acid and acts as hydrogen carrier in the Krebs cycle. This usually function as coenzyme for certain dehydrogenation reactions in the cell. The hydrogens that are removed from compounds by various dehydrogenase enzymes are accepted by NAD^+ (or $NADP^+$). Hydrogens are usually removed in pairs, so that the substrate is oxidised and NAD is reduced to NADH. NADH is oxidised back to NAD by reducing a substrate.

NADP. Nicotinamide adenine dinucleotide phosphate. A coenzyme which is closely related and similar in its action to NAD. Structurally, in NADP, another phosphate group is attached on the left hand ribose unit above. See structure of NAD. A compound which can exist in oxidized or reduced forms. The reduced form is $NADPH_2$. During the light reaction NADP accepts the hydrogen atoms resulting from the splitting of water molecules, giving $NADPH_2$, which is then used in the reduction of CO_2 to carbohydrate in the dark reaction.

naked bud. A bud covered only by rudimentary foliage leaves.

nannandrous. Species of *Oedogonium* with dwarf male.

nano. A prefix meaning dwarfed.

nanoplanktons. Aquatic floating organisms of 2-20 μm diamete.

napiform. Turnip-shaped, referring to spores.

nastic movement. A non-directional plant movement produced by diffuse stimuli causing disproportionate growth or increased turgor pressure in the tissues of one surface e.g., the changes in leaf position which occur in some plants at night, and opening and closing of flowers with a change of light intensity.

nasty. (*pl.* nasties) *See* nastic movements.

native. Grown, produced or originating in a specific region or country.

naturalized. Of a species, having become permanently established after being introduced.

natural selection. Darwin's theory of evolution, also called the survival of the fittest. Natural selection is the selection by the environment of the individuals in a population which are fittest. Only those progency with favourable variations survive; the favourable variations accumulate through subsequent generations, and descendants diverge from their ancestors.

necridium. A dead cell of some filamentous Myxophycean members.

necrology. The study of decomposition, fossilization and other processes affecting plant and animal remains after death.

necrophyte. An organism living on dead material; *cf.* perthophyte; saprophyte.

necrophytophagous. Feeding on dead plant material; nekrophytophagous; necrophytophage, necrophytophagy.

necrosis (*pl.* necroses). Death of plant cells leading to tissue becoming dark in colour; commonly a symptom of fungus infection.

necrotroph. A parasite that derives its food from the dead cells of the host; *cf.* biotroph.

necrotrophic symbiosis. A symbiosis established between two living organisms in which one symbiont continues to use the other as a food source even after complete or partial death has occurred; *cf.* biotrophic symbiosis.

necrotype. A fossil; an extinct organism.

nectar. 1: A liquid containing sugars, amino acids, and other organic compounds. Nectar is secreted by nectaries of many flowers.

2: The sticky, sweet secretion exuded from the pycnia of rusts which contain spores and spermatia and which has an attraction for insects.

nectar guides. A series of marking on the petals of flowers, thought to guide insects to nectar; honey guides.

nectariferous. Nectar producing.

nectarine. A smooth-skinned, fuzzless fruit originating as a spontaneous somatic mutation of the peach, *Prunus persica* and *P. persica* var. *nectarina*.

nectarivorous. Feeding on nectar; nectarivore, nectarivory.

nectarthode. An opening at the base of a flower from which nectar exudes.

nectary. A secretory organ or surface modification of a floral organ in many flowers, occurring on the receptacle, in and around ovaries, on stamens, or on the perianth, secretes nectar. Animals feed on the nectar and at the same time carry pollen from one flower to another. Some plants have extrafloral nectaries, providing food for ants which protect the plant against herbivores.

needle. A slender-pointed leaf, as of the firs and other evergreens.

negative staining. A method in microscopy for demonstrating the form of cells, bacteria, and other small objects by staining the ground rather than the objects.

negative taxis. *See* tactic movements.

negative tropism. *See* tropic movements.

nema. Nematode, eelworm.

nematicide. A chemical lethal to nematodes; nematocide.

nematode. Nema, eelworm; a round or thread-worm of the phylum Nematoda (in some classifications, a class of the phylum Aschelminthes). They are triploblastic, bilaterally symmetrical, unsegmented, invertebrate animals. Many are free-living, whereas others parasitize animals or plants.

nemoral. Pertaining to or inhabiting a grove or wooded area.

neo-Darwinism. The theory of evolution developed in the 20th century, after Darwin's death. It includes Darwin's theory of natural selection and the more recent knowledge of genetics and inheritance through chromosomes.

neopalynology. A field of palynology concerned with extant microorganisms and disassociated microscopic parts of megaorganisms.

neosexual. *See* protosexual.

neoteny. The condition in which some of the characters of an embryonic or young organism are found in the mature organism of reproductive age. Neoteny may have been important in the evolution of flowering plants.

neotype. A specimen selected as type subsequent to the original description when the primary types are known to be destroyed; a nomenclatural type.

nervicolous. Living on or in the veins of leaves; nervicole.

net aerial production. The biomass or biocontent which is incorporated into the aerial parts, that is, the leaf, stem, seed, and associated organs, of a plant community.

net plankton. Plankton that can be removed from sea water by the process of filtration through a fine net.

nett photosynthesis. Apparent photosynthesis measured as the nett uptake of carbon dioxide into the leaf, and equal to gross photosynthesis less respiration.

nett primary production (NPP). The total assimilation of organic matter by an autotrophic individual or population per unit time per unit area or volume, less that consumed by the catabolic processes of respiration; net primary production; *cf.* gross primary production.

nett production. The total assimilation of organic matter by an individual, population or trophic unit per unit time per unit area or volume, less that consumed by the catabolic processes of respiration; often referred to simply as production; net production; visible production; realized production; observable production.

net production rate. The assimilation rate (gross production rate) minus the amount of matter lost through predation, respiration, and decomposition.

net-veined. Having a network of veins, as a leaf or an insect wing.

neuromotor apparatus. A flagellar apparatus of some motile cells in which two basal granules are connected by a paradesmose, and one of them is also connected with the centrosome of nucleus by rhizoplast.

neuter. Neither male nor female.

neutralism. A neutral interaction between two species, that is, one having no evident effect on either species.

niche. The position and activities of an organism in its habitat. Each species has its own niche, and competition occurs when the niches overlap.

nicotine. $C_{10}H_{14}N_2$. An Aromatic alkaloid found in tobacco leaves. It is a colourless, highly toxic liquid used as an insecticide.

nicotinic acid. C_5NH_4COOH. A component of the vitamin B complex; a white, water-soluble powder stable to heat, acid, and alkali; used for the treatment of pellagra; niacin.

nidose. Having an unpleasant smell; nidorose.

nigrescent. Blackish.

nimbospore. A spore having a gelatinous apparently many layered wall, e.g., *Histoplasma capsulatum.*

ninhydrin. An organic chemical used in detecting amino acids, with which it gives blue colour on heating. It is used in several types of chromatography.

nitidous. Smooth and clear; lustrous.

nitrate bacteria. Bacteria, e.g., *Nitrobacter*, which convert nitrite, NO_2^-, to nitrate, NO_3^-; they are found in soil.

nitrification. 1: Formation of nitrous and nitric acids or salts by oxidation of the nitrogen in ammonia.

2: The process in which nitrifying bacteria oxidise ammonium NH_4^+, or ammonia, NH_3 to nitrite NO_2^-; and nitrite to nitrate, NO_3^-.

nitrifying bacteria. Bacteria in the soil which oxidize ammonia (NH_3) to nitrate (NO_3^-). Examples include. *Nitrococcus*, *Nitrosomonas*. This is one of the important stages in the nitrogen cycle, which makes nitrate available for plants; nitrifiers.

nitrite bacteria. Bacteria, e.g., *Nitrococcus* and *Nitrosomonas*, which convert ammonium, NH_4^+, or ammonia, NH_3, to nitrite NO_2^-; they are found in soil.

nitrogenase. An enzyme present in nitrogen fixing micro-organisms which assists in the transformation of atmospheric nitrogen, N_2 to ammonia, NH_3.

nitrogen fixation. The process in which nitrogen in the air (N_2) is reduced by organisms to ammonia. Only prokaryotic organisms such as blue-green algae and bacteria can carry this out. Some nitrogen-fixing organisms are found in symbiotic relationships, e.g., blue-green algae in lichens or *Rhizobium* bacteria in root nodules.

nitrophilous. Plants having a preference for habitats rich in nitrogen.

nitrophobus. Plants having a preference for habitats poor in nitrogen.

nitrophyte, A plant thriving in soil rich in nitrogenous compounds.

nival. 1: Characterized by or living in or under the snow.

2: Of or pertaining to a snowy environment.

nivalis. Latin meaning 'growing near snow.'

niveus. Latin meaning 'snow white'.

node. The point on a stem from which a leaf grows. Nodes are spaced along stems, with internodes between them; nodal.

nodose. Having many or noticeable protuberances; knobby.

nodule. A swelling on the root of a member of the family Leguminosae caused by symbiotic *Rhizobium* bacteria. These bacteria are involved in nitrgen fixation.

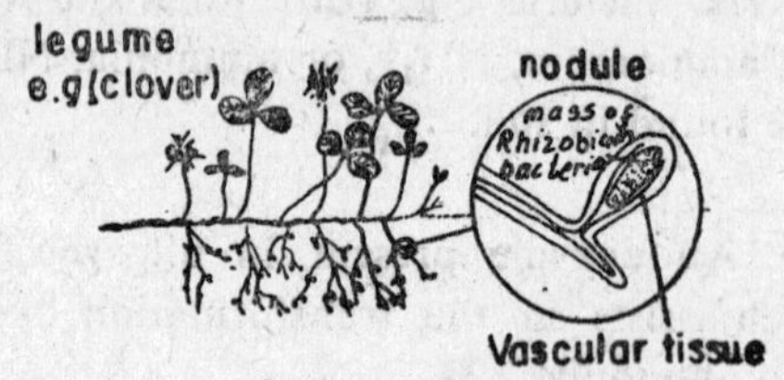

nodular bodies. Pertaining to fungi, in dermatophytes, rounded bodies made up of massed hyphae.

nodulose. 1: Having minute nodules or fine knobs.

2: Pertaining to spores, having broad based, blunt, wart-like excrescences.

nodum (*pl.* noda) A particular well-defined plant community.

nomad. 1: A wandering organism.

2: A pasture plant.

nomenclature. The system of scientific names applied to taxa, or the application of these names; nomenclatural.

nomen ambiguum. One name having different senses.

nomen confusicum. The name of a taxonomic group based on two or more different elements.

nomen conservandum. A name made valid by a decision of the Intervational Botanical Congress.

nomen dubium. A proposed taxonomic name invalid because it is not accompanied by a definition or description of the taxon to which it applies.

nomen nudum. A proposed taxonomic name invalid because the accompanying definition or description of the taxon cannot be interpreted satisfactorily.

nomophyte. A pasture plant; nomophyta.

noncompetitive inhibition. Enzyme inhibition in which the inhibitor can combine with either the free enzyme or the enzyme-substrate complex so that the inhibitor does not compete with the substrate for the enzyme. A common example is the action of heavy metals such as mercury on the active site of enzyme.

noninfectious disease. A disease that is caused by an environmental factor, not by a pathogen.

nonessential amino acid. An amino acid which can be synthesized by an organism and thus need not be supplied in the diet.

nonsaccharine sorghum. *See* grain sorghum.

nonsense codon. A codon that does not code for any amino acid. Only three of the 64 codons in the genetic code are nonsense, and their function is to code for the ends of polypeptide chains.

nonsense mutation. A point mutation within a codon which converts an amino acid specifying codon into a chain terminating codon, resulting in the premature interruption of polypeptide synthesis; *cf.* missense mutation, samesense mutation.

nonvascular. Plants which do not have a vascular system. Nonvascular plants include most bryophytes and all algae.

non-viable. Incapable of normal development, or survival.

Noon pollen unit. The quantity of pollen extract which contains 0.00001 milligram of total nitrogen.

nottate. A surface marked by straight or curved lines.

notch graft. A plant graft in which the scion is inserted in a narrow slit in the stock.

NPP. Nett primary production.

nucellar embryony. Asexual reproduction in plants in which an embryo is produced directly from the nucellus.

nucellus. The oval central mass of tissue of the ovule between the integuments and the embryo sac.

nucivorous. Feeding on nuts; nucivore, nucivory.

nuclear cortex. A fibrous, electron-dense material that coats the nuclear side of the inner membrane of the nuclear envelope.

nuclear membrane. The membrane around the nucleus of a cell. Nuclear membranes have two layers, and many pores connecting the nucleoplasm with the cytoplasm.

nuclease. An enzyme that catalyzes the splitting of nucleic acids to nucleotides, nucleosides, or the components of the latter.

nucleic acid. A large, acidic, chain-like molecule containing phosphoric acid, sugar, and purine and pyrimidine bases, two types are ribonucleic acid and deoxyribonucleic acid.

nuclein. Any of a poorly defined group of nucleic acid protein complexes occurring in cell nuclei.

nucleocytoplasmic ratio. The ratio between the measured cross-sectional area or estimated volume of the nucleus of a cell to the volume of its cytoplasm; karyoplasmic ratio.

nucleolar organizing region (NOR). The sepecific part(s) of the genome containing genes for ribosomal RNA, usually clustered together on one (or more) chromosome.

nucleoid. A region in the cytosol of prokaryotic cells that contains nuclear material but is not segregated by membranes; chromatin body; nuclear equivalent; bacterial chromosome.

nucleolus. Small, dark body inside the nucleus which can only be seen during interphase. Involved in rRNA synthesis and ribosome precursor formation; nucelolar.

nucleolonema. The regions of clustered or segregated granules observable in the electron micrograph of nucleolus.

nucleoplas m. The unstructured portion inside the nucleus of a cell. The nucleoplasm contains the chromosomes and the nucleoli.

nucleoprotein. Any member of a class of conjugated proteins in which molecules of nucleic acid are closely associated with molecules of protein. Plant viruses are nucleoproteens.

nucleoreticulum. Any type of network found within a nucleus.

nucleosidase. An enzyme that catalyzes the hydrolysis of a nucleoside to its component pentose and its purine or pyrimidine base.

nucleoside. The glycoside resulting from removal of the phosphate group from a nucleotide, consists of a pentose sugar linked to a purine or pyrimidine base.

nucleosome. A structural unit of chromatin in eucaryotes observed as spherical particles along decondensed chromatin; each nucleosome consist of the core particle (146 base pairs of DNA plus an octamer of histones) plus the spacer DNA (54 base pairs) and histone HI.

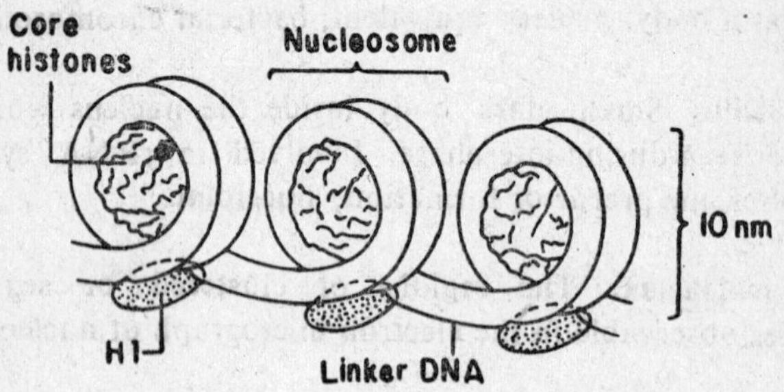

nucleotidase. Any of group of enzymes which split phosphoric acid from nucleotides, leaving nucleosides.

nucleotide. A molecule with a pentose sugar, a phosphate group, and a purine or pyrimidine base containing nitrogen. Nucleotides are the units which form the long chain polymers, nucleic acids.

nucleus. An organelle of a eukaryotic cell, containing the nucleoplasm, nucleoli and chromosomes. Cells usually have only one nucleus, which controls most of the activities of the cell; nuclear.

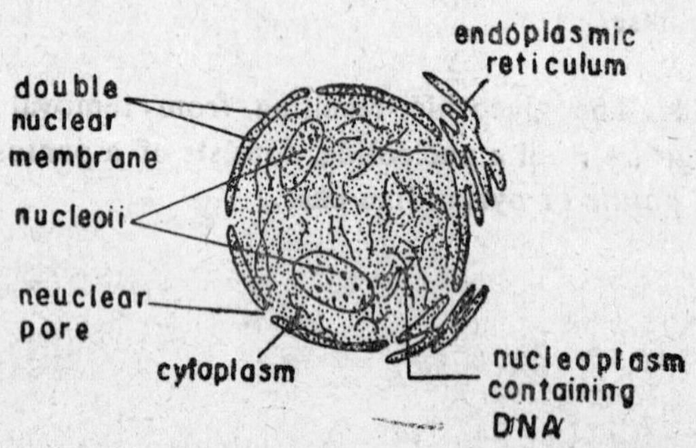

nucule. Female reproductive body of Charales.

nullisomic. A type of aneuploid in which both members of a chromosome pair are missing from the chromosome complement.

numerical taxonomy. Classification based on the numerical comparison of large numbers of equally-weighted characters, scored consistently for all the groups under consideration, and in which individuals are grouped solely on the basis of observable similarities; taxometrics; taximetrics; taxonometrics; numerical phenetics.

nurse graft. A plant graft in which the scion remains united with the stock only until roots develop on the scion.

nut. 1: A fruit which has at maturity a hard dry shell enclosing a kernel consisting of an embryo and nutritive tissue.

2: An indehiscent, one-celled, one-seeded, hard fruit derived from a single, simple, or compound ovary.

nutant. Hanging with the apex downwards; nodding.

nutation. The rhythmic change in the position of growing plant organs caused by variation in the growth rates on different sides of the growing apex.

nutgall. A nut-like gall.

nutmeg. *Myristica fragrans*. A dark-leafed evergreen tree of the family Myristicaceae cultivated for the golden-yellow fruits which resemble apricots, a delicately flavured spice is obtained from the kernels inside the seeds.

nutrient. An inorganic substance which plants require for growth. Nutrients are taken up from the soil by the roots e.g., nitrate, phosphate.

nutriocyte. The inflated portion of the ascogonium of *Pericystis*, which eventually develops into a spore cyst.

nutrition. The process of taking up nutrients and using them in metabolism.

nux vomica. The seed of *Strychnos nuxvomica*, an Indian tree of the family Loganiaceae; contains the alkaloid strychnine, and was formerly used in medicine.

nyctigamous. Used of flowers that open at night and close during the day; nyctigamy.

nyctinasty. Orientation movements of plants during the night; i.e. "sleep movements" of many species, e.g., *Mimosa pudica*, *Phaseolus vulgaris.* Leaflets of nyctinastic plants assume a vertical position; in darkness (i.e. closed) and revert to horizontal position (open) upon illumination.

nyctitropism. An orientation response occurring at night; nyctitropic.

nystatin. An antibiotic from the actinomycete *Streptomyces noursei*; antifungal, widely used against *Candida albicans* infections of man.

O

obclavate. Inversely clavate i.e., widest near the base.

obcordate. Referring to a leaf, heart-shaped with the notch apical.

obligate. Essential; necessary; unable to exist in any other state, mode, or relationship; For instance, the fungi and algae in lichens are obligate symbionts being unable to live without each other, in most cases; *cf.* facultative.

obligate anaerobe. An organism surviving only in the absence of molecular oxygen.

obligate apogamy. Parthenoapogamy.

obligate gamete. A gamete that is unable to develop parthenogenetically.

obligate parasite. An organism which can obtain food only from living protoplasm. Obligate parasites cannot be grown in culture on non-living media.

obligate saprobe. An organism which must obtain its food from dead organic matter, and is incapable of infecting another living organism.

obligate thermophile. An organism requiring a temperature between 65-70°C for optimum growth, and usually unable to grow at temperature below 40C°; *cf.* facultative thermophile.

oblong. Spores twice as long as wide and having somewhat truncate ends.

oblong-ellipsoid. Spores which are rounded oblong; having long sides parallel and ends almost hemispherical.

obovate. Inversely ovate.

obovoid. Solid, egg-shaped, and attached at the narrow end.

obsolete. 1: A part of an organism that is imperfect or indistinct, compared with a corresponding part of similar organisms.

2: Terms which are no longer in use.

obtuse. Of a leaf, having a blunt or rounded free end.

obvolute. Overlapping.

ocellate. Having rounded marks, like eyes.

ocellus. An eyespot functioning as a lens and concentrating the light rays on a photosensitive spot.

ochrosporous. Having yellow or yellow-brown spores.

ochthophyte. A plant living on banks; ochthad; ochthophyta.

ocrea. A tubular stipule or pair of coherent stipules.

ocular micrometer. A glass disk etched with equidistant lines that fits into the eyepiece of a microscope.

odontoid. Tooth-like; dentate.

oecad. Ecad.

oecesis. Ecesis; oecisis.

oecology. Ecology; oecological.

oecophene. Ecophene.

oedema. Intumescence or blister formation due to an increase in inter-cellular water, as in leaves; edema.

oedocephaloid. Having a swelling at the end or lip, as in the conidiophores of *Oedocephalum*.

oidiophore. A hypha that produces oidia from the tip towards the base.

oidium. (*pl.* oidia). 1: One of the small, thin-walled spores with flat ends produced by autofragmentation of the vegetative hyphae or from an oidiophore.

2: It behaves as a male cell i.e., a spermatium.

3: A mildew.

oidization. The union of an oidium with a somatic hypha, resulting in the dikaryotization of the latter.

Okazaki fragments. The discontinuous synthesis of DNA. Short segments, Okazaki fragments, of DNA are synthesized in the 5′→3′ direction. Later the fragments are linked together by ligase; named after their discoverer.

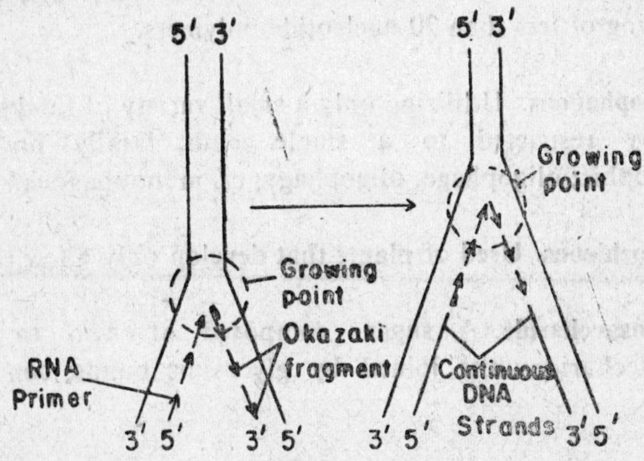

oleoso-locular. Spores having cells like drops of oil.

oleosome. A large fatty inclusion in the cytoplasm of a cell.

oligocene. A geological epoch within the Tertiary period.

oligogene. A gene with a major obvious phenotypic effect; switch gene; key gene; major gene.

oligogenic. Used of characters or traits controlled by only a few genes; *cf.* digenic, monogenic, polygenic, trigenic.

oligogyny. 1: The mating of a single male with a few females.

2: The presence of a few queens within a single colony of social insects; oligogynous; *cf.* monogyny, polygyny.

oligomeric protein. A protein composed of two or more polypeptide chains.

oligomerous. Having one or more whorls with fewer members than other whorls of the flower.

oligomycin. An antibiotic produced by *Streptomyces* sp. and acts as an inhibitor of oxidative phosphorylation. Oligomycin inhibit ATP formation by AT Pase.

oligonucleotide. A polynucleotide of low molecular weight, consisting of less than 20 nucleotide polymers.

oligophagous. Utlilizing only a small variety of food species, typically restricted to a single genus, family or order; oligotrophic; oligophage, oligophagy; *cf.* monophagous.

oligorhizous. Used of plants that develop only a few roots.

oligosaccharide. A sugar composed of two to eight monosaccharide units joined by glycosidic bonds; compound sugar.

oligosaccharins. These are fragments of the cell wall released be enzymes. Indications are that auxins and gibberellns may actually function by activating the enzymes that release these messengers from cell wall. They function as defense against disease, growth, and differentiations in the couse of development: whether to form roots, stems, leaves or flowers and fruits.

oligotaxis. The suppression of a whorl or whorls in a flower.

oligothermic. Tolerating relatively low temperatures; oligothermal; *cf.* polythermic.

oligotokous. Having only a few offspring per brood; oligotocous; *cf.* ditokous, monotokous, polytokous.

oligotrophic. 1: Having low primary productivity; pertaining to waters having low levels of the mineral nutrients required by green plants; used of substrates low in nutrients; oligotropic.

2: Used of any organism requiring only a small nutrient supply, or restricted to a narrow range of nutrients; oligophagous.

3: Used of a lake in which the hypolimnion does not become depleted of oxygen during the summer; *cf.* dystrophic, eutrophic, mesotrophic.

4: Pertaining to insects that visit only a small variety of plant species.

olivaceous. 1: Resembling an oline.

2: Olive coloured.

omnivorous. Feeding on a mixed diet of plant and animal material; pantophagous; omnivore, omnivory.

ontogeny. The course of growth and development of an individual to maturity; ontogenesis; ontogenetic.

onychomycosis. A fungus disease of nails.

oogamete. A large, nonmotile female gamete containing reserve materials.

oogamous. Sexual reproduction involving a small motile male gamete and a large non-motile female gamete, as in algae, fungi, bryophytes and pteridophytes; oogamy.

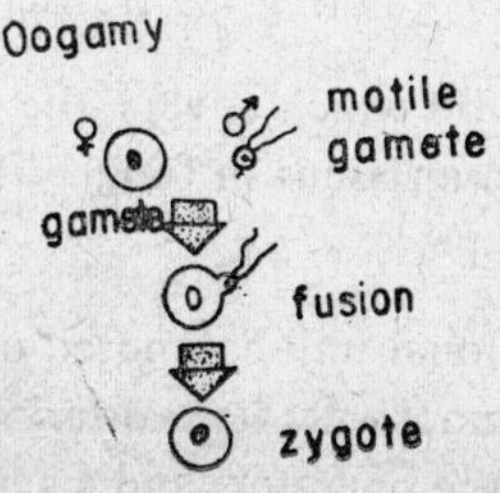

oogonium. Reproductive organ in some fungi and algae, which produces female gametes, or oospheres. Oogonia (*pl.*) are multinucleate.

ooplasm. The protoplasm at the center of the oogonium, which becomes the oosphere in Oomycetes.

oosphere. A large, naked, non-motile female gamete produced in an oogonium.

oospore. A thick walled spore formed after the fertilization of a large, passive female cell, oosphere, by a small, active male cell, or it may develop parthenogenetically.

open-circle deoxyribonucleic acid. See relaxed circular deoxyribonucleic acid.

open community. 1: A plant community in which the niches are unstable or 'empty', allowing entry into the community of new species from outside.

2: A plant community which shows incomplete group cover; typically with the plants or tufts discrete rather than touching, but separated by less than their diameter.

open water. Lake water that is free from emergent vegetation, artificial obstructions, or tangled masses of underwater vegetationat very shallow depths.

operator. A sequence at one end of an operon on which a repressor acts, thus regulating the transcription of the operon.

operculum. The lid which covers the pore at the apex of a moss capsule, or on an ascus in fungi. The operculum opens to allow spores to escape.

operon. A functional unit composed of a number of adjacent cistrons on the chromosome; its transcription is regulated by a receptor sequence, the operator, and a repressor.

opine. A type of amino acid usually not found in nature, such as that secreted by a crown gall.

opisthokont. Having one or more flagella at the posterior end.

opium. 1: A parasite community.

2: A mixture of alkaloids extracted from poppy seeds, (*Papaver sominiferum*, fam. Papaveraceae) containing morphine, codeine, etc. Opium is a pain killer, it soothes and relieves anxiety. It depresses central nervous system thus reducing hunger, thirst, fear or pain. Continued use, leads to delirium and death.

opposite. 1: Located side by side.

2: Of leaves, being in pairs on an axis with each member separated from the other of the pair by half the circumference of the axis

orchid. A monocotyledon of the family Orchidaceae. Most orchids are tropical and epiphytic. The family is one of the largest in the plant kingdom with at least 17000 species.

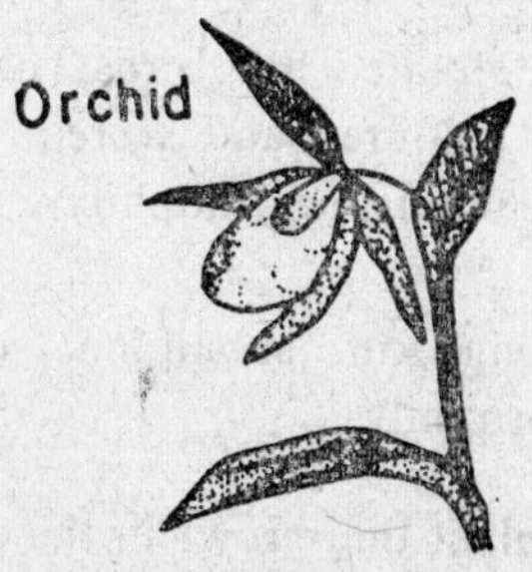

orchil. A purple dye obtained from a lichen *Rocella tinctoria*,

order. A taxon consisting of families. The Latin names of orders usually end with-ales.

orgadophilous. Thriving in open woodlands; orgadophile, orgadophily.

orgadophyte. A plant inhabiting open woodland; orgadophyta.

organ. A group of cells or tissues, forming part of an organism, with a special function, e.g., a leaf, a stamen.

organelle. A body inside a eukaryotic cell, usually with a membrane around it. There are usually several kinds of organelle in any cell, and each kind has a special function. For example, the function of chloroplasts is photosynthesis, the function of mitochondria is respiration.

organogenesis. The formation of organs during development; organogeny.

organotroph. An organism that uses organic compounds as a source of electrons.

ornithocoprophilous. Thriving in habitats rich in bird droppings; ornithoeoprophilic.

ornithophilous. Polinated by birds; ornithophilous flowers are usually brightly coloured and secrete nectar on which the birds feed, ornithogamous; ornithophily.

orophilous. Thriving in subalpine, or in mountainous regions; orophile, orophily.

orophyte. Any plant that grows in the subalpine region.

orthokinesis. A change in the rate of random movement of an organism (kinesis) in which the rate of locomotion (linear velocity) varies with the intensity of the stimulus; *cf.* klinokinesis.

orthotropic. Of axes growing upwards.

orthotropism. The tendency of a plant to grow with the longer axis oriented vertically.

orthotropous. Having a straight ovule with the micropyle at the end opposite the stalk.

orthotropous. Of ovules which are borne erect on the funicle, with the micropyle pointing away from the placenta.

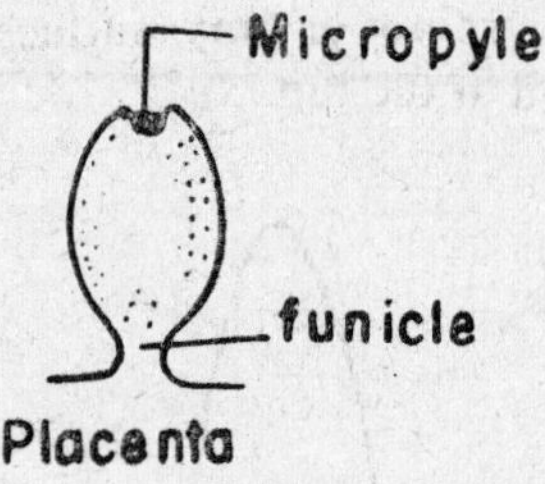

osmophilic. Thriving in a medium of high osmotic concentration; osmophile, osmophily.

osmosis. The process by which water moves across semi-permeable membranes from a hypotonic solution to a hypertonic solution; osmotic.

osmotic potential. A measure of the pressure that needs to be applied to a solution in order to make its water potential equal to that of pure water. When this pressure is applied to a solution, pure water will not pass into it across a semi-permeable membrane. When there is no matric potential or pressure potential, i.e., under experimental conditions, osmotic potential is equal to water potential.

ostiole. A pore; an opening lined with periphyses in a fruiting body through which spores are discharged, as observed in the conceptacle of *Fucus* and in a perithecium and pycnidium in fungi.

otomycosis. A fungal disease of the ear.

outbreeding. Breeding between individuals that are not closely related.

ovariicolous. Living in ovaries.

ovary. The intertior of the carpel of a flower, containing the ovules. The ovary has a thick wall, which develops into the fruit after the ovules inside it have been fertilized by the male gametes carried in the pollen grains.

ovoid. Shape of a spore or any surface like a hen's egg with the narrower end at the top.

ovule. A small body in the ovary, which contains the female gamete, in seed plants. After the gamete is fertilized by one pollen nucleus, the ovule develops into a seed.

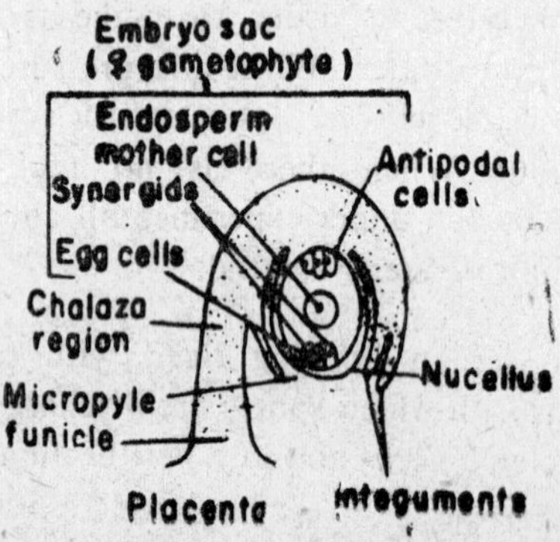

ovum. An egg-cell or a non motile female gamete.

oxidase. An enzyme that catalyzes oxidation reactions by the utilization of molecular oxygen as an electron acceptor.

oxidative phosphorylation. The production of ATP from ADP and orthophosphate, using energy released in the oxidation of organic compounds in the electron transfer chain. Oxidative phosphorylation takes place in the mitochondria and is the main source of ATP in heterotrophic organisms. In the final reaction of the process, molecular oxygen (O_2) is reduced to water.

oxidoreductase. An enzyme catalzing a reaction in which two molecules of a compound interact so that one molecule is oxidized and the other reduced, with a molecule of water entering the reaction.

oxyphilous. Thriving in acidic habitats; acidophilic; oxyphile, oxyphily; *cf.* oxyphobous.

oxyphobous. Thriving in alkaline habitats; intolerant of acidic conditions; basophilous; oxyphobe, oxyphoby; *cf.* oxyphilous.

PABA. *See para*-aminobenzoic acid.

pachycaul. Having fat stems due to massive primary thickening, e.g., in many palms, pachycauly.

palaeobotany. The study of plants of the past, using fossils.

palaeontology. The study of life of the past as recorded by fossils; oryctology; paleontology.

Palaeophytic. The period of geological time between the development of the algae and the first appearance of gymnosperms, during which pteridophytes were abundant; Pteridophytic; Paleophytic.

palaeophytogeography. The study of plant distributions through geological time; paleophytogeography.

Palaeozoic. The geological era which ended about 2.5 million years ago, before the evolution of the angiosperms. Pteridophytes were the dominant plants on land during this time.

palea. 1: One of the pair of inner bracts at the base of a grass spikelet.

2: A chaffy scale found on the receptacle of the disk flowers of some compositae members.

paleate. Having a covering of chaffy scales, as some rhizomes.

paleoagrostology. The study of fossil grasses.

paleoalgology. The study of fossil algae; paleophycology.

paleobiochemistry. The study of chemical processes used by organisms that lived in the geologic past.

paleoecology. The ecology of the geologic past.

paleomycology. The study of fossil fungi.

paleopalynology. A field of palynology concerned with fossils of microorganisms and of dissociated microscopic parts of megaorganisms.

paleophycology. *See* paleoalgology.

palindrome. A nucleic acid sequence that is self-complementary, i.e., the base sequence in a stretch of DNA reads the same from 3′ or 5′ end.

palisade cell. One of the columnar cells of the palisade mesophyll which contain numerous chloroplasts.

palisade mesophyll. A tissue system of the chlorenchyma in well-differentiated broad leaves composed of closely spaced palisade cells oriented parallel to one another, but with their long axes perpendicular to the surface of the blade.

palm. A monocotyledon of the family Palmae. Palms are the largest monocotyledons, and are found mostly in tropical forests. They usually have a thick unbranched pachycaul trunk, with rings on the surface. These ring mark the position of fallen leaves. Most palms have compound leaves, borne in a thick crown at the top of the trunk.

palmate. Of compound leaves with leaflets arising from a central point at the end of the petiole, or, of simple leaves with lobes, in which the main veins arise in the same way.

palmella-stage. Temporary stage in the life-history of some algae, consisting of non-motile cells embedded in gelatinous matrix.

palmelloid. A habit like *Palmella*.

palmetto fiber. Brush or broom fiber obtained from young leafstalks of the cabbage palm tree (*Sabal palmetto*).

paludification. Bog expansion resulting from the gradual rising of the water table as accumulation of peat impedes water drainage.

palustrine. Being, living, or thriving in a marsh.

palynology. The study of spores and pollen, both living and fossil. This is an important way of studying the history of vegetation. Pollen grains can last for thousands or millions of years, and it is possible to identify their genus or family by their shape and surface patterns; palynologic.

palynomorph. A microscopic feature such as a spore or pollen that is of interest in palynological studies.

pandurate. Of a leaf, having the outline of a fiddle.

pangenesis. Theory of heredity, that somatic cells contain particles (gemmules) that can be influenced by the environment and the activity of the organs containing them; these particles can move to the sex cells to influence the course of heredity; gemmule theory.

panicle. A branched inflorescence, consisting of a number of racemes, as in many grasses.

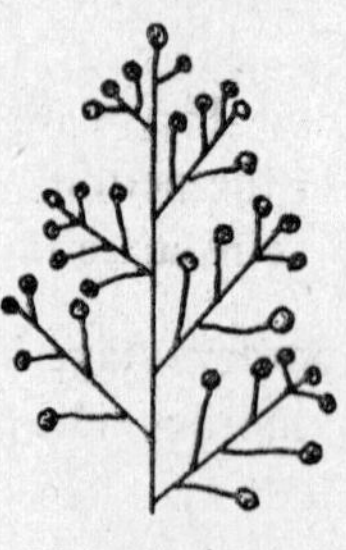

Panicle

pannose. Having a felty or woolly texture.

pantacronematic. A flagellum having two rows of fibrils and a terminal fibril.

pantonematic. A pantacronematic flagellum devoid of terminal fibril.

papain. An enzyme preparation obtained from the juice of the fruit and leaves of the papaya (*Carica papaya,* fam. Caricaceae); contains proteolytic enzymes.

papilla. A small, nipplelike eminence.

papillate. 1: Having or covered with papillae.
2: Resembling a papilla.

pappus. An appendage or group of appendages consisting of a modified perianth on the ovary or fruit of various seed plants; adapted to dispersal by wind and other means, e.g., a group of fine hairs on a small dry fruit in the family compositae.

papula. A small papilla.

paracentric inversion. A type of chromosomal alteration that occurs within one arm of a chromosome and does not span the centromere.

parachoric. Used of organisms that live within the body of another organism without being parasitic.

parallel key. Bracketed key.

parallel-veined. Of a leaf, having the veins parallel, or nearly parallel, to each other.

paramylum. A reserve, starch-like carbohydrate of various protozoans and algae. It is stored in grains.

parapatric. Used of populations whose geographical ranges are contiguous but not overlapping, so that gene flow between them is possible; macroparapatric; parapatry; *cf.* allopatric, dichopatric, sympatric.

parapatric speciation. Speciation in which initial segregation and differentiation of diverging populations take place in disjunction but complete reproductive isolation is attained after range adjustment so that populations become separate but contiguous.

paraphyses. (sing. paraphysis) 1: Sterile, basally attached structures in a hymenium in some algae like *Ficus and* in fungi classes Ascomycetes and Basidiomycetes.

2: Small hairs one cell thick, often with a large round cell at the top, which grow among the antheridia in mosses. They protect the antheridia and may provide them with photosynthetic products.

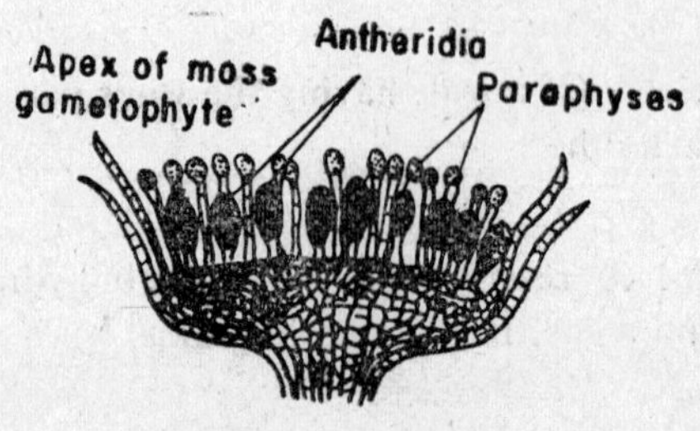

parasexual. Pertaining to organisms that achieve genetic recombination by means other than the regular alternation of meiosis and karyogamy; parasexual reproduction; *cf.* eusexual, protosexual.

parasexual cycle. A series of events leading to genetic recombination in vegetative or somatic cells; it was first described in filamentous fungi; there are three essential steps; heterokaryosis; fusion of unlike haploid nuclei in the heterokaryon to yield heterozygous diploid nuclei; and recombination and segregation at mitosis by two independent processes, mitotic crossing-over and vegetative haploidization of diploid nuclei. Starting from a culture containing a mixture of genetically different haploid nuclei the result of the parasexual cycle is a much more varied mixture of nuclei consisting of haploids like the originals, haploid recombinants, homozygous and heterozygous diploid. The implications of this cycle are; it is possible to study (i) the genetics of a sexual organisms; (ii) the appearance of new strains of pathogenic fungi.

parasite. Any organism that is intimately associated with, and metabolically dependent upon another living organism (the host) for completion of its life cycle, and which is typically detrimental to the host to a greater or lesser extent. Many fungi and bacteria are parasites, and so are a few flowering plants, e.g., dodder, broomrape; parasitic.

parasite-saprophyte. A plant parasite that kills its host then lives saprophytically on its decomposing remains.

parasitism. A symbiotic relationship in which the host is harmed, but not killed immediately, and the species feeding on it is benefitted.

paratype. A specimen other than the holotype which is before the author at the time of original description and which is designated as such or is clearly indicated as being one of the specimens upon which the original description was based.

parenchyma. A tissue of higher plants consisting of living cells with thin walls and often with inter cellular spaces, that are agents of photosynthesis and storage; abundant in leaves, roots, and the pulp of fruit, and found also in leaves and stems.

parietal. 1: Situated against inner wall of the cell.

2: Of a kind of placentation in which the ovules are arranged in rows down the wall of the ovary. The rows mark the lines where the margins of the carpels are joined together.

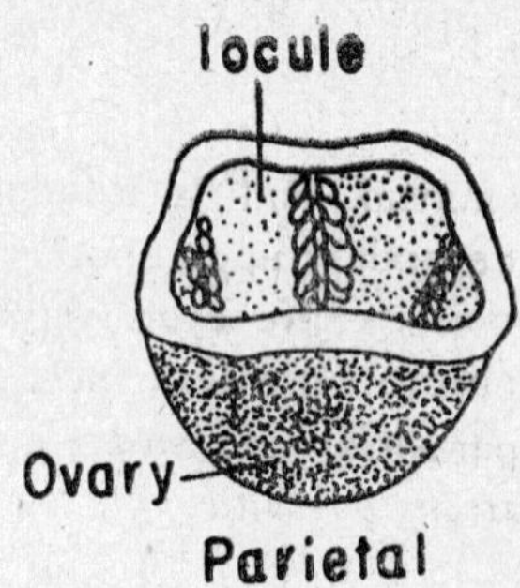

Parietal

parthenocarpy. The development of fruit without seeds, typically resulting from lack of fertilization; parthenocarpic.

parthenogamete. A gamete capable of parthenogenetic development.

parthenogamy. The parthenogenetic development of a diploid organism, arising from the fusion of two nuclei within a single gamete; automictic parthenogenesis.

parthenogenesis. The development of an individual from a female gamete without fertilization by a male gamete; gene-agenesis; gynecogeny.

parthenokaryogamy. The fusion of two female haploid nuclei; parthenocaryogamy.

parthenomixis. The fusion of two female nuclei produced within a single gamete or gametangium.

parthenospore. A spore produced from an unfertilized famale gamete; aboospore; azygospore.

passive uptake. The absorption of ions by diffusion or other physical process not involving the expenditure of metabolic energy; *cf.* active uptake.

Pasteur-Chamberland filter. A porcelain filter with small pores used in filtration sterilization.

Pasteur effect. Inhibition of fermentation by supplying an abundance of oxygen to replace anaerobic conditions.

pasteur point. The transition point at which a facultative anaerobe no longer utilizes oxygen for respiration but obtains energy through fermentation.

patch budding. Budding in which a small rectangular patch of bark bearing the scion (a bud) is fitted into a corresponding opening in the bark of the stock.

patho-, pathy. Prefix meaning suffering, feeling, associated with disease.

pathogen. An organism which causes disease or illness in another organism. Many viruses, fungi, bacteria and pathogenicity, mycoPlasmas are pathogens; pathogenic.

Bacteria causing galls
on apple tree stem

pathogenesis. That portion of the life cycle of a pathogen during which it becomes, and continues to be, associated with its suscept.

pathogenic. Producing or capable of producing disease; nosogenic; pathogen, pathogenicity.

pathology. The study of disease; nosology.

patobiontic. Used of an organism that spends its entire life cycle in the forest litter; cryptozoic; patabiontic; patobiont.

pathovar (*abbr.* pv.). Of bacteria, a strain of a species differing from other strains in its pathogenicity to species of suscepts; *cf.* forma speciais.

patoxenous. Used of an organism that occurs in the forest litter only by accident; patoxene.

peat ball. A lake ball containing an abundance of peaty fragments.

peat flow. A mudflow of peat produced in a peat bog by a bog burst.

peat moss. Moss, especially sphagnum moss, from which peat has been produced.

pectic acid. A complex acid, partially demethylated, obtained from the pectin of fruits.

pectin. A purified carbohydrate obtained from the inner portion of the rind of citrus fruits, or from apple pomace; consists chiefly of partially methoxylated polygalacturonic acids.

pectinase. An enzyme that catalyzes the transformation of pectin into sugars and galacturonic acid.

pectinesterase. An enzyme that catalyzes the hydrolytic breakdown of pectins to pectic acids.

pedicel. 1: The stalk of a single flower in an inflorescence.
2: A small stalk.

pedicellate. Having a pedicel.

pedology. Study of the structure and formation of soils; soil science.

peduncle. The stalk of a whole inflorescence.

peg graft. A graft made by driving a scion of leafless dormant wood with wedge-shaped base into an opening in the stock and sealing with wax or other material.

peitschengeissel. Whiplash flagellum lacking lateral hairs.

pellicle. 1: A skin-like aggregation of bacteria or yeasts on the surface of liquid media.
2: Any surface, skin-like growth.

peltate. Of leaves, having the petiole attached to the lower surface instead of the base.

penicillate. Having a tuft of fine hairs.

penicillus (*pl.* penicilli). The conidiophore of the genus *Penicillium.*

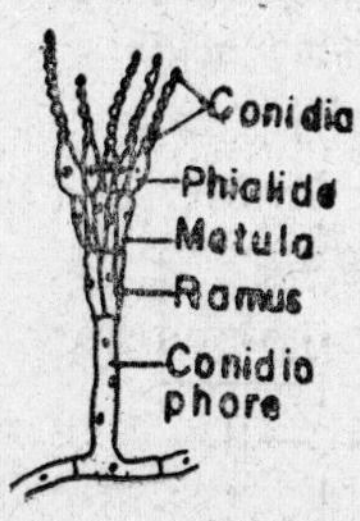

pentadelphous. Having the stamens in five sets with the filaments more or less united within each set.

pentamerous. Having each whorl of the flower consisting of five members, or a multiple of five.

pentandrous. Having five stamens.

pentosan. A hemicellulose present in cereal, straws, brans, and other woody plant tissues; yields five-carbon-atom sugars.

pentose. Any monosaccharide with five carbon atoms. Important pentoses are ribose and deoxyribose, which are found in RNA and DNA, and ribulose, which, as ribulose-diphosphate, is used for CO_2 fixation in photosynthesis.

pentose phosphate pathway. A pathway by which glucose is metabolized or transformed in plants and microorganisms; glucose-6-phosphate is oxidized to-6-phosphogluconic acid, which then undergoes oxidative decarboxylation to form ribulose-5-phosphate, which is ultimately transformed to fructose-6-phosphate.

pepo. A fleshy indehiscent berry with many seeds and a hard rind; characteristic of the family Cucurbitaceae.

peptide. A compound of two or more amino acids joined by peptide bonds.

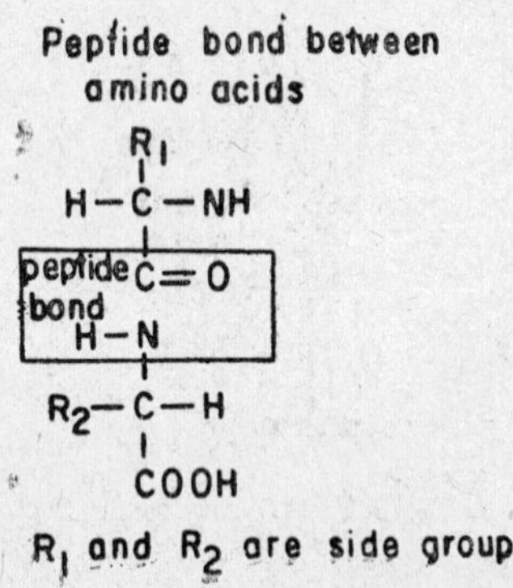

peptidoglycan. Macromolecule formed from both protein and carbohydrate parts, i.e., N-acetyl glucosamine, abundant in the cell walls of bacteria.

perennation. The survival of an individual over successive years, or of a dormant organ during unfavourable seasons; perennate.

perennial. 1: Used of plants that persist for several years with a period of growth each year; restibilic.

2: Occurring throughout the year; *cf.* annual, biennial.

perfect. 1: Pertaining to a hermaphrodite or bisexual flower; monoclinous.

2: Pertaining to fungi or fungal life cycle stages that produce sexual spores.

perfoliate. Pertaining to the form of a leaf having its base united around the stem.

perianth. The outer whorls of the flower, i.e., the calyx and corolla, which are the parts of the flower not concerned with the production of gametes. The function of the perianth is to protect the reproductive organs, and to attract pollinators to the flower.

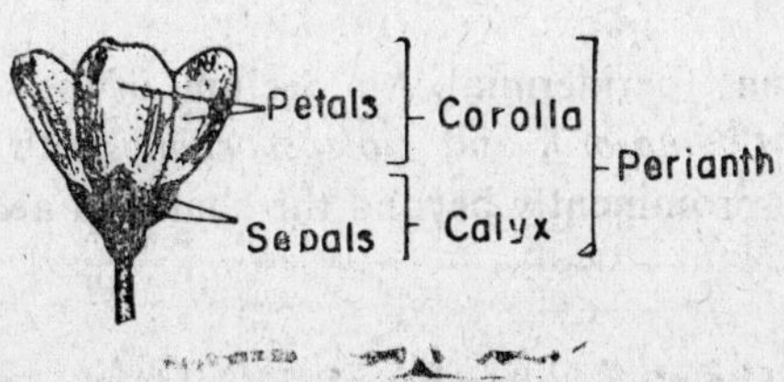

periblem. A layer of primary meristem which produces the cortical cells.

pericarp. 1: The whole wall of the ripe ovary or fruit, usually consisting of exocarp, mesocarp and endocarp.

2: Covering of the vegetative sterile filaments around carposporophyte.

pericentric inversion. A type of chromosome aberration in which chromosome material involving both arms of the chromosome is inverted, thus spanning the centromere.

periclinal. Of cell divisions which occur parallel to the surface of the plant; *cf*. anticlinal.

pericycle. A layer of cells lying inside the endodermis, on the surface of the vascular cylinder of plants.

periderm. Tissue forming part of the bark, consisting of phelloderm, phellogen and phellem.

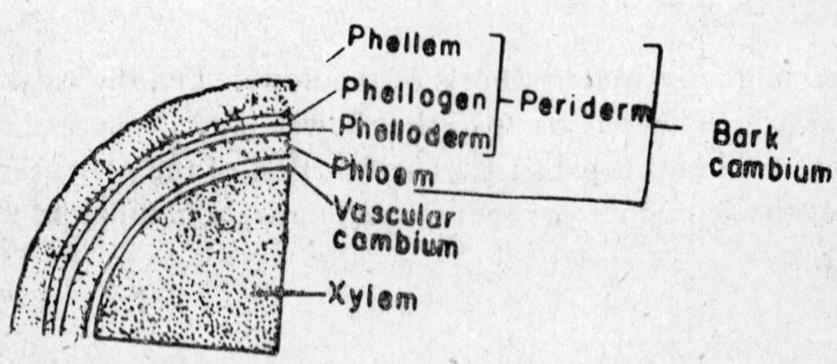

peridermium (peridermia). An aecium, of the sort found in *Cronartium*, *Melampsora*, and *Coleosporium*, with a peridium that extends prominently beyond the chains of aeciospores; *cf*. roestelium.

peridiole. The glebal chamber of the Nidulariales, which has a hard, waxy wall of its own; contains the basidiospores, but acts as a propagating unit as a whole.

peridium (*pl.* peridia). A membrane of sterile cells that forms around the aecia produced in aecidia or in peridermia or roestelia, and around sporangium or any other kind of fruiting body in fungi.

perigonium. The perianth of a liverwort.

perigynium. A fleshy cup-or tubelike structure surrounding the archegonium of various bryophytes.

perigynous. Bearing the floral organs on the rim of an expanded saucer-or cup-shaped receptacle or hypanthium.

peripheral protein. Term designating a protein that interacts only with the hydrophilic surface of the lipid bilayer of a membrane, hence is easily extractable from it.

periphyses (sing. periphysis). Short, hair-like growths in the form of a fringe the lining inside of an ostiole or of a pore in a stroma.

periplasm. A layer of protoplasm surrounding the oosphere of certain Oomycetes.

perisperm. In a seed, the nutritive tissue that is derived from the nucellus and deposited on the outside of the embryo sac.

peristome. A set of tooth-like plates under the operculum of the capsule of a moss. These plates are called peristome teeth. They control the release of spores into the air, by altering their position with changes in humidity.

perithecium (*pl.* perithecia). A spherical or flask-shaped, thick-walled fruiting body of Ascomycetes, consisting of an internal hymeniumof asci and paraphyses and an ostiole through which ascospores are discharged.

peritrichous. Of bacteria, having a uniform distribution of flagella on the body surface.

permanent wilting. The condition in which the leaves of a plant are unable to recover turgidity even in saturated air without the addition of water to the soil.

permeable. Of membranes which allow the movement of substances from one side to the other; permeability.

permissive cell. A cell in which viral genome replication can occur after integration of the genome into one or more chromosomes of the cell

peroxidase. An enzyme that catalyzes reactions in which hydrogen peroxide is an electron acceptor.

peroxisome. A small organelle surrounded by a membrane, containing the enzyme catalase, which catalyses the breakdown of hydrogen peroxide (H_2O_2) to water and oxygen. Catalase prevents the build-up of H_2O_2 in cells; H_2O_2 is toxic, and is known to be a product of some metabolic reactions. Peroxisomes also contain enzymes involved in the oxidation of glycolic acid ($COOHCH_2OH$).

persistent. Of a leaf, withering but remaining attached to the plant during the winter.

perthophyte. A plant living on dead, or partially decayed, tissues of a living plant; perthophytic.

petal. An often brightly coloured leaf-like organ. A whorl of petals forms the corolla of a flower. The function of coloured petals is often to attract pollinators to the flower.

petiole. The stalk of a leaf, which joins it to a node on the stem.

petri dish. A shallow glass or plastic dish with a loosely fitting overlapping cover used for bacterial plate cultures and plant and animal tissue cultures; named after R.J. Petri, a German scientist.

petroleum microbiology. Those aspects of microbiological science and engineering of interest to the petroleum industry, including the role of microbes in petroleum formation, and the exploration, production, manufacturing, storage, and food synthesis from petroleum.

phage. Bacteriophage.

phallotoxin. One of a group of toxic peptides produced by the mushroom *Amanita phalloides.*

phanerogam. A general name for all seed plants in an old classification of the plant kingdom, so called because the organs of reproduction are clearly visible. This name has been replaced by spermatophyte.

phanerophytes. Tall aerial perennial plants, mostly trees or shrubs with their renewal buds at least 250 mm above ground level; subdivided on the basis of bud height above the ground into—nanophanerophyte (up to 2 m), microphanerophyte (2—8 m), mesophanerophyte (8—30 m), megaphanerophyte (greater than 30 m); *cf.* Raunkiaerian life forms.

phaneroplasmodium (*pl.* phaneroplasmodia). A plasmodium consisting of a well-differentiated advancing fan and conspicuous thick strands in which ecto—and endoplasmic regions are well differentiated, and in which the protoplasm is coarsely granular and show reversible streaming (shuttle streaming). Characteristic of the Physarales.

phellem. Cork; the outer tissue layer of the periderm.

phelloderm. The inner parenchymatous cells layer of the periderm, inside the cork.

phellogen. The meristematic portion of the periderm, consisting of one layer of cells that initiate formation of the cork and secondary cortex tissue. It is sometimes called the cork cambium.

phenocopy. The nonhereditary alteration of a phenotype to a form imitating a mutant trait; caused by external conditions during development.

phenology. The study of organisms and their activities in relation to the seasons of the year.

phenotype. The observable characteristics of an organism, regarded as the result of the interaction between the genotype and the environment; phenotypic.

phenotypic lag. Delay in the expression of a newly acquired character.

phialide. A small bottle-shaped structure from which phialospores are produced. The latter are characteristically formed inside the phialide and extruded.

phialopore. A space in the young daughter colony of Volvocales through which inversion takes place.

phialospore. One of a chain of spores produced successively on phialides. For example, in *Aspergillus*; conidia are formed on phialides borne on a swollen vesicle which is, itself, part of the conidiophore.

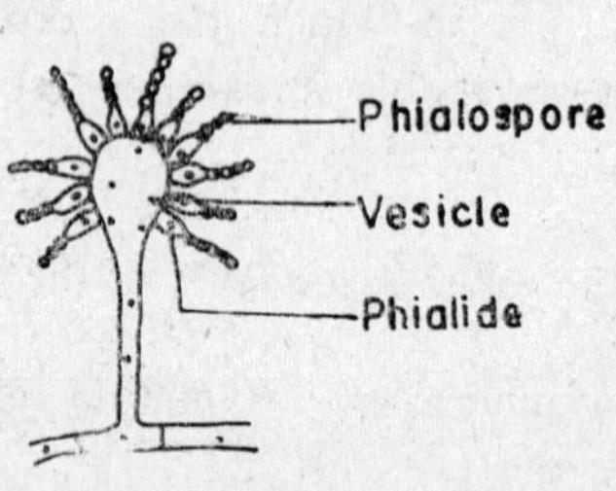

phloem. One of the conducting tissues in the vascular system. Phloem, unlike xylem, is mainly a living tissue, whose cells contain cytoplasm. It is made of sieve elements and companion cells. The phloem can translocate substances in both directions, and its main function is to translocate the products of photosynthesis from leaves to other parts of the plant; bast; sieve tissue.

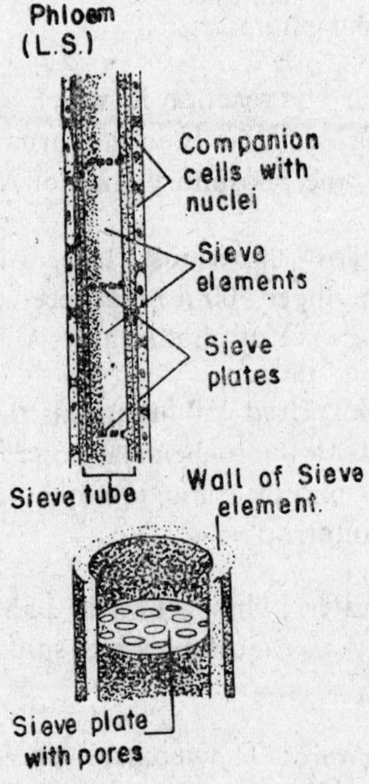

phosphoglyceric acid (PGA). A compound with three atoms of carbon, which is the first product of the reaction between CO_2 and ribulose-diphosphate in the reductive pentose pathway of photosynthesis.

phospholipid. Any of a class of esters of phosphoric acid containing one or two molecules of fatty acid, an alcohol, and a nitrogenous base, phosphatide.

phosphorylase. An enzyme that catalyzes the formation of glucose-1-hposphate from glycogen and inorganic phosphate; it is widely distributed in animals, plants, and microorganisms.

phosphorescence. The emission of light in darkness by the release of absorbed radiation; also the light so produced. Phosphorescence takes place milliseconds after the trapping of the light energy, at a longer wavelength than fluorescence, and may last from a fraction of a second to an hour or more; phosphorescent; phosphoresce.

phosphorylation. The reaction in which a phosphate group is added to a molecule, usually in the form of a trivalent phosphoryl group, e.g., the phosphorylation of ADP to give ATP.

photic zone. The uppermost layer of a body of water (approximately the upper 100 m) that receives enough sunlight to permit the occurrence of photosynthesis.

photoautotrophic. Used of organisms that obtain metabolic energy from light by a photochemical process and manufacture their own food, phototrophic; photoautotroph, photoautotrophy; *cf.* chemoautotrophic.

photolysis of water. The part of the light reaction of photosynthesis in which water molecules are split to give hydrogen and oxygen.

photomorphogenesis. The control exerted by light over growth, development, and differentiation of plants that is independent of photosynthesis.

photoperiod. The number of hours of daylight needed by a plant before it will begin to flower.

photophillic. Thriving in full light.

photophobic. 1: Avoiding light.

2: Exhibiting negative phototropism.

photophosphorylase. An enzyme that is associated with the surface of a thylakoid membrane and is involved in the final stages of adenosinetriphosphate production by photosynthetic phosphorylation.

photophosphorylation. The part of the light reaction, in which ADP is phosphorylated to ATP using energy from light.

photoreactive chlorophyll. Chlorophyll molecules which receive light quanta from antenna chlorophyll and constitute a photo reaction center where light energy conversion occurs.

photorespiration. The process in which plants, in the presence of light, high oxygen concentration and low carbon dioxide concentration, will take up oxygen and give off carbon dioxide, due to the oxidation of organic compounds produced by CO_2 fixation. Photorespiration involves the chloroplasts, mitochondria and peroxisomes. Its function is unclear.

photosynthesis. The process by which plants use energy from sunlight to produce carbohydrates from carbon dioxide (CO_2) and water (H_2O), i.e., the conversion of simple inorganic compounds to complex organic compounds. Sunlight energy is captured by molecules of chlorophyll in the chloroplasts of cells in green leaves. The general equation for photosynthesis is $CO_2 + 4H_2O \rightarrow (CH_2O) + 3H_2O + O_2$. Some bacteria, also use this process; photosynthetic; photosynthesize.

photosystem. 1: One of two reaction sequences of the light phase of photosynthesis in green plants that involves a pigment system which is excited by wavelengths shorter than 700 nanometers and which transfers this energy to energy carriers such as NADPH that are subsequently utilized in carbon dioxide fixation.

2: One of two reaction sequences of the light phase of photosynthesis in green plants which involves a pigment system excited by wavelengths shorter than 685 nanometers and which is directly involved in the splitting or photolysis of water.

phototaxis. Movement of a motile organism or free plant part in response to light stimulation.

phototroph. An organism that utilizes light as a source of metabolic energy.

phototrophic bacteria. Primarily aquatic bacteria comprising two principal groups: purple bacteria and green sulfur bacteria; all contain bacteriochlorophylls.

phragmoid. Having septae perpendicular to the long axis, as the conidia of certain fungi.

phragmoplast. A thin barrier which is formed across the spindle equator in late cytokinesis in plant cells and within which the cell wall is laid down.

phragmosome. A differentiated cytoplasmic partition in which the phragmoplast and cell plate develop during cell division in plant cells.

phycobilin. Any of various proteinbound pigments which are open-chain tetrapyrroles and occur in Myxophyceae and Rhodophyceae.

phycobilisomes. Spherical biliprotein-containing bodies on the thylakoids in blue-green algae and red algae.

phycobiont. The algal partner of a lichen (an algal/fungal symbiosis); *cf.* mycobiont.

phycocyanin. A blue phycobilin pigment of Myxophyceae and Rhodophyceae.

phycocyanobilin. $C_{31}H_{38}O_2N_4$. Phycobilin with an ethylidene side chain ($=CH—CH_3$) and only one asymmetric carbon atom (C_1).

phycoerythrin. A red phycobilin pigment of Myxophyceae and Rhodophyceae.

phycology. Study of algae.

phycoplast. Arrangement of microtubules perpendicular to the spindle axis and at equator of the cell at telophase stage.

phycovirus. Algal virus.

phyllary. A bract of the involucre of a composite plant.

phyllobiology. The study of leaves.

phylloclade. A flattened stem that fulfills the same functions as a leaf.

phyllode. A broad, flat petiole that replaces the blade of a foliage leaf.

phyllody. A change of floral leaves (petals) to foliage leaves.

phyllosphere. The microhabitat of a leaf; the immediate surroundings of plant leaves that are influenced by the presence of the leaves; *cf*. rhizosphere.

phyllotaxy. The arrangement of leaves on a stem, e.g., opposite phyllotaxy, alternate phyllotaxy, spiral phyllotaxy, whorled phyllotaxy. This is an important character in classification.

phylogeny. The evolutionary history of an organism or taxonomic group of organisms; phylogenetic.

physiological anisogamy. A type of isogamy in which the fusing isogametes differ physiologically.

physiologic race. A biotype, or group of closely related biotypes, that differ from other biotypes of that species in the ability to infect particular varieties of the susceptible plant species; *cf*. forma specialis.

physiologic specialization. The existence of a number of races or form of one species of pathogen.

physiology. The study of the internal processes of organisms.

phytal zone. The part of a lake bottom covered by water shallow enough to permit the growth of rooted plants.

phytase. An enzyme occurring in plants, especially cereals, which catalyzes hydrolysis of phytic acid to inositol and phosphoric acid.

phytoalexin. In hypersensitive hosts, a metabolic by-product, toxic both to parasite and host, that is formed only in response to attack by the parasite.

phytochemistry. The chemistry of plants; phytochemical.

phytochrome. A protein plant pigment which serves to direct the course of plant growth and development in response to the presence or absence of light, to photoperiod, and to light quality. Phytochrome absorbs the red and far-red wavelengths of light.

phytocoenosis. The entire plant population of a particular habitat.

phytogenic dam. A natural dam consisting of plants and plant remains.

phytogenic dune. Any dune in which the growth of vegetation influences the form of the dune, for example, by arresting the drifting of sand.

phytogeography. Study of the biogeography of plants; geobotany.

phytography. Descriptive botany.

phytolectin. A lectin found in plants, phytohemagglutinin.

phytolith. A fossilized part of a living plant that secreted mineral matter.

phytopathogen. An organism that causes a disease in a plant.

phytopathology. The study of the diseases of plants.

phytoplankton. The small plants, mostly diatoms and other unicellular algae, which are found near the surface of oceans and lakes. Phytoplankton is one of the world's most important primary producers.

phytosociology. A broad study of plants that includes the study of all phenomena affecting their lives as social units.

phytosterol. Any of various sterols obtained from plants, including ergosterol and stigmasterol.

phytotoxin. 1: A substance toxic to plants.
2: A toxin produced by plants.

phytotron. A laboratory for large-scale culture of plants under controlled conditions.

piece-root grafting. Grafting in which each piece of a cut seedling root is used as a stock.

pigment. Any coloured substance present in the tissues of an organism. Pigments absorb energy from light. Some of them, like chlorophyll, are important in photosynthesis. Others, e.g., phytochromes, help to control growth.

pigment cell. Any cell containing deposits of pigment.

pileus (*pl.* pilei). The cap of a mushroom, bearing the hymenium on its lower surface.

pilus. 1: A hair.
2: A fine, slender, hair-like body.
3: Any filamentous appendage other than flagella on certain Gram-negative bacteria; fimbria.

piliferous layer. The layer of cells, in the epidermis of the root, which bears root hairs.

pina. A fiber obtained from the large leaves of the pineapple plant; pineapple fiber.

pinna (*pl.* pinnae). The leaflet of a pinnately compound leaf.

pinnate. 1: Having parts arranged like a feather, branching from a central axis.

2: Of a compound leaf with a central axis and leaflets-pinnae-on either side of it.

pinnule. The leaflet on the pinna of a bipinnate leaf, as in many members of the order Filicales, the ferns.

pioneer. An organism that is able to establish itself in a barren area and begin an ecological cycle.

pistil. The female reproductive organ of a flower, consisting of the ovary, style and stigma.

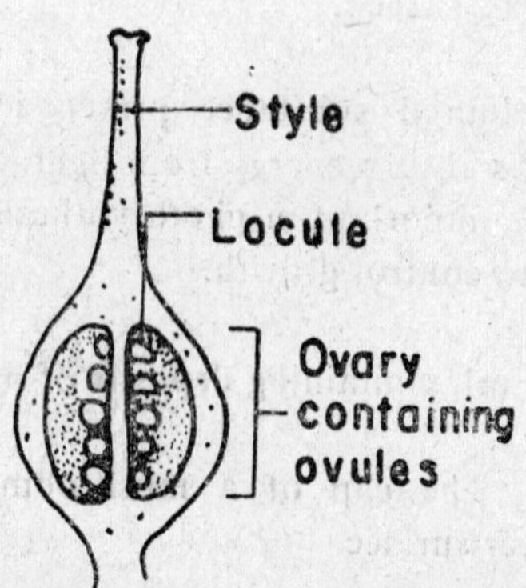

pistillate. 1: Having a pistil.

2: Of flowers which have pistils but no stamens, i.e., female flowers.

pit. 1: A cavity in the secondary wall of a plant cell, formed where secondary deposition has failed to occur, and the primary wall remains uncovered; two main types are simple pits and bordered pits.

2: The stone of a drupaceous fruit. Pits enable the easy passage of substances from one cell to the next. They are common in the tracheids of the xylem.

pitch. The distance between two adjacent turns of double-stranded deoxyribonucleic acid.

pit-connection. A lens-shaped plug of many red algae held within the aperture of the cross-wall situated between two adjacent cells.

pith. A central zone of parenchymatous tissue that occurs in most vascular plants and is surrounded by vascular tissue. Its function is to store food.

placenta 1: The margin of a carpel, where the ovules are attached.

2: A plant surface bearing a sporangium.

placental **cell.** A common cell formed by the fusion of auxiliary cell with some nearby cells; found in many members of red algae.

placentation. The arrangement of ovules in the ovary. Because the ovules are attached to the margins of the carpels, placentation depends on the way in which the carpels are joined together. Common types of placentation are axile, parietal and free central. This is an important character in the classification of angiosperms.

plagiogeotropism. Growth at an angle, in response to gravity, e.g., in lateral branches; plagiogeotropic.

plaglotropic. O branches which grow more or less parallel to the ground.

planogamete. A motile gamete.

planogametic copulation. Fusion of naked gametes, one or both of which are motile.

plant. 1: Any member of the kingdom Plantae.

2: A multicellular eukaryote organism typically exhibiting holophytic nutrition, lacking locomotion, lacking obvious nervous or sensory organs, and possessing cellulose cell walls.

plant fermentation. A form of plant metabolism in which carbohydrates are partially degraded without the consumption of molecular oxygen.

p'ant geography. A major division of botany; concerned with all aspects of the spatial distribution of plants, geographical botany; phytochorology; phytogeography.

plant key. An analytical guide to the identification of plants, based on the use of contrasting characters to subdivide a group under study into branches.

plant pathology. The branch of botany concerned with diseases of plants.

plant physlology. The branch of botany concerned with the processes which occur in plants.

plant societies. Assemblages of plants which constitute structural parts of plant communities.

plant virus. A virus that replicates only within plant cells.

plaque A clear area representing a colony of viruses on a plate culture formed by lysis of the host cell.

plasmagene. Any gene that is not contained in the nucleus, i.e., genes that are found in the cytoplasm in bodies such as plastids or mitochondria. These genes are passed from one generation to the next by cytoplasmic inheritance. The inheritance of plasmagenes does not usually obey Mendel's laws, since they are not organized into chromosomes.

plasmalemma. The cell membrane.

plasma membrane. The cell membrane.

plasmid. An extrachromosomal genetic element found among various strains of *Escherichia coli* and in *Agrobacterium tumefaciens*—a bacterium causing crown gall of plants.

plasmodesmata (sing. plasmodesma). Threads of protoplasm which run through the cell wall between cells, along which substances can be passed.

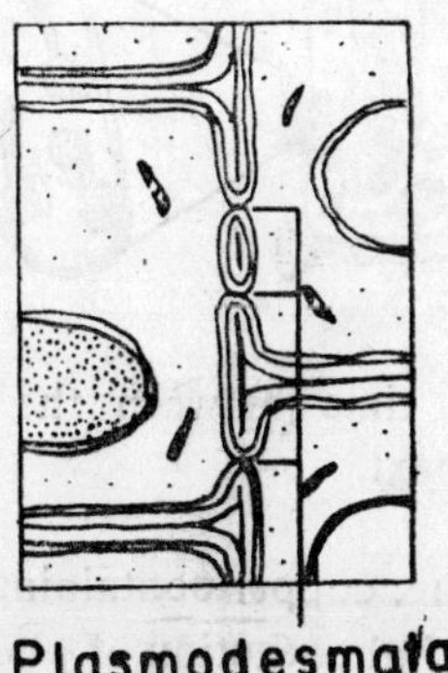

Plasmodesmata

plasmodiocarp. A curved or branched, vein-like fruiting structure of some of the Myxomycetes.

plasmodium. The noncellular, multinucleate, jellylike, assimilative somatic phase of the Myxomycetes, and the Plasmodiophoromycetes, which move and feed in amoeboid fashion.

plasmogamy. Fusion of the cytoplasm of two cells from different parents. This is the beginning of sexual reproduction in fungi.

plastid. The general name for the type of organelle in plant cells which is surrounded by a double membrane and contains plastoglobuli and a network of internal membranes and vesicles. There are several different kinds of plastids, each with a special function, e.g., chloroplasts, chromoplasts, amyloplasts.

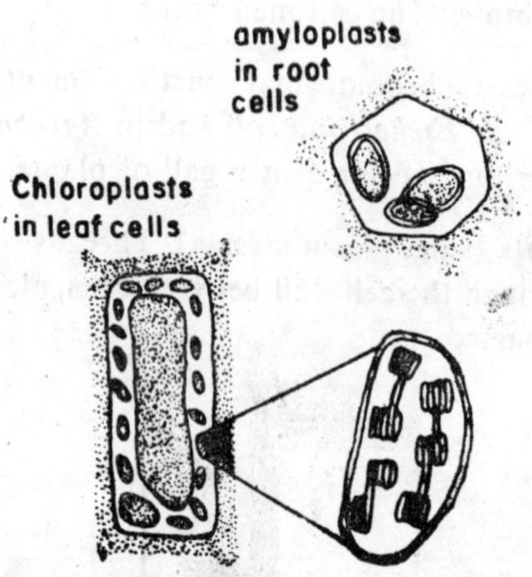

plastochrone. The time between the formation of one leaf primordium and the next.

plastocyanin. A blue, copper-containing protein, which is an electron carrier in the light reaction of photosynthesis.

plastogene. A cytoplasmic factor, controlled by or interacting with the nucleus, which determines differentiation of a plastid.

plastoglobuli. Small, round droplets of lipid, found in plastids.

plastoquinone. A non-protein electron carrier in the light reaction of photosynthesis.

plate budding. Plant budding by inserting a rectangular scion with a bud under a flap of bark on the stock in such a manner that the exposed wood on the stock is covered.

plectenchyma. The general term employed to designate all types of fungal tissues. The two most common types of tissues are prosenchyma and pseudoparenchyma.

plectostele. A protostele that has the xylem divided into plates.

pleiomorphic. Exhibiting polymorphism at different stages of the life cycle; pleomorphic, e.g., mycoplasma.

pleiotropic. Of genes which appear to control several different characters in an organism.

pleistocene. The geological epoch which began about 2 million years ago and ended 10 thousand years ago with the last Ice Age.

pleomorphism. The occurrence of several forms in the life cycle for example, many rust fungi are pleomorphic in that they produce as many as five different spore forms in their complete life cycles; polymorphism.

hlerome. Central core of primary meristem which gives rise to all cells of the stele from the pericycle inward.

plesionecrosis (*pl.* plesionecroses). A symptom exhibited by tissues not yet dead but in the process of dying e.g, wilting,

pleurocarpous. Of mosses which have a stem with many branches, spreading across the ground. The reproductive organs are borne on short side-branches, or in leaf axils along the side of the stem.

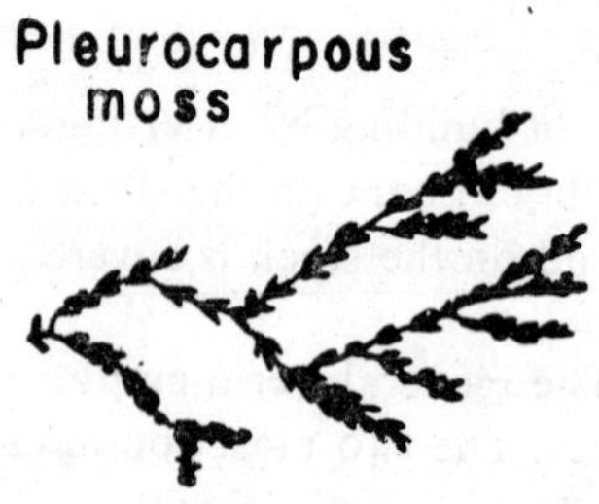

pleuronematic. A flagellum with hairs; tinsel; stichonematic; pantonematic, pantacronematic; *cf.* acronematic; whiplash flagellum.

pleuropneumonialike organism (PPLO). Any of a poorly defined group of microorganisms classified in the order Mycoplasmatales, including the smallest organisms capable of independent life, and comparable in size to the large filterable viruses.

ploidy. Number of complete chromosome sets in a nucleus.

plotophyte. A floating plant, usually having special flotation structures.

plumule. The apical part of the epicotyl of an embryo, from which the first true leaves of the seedling develop.

plurilocular. A sporangium or gametangium having many locules, characteristics of *Ectoca'pus*, a member of Phaeophyceae.

plus strand. A parental ribonucleic acid strand in RNA bacteriophages that is used as a template for a complementary RNA strand (minus strand) produced with the formation of a double-stranded (double-helix) RNA.

pneumatocyst. Gas bladders found on the stipe in some brown algae.

pneumatophore. 1: An air bladder in marsh plants.

2: A submerged or exposed erect root that functions in the respiration of certain marsh plants.

pod. A long, thin, dry fruit, developed from a single carpel, splitting down the side where the margins of the carpel were joined; a legume.

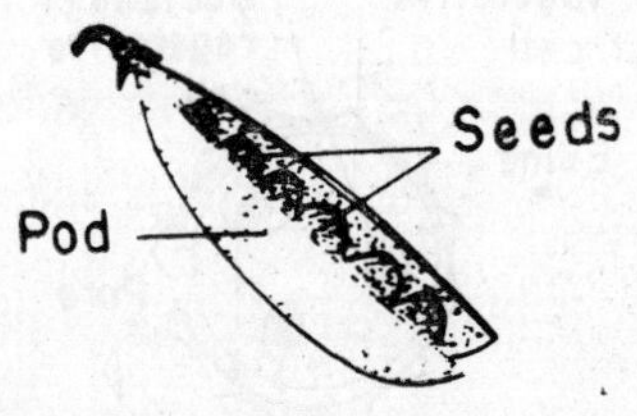

pogonochore. A type of plant that produces plumed disseminucles.

point mutation. Mutation of a single gene due to addition, loss, replacement, or change of sequence in one or more base pairs of the deoxyribonucleic acid of that gene.

polar. Of the very cold regions of the world close to the north and south poles, where the sun does not rise in midwinter and does not set in midsummer. Hardly any plants survive in polar regions.

polar nodule. Thickened regions of the diatom frustule situated on two poles. In blue-green algae the thickened portion of heterocyst is also called a polar nodule.

polar nucleus. One of the two nuclei in the center of the embryo sac of a seed plant which fuse to form the endosperm nucleus.

pollen. Small grains containing the male gametophyte, in seed plants. A pollen grain has a hard wall, or exine. The gametophyte consists of just three cells in angiosperms, and between four and forty cells in gymnosperms; in both cases, only two cells are gametes. The pollen grain protects the male gametophyte during its journey to the female reproductive organs. In angiosperms, pollen is produced in the anthers; in gymnosperms, it is produced in male cones.

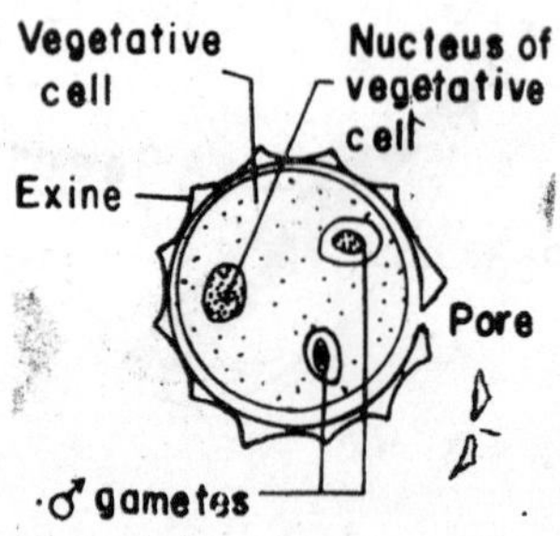

pollen count. The number of grains of pollen that collect on a specified area (often taken as 1 square centimeter) in a specified time.

pollen diagram. A diagram showing the variation in the pollen content of sediments over long periods of time (often tens of thousands of years). Pollen diagrams are used to determine the history of vegetation.

pollen sac. The hollow space inside an anther, where pollen grains are produced.

pollen tube. A thread of cytoplasm, covered by a membrane, which grows from the pollen grain into the micropyle of the ovule, through the tissues of the style. In angiosperms, the pollen tube carries two haploid nuclei into the ovule, one to fertilize the ovum, and the other to fertilize the endosperm nucleus. The pollen tube will only grow if the pollen grain has landed on the stigma.

pollination. The process in which pollen is carried from the anther to the surface of the stigma, in angiosperms, or from the male cone to the female cone in gymnosperms. This can happen in several ways, e.g., by wind, water, insects, birds, bats, or even non-flying mammals, depending on the species of the plant; pollinate; pollinator.

pollinium (*pl.* pollinia). A large group of pollen grains, which are carried together during pollination, as in the orchid family Orchidaceae.

pollution. Destruction or impairment of the purity of the environment.

polyacrylamide Gel Electrophoresis (PAGE). A type of electrophoresis using polyacrylamide gel as the support through which molecules possessing charge move and are separated.

polyadelphous. Of stamens which are united by their filaments into three or more groups in a flower.

polycarpic. Producing fruit or spores more than once during a life cycle; polytokous; *cf.* monocarpic.

polycentric. A thallus radiating from many centers at which reproductive organs (sporangia or resting spores) are formed.

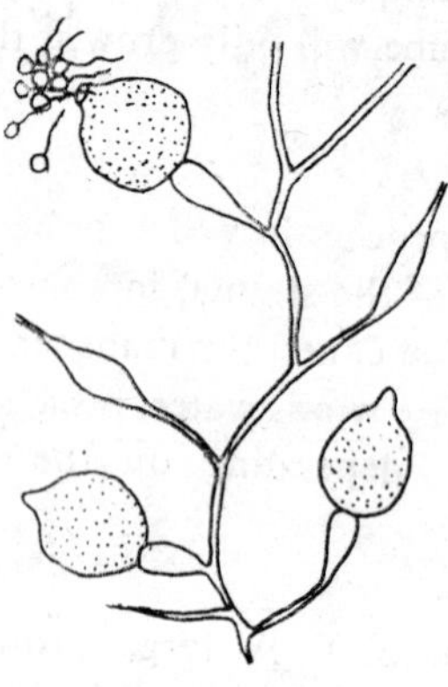

polyembryony. The formation of multiple embryos from a single zygote, ovule or other cell.

polygamous. 1: Pertaining to the condition in which a single male has many female mates at any one time; monarsenous; polygamy; *cf.* monogamous.

2: Used of a plant having perfect and imperfect flowers together on the same individual.

polygene. One of a group of nonallelic genes that collectively control a quantitative character.

polygenic. Used of characters or traits controlled by the integrated action of multiple independent genes; multifactorial; multigenic; polyergistic; polyfactorial; polygenetic; *cf.* monogenic.

polymer. A chemical substance formed by the joining together of many molecules of the same kind, e.g., polysaccharides, polypeptides and nucleic acids.

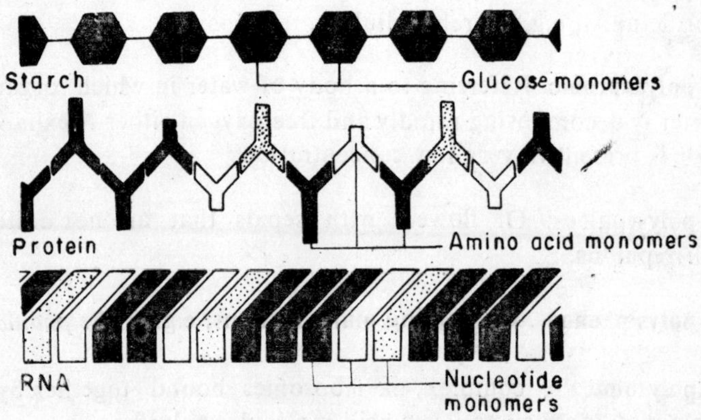

polymorphism. The occurrence of two or more forms of a species in the same population or habitat; polymorphic.

polynucleotide. A linear sequence of nucleotides.

polypeptide. A peptide with a large number of amino acids. Polypeptide chains become folded to form proteins.

polypetalous. Having distinct petals, in reference to a flower or a corolia; choripetalous.

polyphenol oxidase. A copper-containing enzyme that catalyzes the oxidation of phenol derivatives to quinones.

polyphyletic. Derived from two or more distinct ancestral lineages; used of a group comprising taxa derived from two or more different ancestors; pleophyletic; polyphylesis; polyphyly; *cf*. monophyletic.

polyplanetic. Having several motile stages and intervening resting phases during the life cycle; polyplanetism; *cf*. diplanetic, monoplanetic.

polyploid. Of cells which have three or more sets of homologous chromosomes in their nuclei; polyploidy.

polyribosome. *See* polysome.

polysaccharide. Any polymer consisting of many monosaccharide units, e.g., starch, cellulose.

polysaprobic Referring to a body of water in which organic matter is decomposing rapidly and free oxygen either is exhausted or is present in very low concentrations.

polysepalous. Of flowers with sepals that are not united; chorisepalous.

polysiphonous. Containing many siphons, e.g., *Polysiphonia.*

polysome. A complex of ribosomes bound together by a single messenger ribonueleic acid molecule; polyribosome.

polysomy. The occurrence in a nucleus of one or more individual chromosomes in a number higher than that of the remainder.

polystele. A stele consisting of vascular units in the parenchyma.

polytypic. A taxon that contains two or more taxa in the immediately subordinate category.

polytokous. Having many offspring per brood; fruiting many times during the life cycle; caulocarpous; multiparous; polytocous *cf.* monotokous, oligotokous.

population. A group of individuals of the same species living in the same place or area close enough to breed together.

population dispersion. The spatial distribution at any particular moment of the individuals of a species of plant or animal.

pore. Any small hole in a surface or membrane, which allows substances to pass through it, e.g., pores in nuclear membranes.

poricidal dehiscence. Spontaneous release of the contents of a ripe fruit through pores.

porins. Channel—containing proteins that span the outer membrane of Gram-negative bacteria.

porogamy. Passage of the pollen tube through the micropyle of an ovule in a seed plant.

porospore. A spore produced from pores of a conidiophore.

porphyrin. A class of red-pigmented compounds with a cyclic tetrapyrrolic structure in which the four pyrrole rings are joined through their α-carbon atoms by four methene bridges (=C—) in a ring around a central metal atom, the porphyrins form the active nucleus of chlorophylls and hemoglobin.

potamoplankton. Plankton found in rivers.

potometer. An instrument for measuring the rate of transpiration or rate of absorption of water by a cut leafy shoot. It is normally used to compare the rates of transpiration of the same shoot under different conditions, by timing the movement of a water meniscus or air bubble over a given distance on the scale.

pour-plate culture. A technique for pure-culture isolation of bacteria; liquid, cooled agar in a test tube is inoculated with one loopful of bacterial suspension and mixed by rolling the tube between the hands; subsequent transfers are made from this to a second test tube, and from the second to a third; contents of each tube are poured into separate petri dishes; pure cultures can be isolated from isolated colonies appearing on the plates after incubation.

prairie. A North American grassland.

pratincolous. 1: Living in meadows.
2: Living in low grass.

preadaptation Possession by an organism or group of organisms, specialized to one mode of life, of characters which favour easy adaptation to a new environment.

predisposition. An increase in susceptibility resulting from the influence of environment on the suscept.

pressure potential. Symbol ψp, a measure, in units of pressure, of the squeezing effect that the cell wall has on the contents of the cell, in turgid plant cells.

primary cycle. The first cycle of a pathogen initiated after a period of pathological inactivity, usually after rest or seasonal inactivity.

primary meristem. Meristem which is derived directly from embryonic tissue and which gives rise to epidermis, vascular tissue, and the cortex.

primary phloem. Phloem derived from apical meristem.

primary production. The total amount of new organic matter produced by photosynthesis.

primary productivity. The amount of matter that can be synthesized by all the autotrophic organisms in a given area in given time.

primary root. The first plant root to develop; derived from the radicle.

primary succession. *See* prisere.

primary thickening. The thickening of a stem or root which occurs near the growing apex.

primary tissue. Plant tissue formed during primary growth.

primary vegetation. vegetation which has not been disturbed or changed by man.

primary xylem. Xylem derived from apical meristem.

primitive. Of organisms which have the characteristics of early stages in evolutionary history. This can also apply to characters and organs.

primordium. An undeveloped organ, e.g., a leaf bud contains leaf primordia (*pl.*) a young flower bud contains the primordia of reproductive organs.

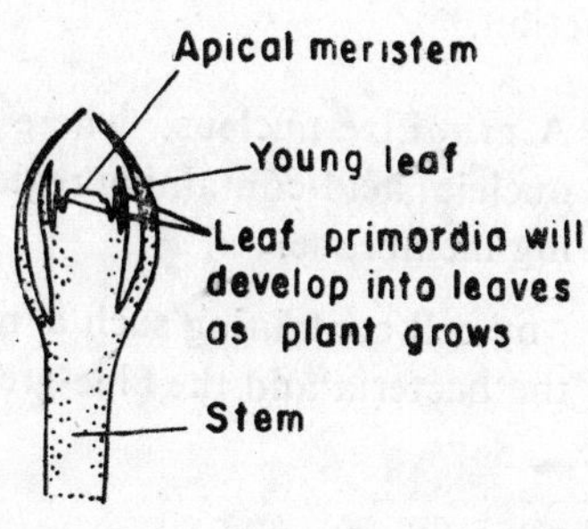

prions. These are the smallest known agents of disease that contain little or no nucleic acid and consisting only of protein, of moleculer weight between 50,000—100,000 and the particle size that is 100 times shorter than smallest viruses. Till now, they have been found to be associated with a disease of goats and sheep (*Scrapie*) and with a disease of the human nervous system—Creutzfeldt—Jakob disease.

prisere. The ecological succession of vegetation that occurs in passing from barren earth or water to a climax community; primary succession.

probasidium (*pl.* probasidia). A binucleate hyphal cell which may undergo morphological changes before developing into a mature basidium.

procarp. Multicellular female sex organ of many Rhodophyceae; carpogonium and auxiliary cell.

procumbent. Having stems that lie flat on the ground but do not root at the nodes.

producer. An autotrophic organism in an ecosystem, which produces organic matter using chemical energy from light. Plants are the main producers in the biosphere.

progametangium (*pl.* progametangia). A cell which gives rise to a gametangium.

progeny. The offspring or young one produced by an organ ism during reproduction.

prokaryote. 1: A primitive nucleus, where the deoxyribonucleic acid-containing region lacks a limiting membrane.

2: Any cell containing such a nucleus, such as the bacteria and the blue-green algae.

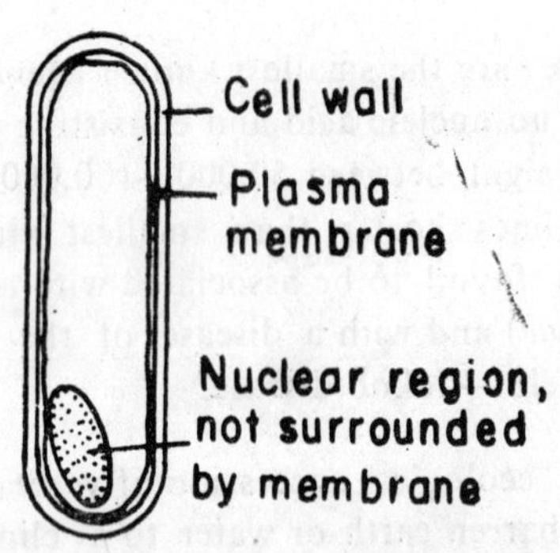

prolamin. Any of the simple proteins, such as zein, found in plants; soluble in strong alcohol, insoluble in absolute alcohol and water.

prolepsis (*pl.* prolepses). A hyperplastic symptom of disease in which organs appear before the natural time, e.g., the sprouting of shoots from adventitious buds after disease has impaired the metabolism of the organ in question.

proliferation. A rapid and repeated production of new cells, tissues, or organs; specifically, a hyperplastic symptom of plant disease in which organs continue to develop after they have reached the point beyond which they normally do not grow.

prometaphase. A stage between prophase and metaphase in mitosis in which the nuclear membrane disappears and the spindle forms.

promoter. The site on deoxyribonucleic acid to which ribonucleic acid polymerase binds preparatory to initiating transcription of an operon.

promycelium (*pl.* promycelia). An organ of meiosis of smuts and rusts that bears spores exogenously; analogous to, and often termed, a basidium.

propagation. The process of reproduction, either by natural or artificial means; propagate.

propagule. 1: Any part of an organism, produced sexually or asexually, that is capable of giving rise to a new individual; diaspore; propagulum.

2: The minimum number of individuals of a species required for colonization of a new or isolated habitat.

3: A reproductive structure of brown algae.

prophage. Integrated unit formed by union of the provirus into the bacterial genome.

prophase. The initial stage of mitotic or meiotic cell division in which chromosomes are condensed from the nuclear material and split logitudinally to form pairs.

proplastid. Precursor body of a cell plastid.

prop root. A root that serves to support or brace the plant; brace root.

prosenchyma. Pertaining to fungal tissue, a type of plectenchyma in which the component hyphae lie parallel to one another and are easily recognized as such.

prosthetic group. A characteristic nonamino acid substance that is strongly bound to a protein and necessary for the protein portion of an enzyme to function; often used to describe the function, as in haemeprotein for haemoglobin.

protandrous. Of flowers whose anthers produce pollen before the ovules or stigma of the same flower are functional. This is a way of avoiding self-fertilization; protandry.

protection. As a principle of plant-disease control, the placing of a barrier between suscept and pathogen, e.g., the use of protective chemical dusts or sprays.

protein. A substance made of one or more polypeptides, which themselves are made of amino acids. There are very many different kinds of protein, each with its own sequence of amino acids. Some are structural, e.g., in membranes, and others are enzymes which catalyze reactions in cells.

protein structure. The structure of proteins can be studied at four levels—primary, secondary, tertiary and quaternary. The primary structure of a protein is the sequence of amino acids in a polypeptide. The secondary structure of a protein is the coiling of the polypeptide into a helix or pleated sheet. The tertiary structure of a protein is the twisting and folding of the polypeptide helix or pleated sheet to form a three-dimensional protein molecule. The quaternary structure of a protein is the structure of several protein molecules when bonded together.

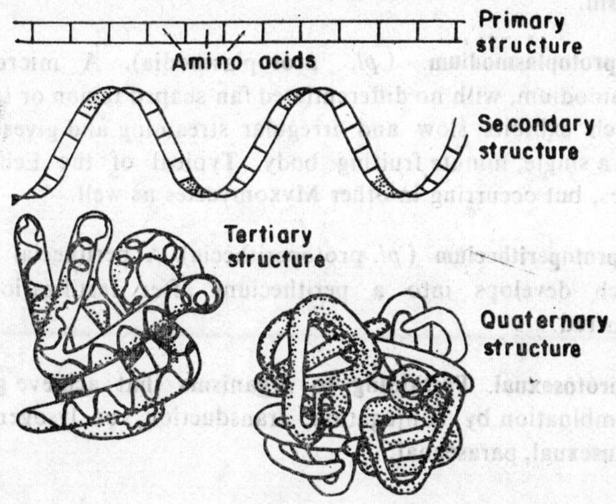

prothallus (*pl.* prothalli). The gametophyte of mosses, liverworts and pteridophytes.

Protista. A kingdom proposed by Haeckel in an attempt to classify organisms showing characteristics of both plants and animals.

protogynous. Of flowers in which the female parts become functional before the male parts. This is a way of avoiding self-fertilization; protogyny.

protonema. The young gametophyte of a moss, in the early stages after the germination of the spore.

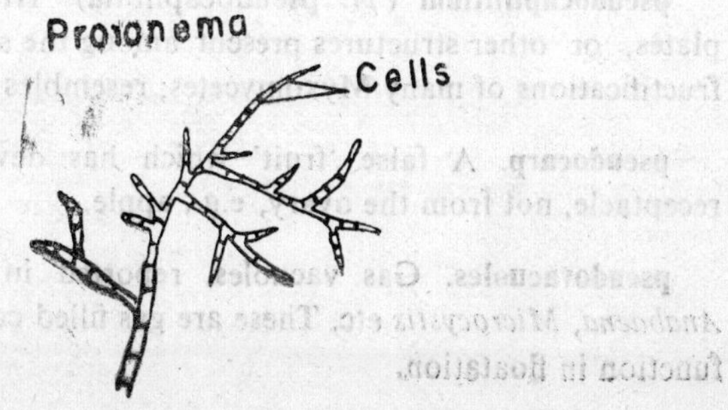

protoplasm. The general name for all the substances and bodies inside a cell. All living organisms are made of protoplasm.

protoplasmodium. (*pl.* protoplasmodia). A microscopic plasmodium, with no differentiated fan-shaped region or strands, which exhibits slow and irregular streaming and gives rise to but a single, minute fruiting body. Typical of the Echinosteliales, but occurring in other Myxomycetes as well.

protoperithecium (*pl.* protoperithecia). A perithecial initial which develops into a perithecium after fertilization has occurred.

protosexual. Pertaining to organisms that achieve genetic recombination by conjugation; transduction, or lysogenation; *cf.* eusexual, parasexual,

protospecies. An ancestral species.

prototrophic. 1: Autotrophic; protrophic.

2: Obtaining nourishment from a single source only; prototroph, prototrophism.

3: Used of microorganisms capable of growth on simple sugars and salts without specific additional nutrient requirements.

pruning. The cutting away of dead or overgrown branches of a tree in order to promote growth.

pseudocapillitium (*pl.* pseudocapillitia) Irregular threads, plates, or other structures present among the spores within the fructifications of many Myxomycetes; resembles capillitium.

pseudocarp. A false 'fruit' which has developed from the receptacle, not from the ovary, e.g., apple.

pseudovacuoles. Gas vacuoles, reported in spp. of *Nostoc*, *Anabaena*, *Microcystis* etc. These are gas filled cavities serving to function in floatation.

pseudogamy. The activation of an egg by a male gamete that degenerates without its nucleus fusing with that of the egg; gynogenesis; pseudomixis; somatogamy.

pseudometollaphyte. A plant able to tolerate high levels of heavy metals but one not confined to such substrates; facultative metallophyte; *cf.* metallophyte.

pseudomycelium (*pl.* pseudomycelia). A series of cells adhering end to form a chain. Produced by some yeasts.

pseudoparaphyses (sing. pseudoparaphysis). Sterile threads attached both to the roof and to the base of an ascocarp.

pseudoparenchyma (*pl.* pseudoparenchymata). A type of plectenchyma consisting of oval or isodiametric cells, the component hyphae having lost their individuality.

pseudoperithecium (*pl.* pseudoperithecia). A uniloculate ascostroma embedded within fungal stroma characteristics of Loculoascomycetes.

pseudoplasmodium (*pl.* pseudoplasmodia). An aggregation of amoeboid cells constituting the initial stage of fruiting of the Acrasiomycetes (cellular slime moulds).

pseudoseptum (*pl.* pseudosepta). A plug-like partition of cellulin or other substance in a hypha, resembling a septum.

pseudospore. A non-motile maked spore; found in some Acrasiomycetes.

psychrotroph. An organism which is able to grow at 0°C but which grow best at a temperature of 20°C; facultative psychrophile.

pteridophyte. A member of the division Pteridophyta, which includes the ferns, clubmosses and horsetails. In pteridophytes the sporophyte is the main vegetative stage of the life cyle. The gametophyte is small, and is independent of the sporophyte, as in bryophytes. Large tree-like forms of pteridophytes were common during the Carboniferous period, and were fossilized to form coal.

pubescent. Hairy.

pulp. The succulent part of a fruit.

pure line. A series of generations that are homozygous for all characters.

purine. One of the two kinds of nitrogen-containing base in nucleic acids. A purine molecule consists of two rings of carbon and nitrogen atoms. The main purines in nucleic acids are adenine and guanine.

pustule. A pimplelike, eruptive fruiting structure, such as a uredinium of a rust fungus.

pycnidiospore. A conidium borne in a pycnidium.

pycnidium (*pl.* pycnidia). A hard-walled, flask-shaped, fungal fruiting body that contains conidia.

pycnium (*pl.* pycnia). A spermagonium, a flask-shaped fruiting body of a rust fungus; it contains haploid pycniospores (spermatia) and filaments (receptive hyphae) that extend through the ostiole. A pycnium and its contents are of one mating type.

pycnosclerotium (*pl.* pycnosclerotia). A more or less hard-walled structure resembling a pycnidium but containing no spores.

pygmism. The state of being dwarfed or reduced in size.

pyramid of numbers. The number of organisms at each trophic level in a food web or food chain. At each level, energy is lost through respiration and other metabolic processes, and there is less energy available for the next trophic level. For this reason, the number and biomass of consumers in an ecosystem are less than the number and biomass of producers.

pyrene. A singe stone in a small drupe.

pyrenoid. A small grain of protein in the chloroplast of an algal cell, around which starch is deposited.

pyrimidine. One of the two kinds of nitrogen-containing base found in nucleic acids. The molecules consist of a single ring of carbon and nitrogen atoms. The main pyrimidines in nucleic acids are thymine, cytosine and uracil.

pyrrole. An organic compound with one atom of nitrogen and four atoms of carbon, each bonded to a hydrogen atom, arranged in a ring. Four pyrrole groups make up the porphyrin structure in chlorophyll and cytochromes.

pyruvic acid. $CH_3COCOOH$. The end product of glycolysis. Pyruvic acid is the fuel for the Krebs cycle in aerobic organisms.

Q

Q_{10}. Temperature coefficient. The increase in rate of a process (expressed as a multiple of initial rate) produced by raising temperature 10°C. Often between two and three for biological as for many chemical processes.

Q-enzyme. This enzyme was first isolated from potato extracts and is thought to catalyze the transfer of small chains of glucose units from an amylose type molecule onto the carbon – 6 of one of the glucose units of the acceptor molecule to form α-1, 6 glycosidic linkages.

donor (optimum 40 units)

Acceptor

Q-enzyme

QO_2. A Term used in measurement of respiration. The oxygen uptake of an organism, expressed in microlitres per milligram (dry weight) per hour.

quadrat. 1: A delimited area for sampling flora or fauna; usually taken randomly within the study area and typically consisting of a 1 metre square frame; a laiger area of 4 square metres is often referred to as a major quadrat.

2: The sampling frame itself.

quadrat chart. A diagram showing the position of each organism within a quadrat.

quadriflagellate zoospores. Pertaining to the zoospores of *Ulolhrix zonate* a member of Chlorophyceae which, have four flagella. Depending on the size these zoospores may be macro-zoospores, or microzoospores. All these zoospores are a means of asexual reproduction.

quadripartition. The division of a spore-mother-cell into four spores.

quadriplex. Having four dominant genes.

quadriscutate. Pertaining to *Chara zeylanica*, a member of class Chlorophyceae, when the globular woll consists of four shield cells.

quadrivalent. 1: An association of four chromosomes synapsed during the first meiotic division.

2: A nucleus having 2 pairs of homologous chromosomes, or an individual containing such nuclei.

qualitative character. A character having discrete states that are not numerical or morphometric; discrete character; qualitative trait; *cf.* quantitative character.

qualitative inheritance: The inheritance of phenotypic characters showing discontinuous variation, that are mainly controlled by only one or a few genes; *cf.* quantitative inheritance.

quantasome. One of the highly ordered array of units that has a "cobblestone" appearance in electron micrographs of the lamella of chloroplasts, and thought to be the most probable site of the light reaction in photosynthesis. Each quantasome is estimated to contain approximately 300 chlorophyll molecules which are believed even now by some to function as a photosynthetic unit in absorption of light quanta. However, this entity is not considered now by some scientists to be a functional photosynthetic unit. It does not seem to be large enough to contain the necessary components for complete photosynthesis.

quantitative biology. The application of statistical methods to biological problems and the mathematical analysis of biological data; biometry; biometrics.

quantitative character. 1: A character based on counts, measurements, ratios or other numerical values; numerical character; quantitative trait; *cf.* qualitative character.

2: A character showing continuous variation and controlled by several interacting genes.

quantitative inheritance. The inheritance of phenotypic characters showing continuous variation, that are controlled by several interacting genes; *cf.* qualitative inheritance.

quantum yield (ϕ). Of photosynthesis, the photochemical work performed by the absorption of light quanta, expressed in mole oxygen liberated per mole absorbed photons.

quaquaversal. Bending every way.

quarantine. The isolation of an individual who is suffering from or has been exposed to a communicable disease, in order to protect others against infection. The incubation period of the disease may determine the length of quarantine. The term also applies to imported animals.

quasi symbiosis. It is a kind of symbiosis between two completely independent plants like rice and *Aulosira* or *Anabaena* where each seems to do better than if grown separately.

Quaternary. The second geological sub-era of the Cenozoic, It began with the Pleistocene and has lasted until the present, characterized by the rise of man and modern mammals.

quaternary structure. The manner in which the separate polypeptide chains of a protein are held together and oriented in space with respect to one another. The individual polypeptide chains of a protein are usually referred to as its subunits. Ionic bonds, hydrogen bonds, hydrophobic bonds and disulphide bridges serve to stabilize quaternary association of subunits.

quebracho. Any of a number of South American trees in different genera in the order Sapindales, e.g., *Schinopsis spp.*, all being a valuable source of wood, bark, and tannin.

quercetin. $C_{15}H_5O_2(OH)_5$. A yellow, crystalline flavonol obtained from bark of *Quercus velutina* family fagaceae and Douglas fir bark; used as an antioxidant and absorber of ultraviolet rays, and in rubber, plastics, and vegetable oils.

quiescence. 1: A temporary resting phase characterized by reduced activity, inactivity or cessation of development; quiescent.

2: The condition of seeds that have not germinated because one or more environmental variables is unfavourable; *cf.* dormancy.

3: That period of the prepenetration stage during which a pathogen may be inactive because environmental conditions are unfavorable for its growth.

quiescent centre. A region of cells in the root tip, at the end of the stele where the cells remain inert i.e., no cell divisin takes place. The actively dividing cells occur at the boundary or surface of the quiescent centre.

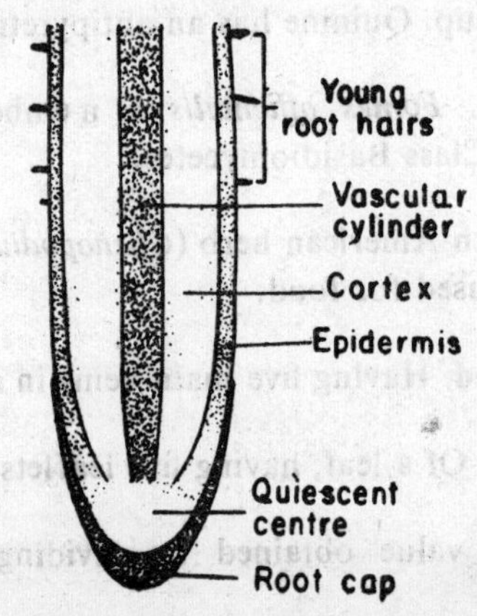

quilled. Designating a ligulate floret which has become tubular.

quillwort. The common name for plants of the genus *Isoetes.*

quinary. In fives; quintuple.

quinate. Arranged in groups or sets of five.

quince. A rosaceous tree (*Cydonia oblonga*) or the large, round, acid fruit it produces, which are commonly used in the preparation of jams, jellies, marmalades. In India it is grown in Kashmir on a small scale.

quincuncial aestivation. A particular type of imbricate aestivation in a five-petalled corolla. Two petals overlap their neighbours by both edges, two are overlapped on both their edges, and overlaps one neighbour, and is overlapped by the other.

quinine. A white or colourless crystalline substance obtained from the bark of the cinchona tree (*Cinchona officinalis*, fam. Rubiaceae). It has a very better taste. About twenty or more alkaloides have been extracted, out of which cinchonidine, cinchonine quinidine, are a few most useful in medicine. Europeans used natural quinine to relieve symptoms of malaria as early as 1630. Now-a-days other less toxic and more effective drugs have come up. Quinine has an antipyretic effect.

quinine-fungus. *Fomes officinalis* a member of the family Polyporaceae of Class Basidiomycetes.

quinoa. A south American herb (*Chenopodium quinoa*) whose seeds are widely used for food.

quintriplinerved. Having five main veins in a leaf.

quinquefoliate. Of a leaf, having five leaflets.

quotient. The value obtained by dividing one quantity by another.

R

race. 1: An intraspecific category charactericed by conspicuous physiological (physiological race), biological (biological race), geographical (geographical race) or ecological (ecological race) properties.

2: Pertaining to a pathogen, a physiological race or a form is one of a group of forms like in morphology but unlike in certain cultural, physiological, biochemical, pathological or other characters. For example *Puccinia graminis* is the cause of stem rust of cereals. Some individuals of this species, attack only wheat or only barley etc., and these individuals comprise groups that are called varieties or special forms such as *Puccinia graminis tritici*, *P.g. hordei* etc. But even within each special form, some individuals attack some of the varieties of the host plant but not the others, each group of such individuals comprising a race. Thus there are more than 200 races of *Puccinia graminis tritici* (race 15, 222, etc).

raceme An inflorescence on which stalked flowers are borne in acropetal succession on an unbranched main stalk that continues to grow during flowering.

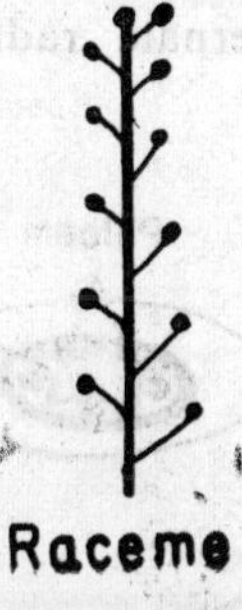

Raceme

racemose. Having flowers in raceme like inflorescences that may or may not be true racemes.

racemule. A small raceme.

race non-specific resistance. Host-plant resistance which is operational against all races of a pathogen species. It is variable, sensitive to environmental changes and is usually polygenically controlled, often referred to as horizontal resistance.

race-specific resistance. Host-plant resistance which is operational against one of a few races of a pathogen species; generally produces an immune or hypersensitive reaction and is controlled by one or few genes, often referred to as vertical resistance.

rachilla. The secondary axis, or rachis; in particular in the grasses and sedges the axis that bears the florets.

rachis. 1: The main axis of a pinnately compound leaf. The rachis is a continuation of the petiole.

2: The main axis of an inflorescence.

3: In fungi, a spore bearing extension of the sporogenous cell.

racket cell. Of dematophytes, a hyphal cell having a swelling at one end; racquette cell.

radial bundle. A vascular bundle having the primary xylem and phloem lying on alternate radii. They are usually found in roots.

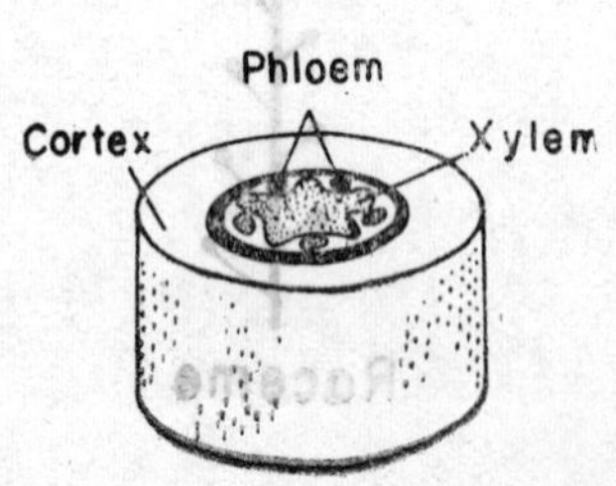

radial symmetry. If an organ or organism can be split down any radius longitudinally to give two identical halves, it is radially symmetrical, e.g., *Alectoria, Usnea* (lichens).

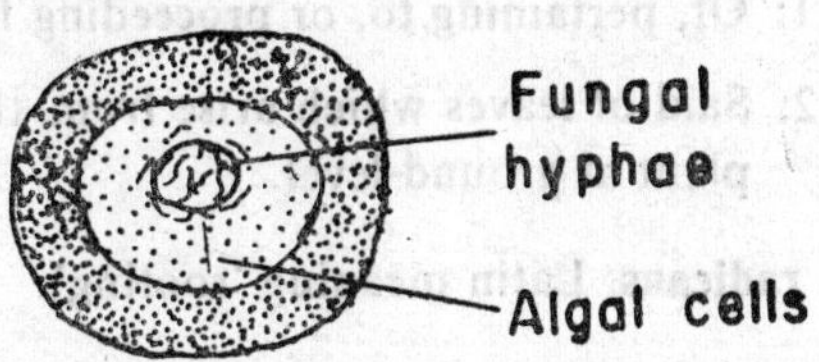

radial wall. An anticlinal wall placed on or across the radius of an organ.

radiate. 1: Spreading from a centre.

2: Said of a capitulum which has ray-florets.

3: Said of a flower which spreads like a ray from the periphery of any densely packed inflorescence.

4: Said of a stigma in which the receptive surface radiate outwards from the centre.

radiation biochemistry. The study of the response of the constituents of living matter to radiation.

radiation biology. *See* radiobiology.

radiation cytology. An aspect of biology that deals with the effects of radiations on living cells.

radiation effects. The harmful effects of ionizing radiation on humans and other animals, such as production of cancers, cataracts, and radiation ulcers, loss of hair, reddening of skin, sterilization, nausea, vomiting, mucous or bloody diarrhea, purpura, epilation, and agranulocytic infections.

radiation genetics. The study of the genetic effects of radiation; radiogenetics.

radiation microbiology. A field of basic and applied radiobiology concerned chiefly with the damaging effects of radiation on micro-organisms.

radical. 1: Of, pertaining to, or proceeding from the root.

2: Said of leaves which arise from the base of the plant at ground-level.

radicant, radicans. Latin meaning 'rooting'.

radicating. 1: Rooting.

2: Said of a mushroom stipe which is like a root.

radicicolous. A parasite growing on or in roots; radicolous; radicicole.

radiciform. Shaped like a root.

radicivorous. Feeding on roots, radicivore, radicivory.

radicle. 1: The part of the embryo that develops into the root of the seedling.

2: Any very small root.

3: A rhizoid of a moss.

radiculose. Said of the 'stem' of a moss which bears many rhizoids at the base.

radioautography. *See* autoradiography.

radiobiology. Study of the scientific principles, mechanisms, and effects of the interaction of ionizing radiation with living matter; radiation biology.

radioecology. The interdisciplinary study of organisms, radionuclides, ionizing radiation, and the environment.

radiogenetics. *See* radiation genetics.

radiomutation. A mutation which is the result of exposure of living-tissue chromosomes to ionizing radiation.

radiotropism. A growth effect produced by radiation; radiotropic.

radula spore. One of the slimy spores borne over the surface of ascospores of *Nectria coryli* while still in the ascus; radulaspore; radulospore.

raffinase. An enzyme that hydrolyzes raffinose, yielding fructose in the reaction.

raffinose. $C_{18}H_{32}O_{16}5H_2O$. A white, crystalline trisaccharide found in sugarbeets, plantains, cottonseed meal, and molasses, yields glucose, fructose, and galactose on complete hydrolysis. gossypose; melitose; melitriose.

rain desert. A desert in which rainfall is sufficient to maintain a sparse general vegetation.

rain factor. A measure of humidity expressed as the ratio of mean annual rainfall (in millimetres) to mean annual temperature (in degrees Celsius).

rain forest. A tropical forest having an annual rainfall of at least 254 centimetres (100 inches), consist of a forest of broad-leaved, mainly evergreen trees and a very large number of plant species.

raised bog. An area of acid, peaty soil, especially that developed from moss, in which the center is relatively higher than the margins.

ramate. Having branches.

rambler. A weak-stemmed plant that leans on and scrambles over the surrounding vegetation.

ramentaceous. Covered with ramenta.

ramentum (*pl.* ramenta). A thin brownish scale consisting of a single layer of cells and occurring on the leaves and young shoots of many ferns.

ramicolous. Living on twigs and branches; ramicole.

ramie. *Boehmeria nivea.* A shrub or half-shrub of the nettle family (Urticaceae) cultivated as a source of a tough, strong, durable, lustrous natural woody fiber resembling flax, obtained from the phloem of the plant; used for high-quality papers and fabrics; China grass; rhea.

ramo-conidium. A conidium formed from the apical branch of a conidiophore which gets reduced in size and functions as a conidium, as in *Cladosporium*.

ramulus. A very small branch of a stem or leaf.

random mating. The situation within a population where each individual has an equal chance of mating with any other individual, including itself, irrespective of genotype or phenotype; panmixis.

range. 1: The area over which a species is spread in the wild.

2: Of a plant pathogen, the geographical region or regions in which it is known to occur.

rangiferoid. Branched like a reindeer's horn.

rapaceous. Latin meaning 'turnip-shaped'.

raphe. 1: A long ridge on the coat of a seed which has developed from an anatropous ovule. The raphe marks the position where the funicle of the ovule used to be.

2: A longitudinal median line or fissure in the wall of diatom valve; it bears a nodule at each end and one in the middle.

raphide. One of the long, needle-shaped crystals, usually consisting of calcium oxalate, occurring as a metabolic by-product; present in the form of bundles, rounded masses or singly in aroids and aquatic plants like *Pistia* and water hyacinth, and in lichens.

Raunkiaerian leaf size classes. A system of classifying leaf sizes according to surface area, as a physiognomic character of vegetation; the six categories are leptophyll (up to 25 mm^2), nanophyll (25—225 mm^2), microphyll (225—2025 mm^2), mesophyll (2025—18 225 mm^2), macrophyll (18225—164 025 mm^2). megaphyll (over 164 025 mm^2); the category notophyll has been added for a small mesophyll (2025—4500 mm^3).

Raunkiaerian life forms. A system of classification of the life forms of plants based on the type and position of the renewal buds with respect to ground level; categories include chamaephyte, cryptophyte, geophyte, helophyte, hemicryptophyte, hydrophyte, phanerophyte, therophyte.

ray. A band of parenchyma and/or sclerenchyma cells, running from the cortex towards the centre of a stem. That between the primary vascular bundles is the interfascicular ray, and that in the secondary vascular tissue, by division of the cambium is a vascular ray.

ray-floret. A flower at the edge of a composite inflorescence. Most ray-florets have a single strap-shaped petal called a ligule; ray flower.

ray fungi. *See* Actinomycetes.

ray initial. One of the cells of the cambium which divide to produce new phloem and xylem ray cells.

ray shake. A radial crack in wood caused by wounds in a tree along the barrier zone.

ray tracheid. A thick-walled cell which occurs in the vascular rays of pines. It has bordered pits and conducts aqueous solution horizontally.

RDP. Ribulose diphosphate.

reaction. Any change in the activity of an organism in response to a stimulus.

reaction time. The time interval between a stimulus and the response to that stimulus; latent period.

reaction wood. An abnormal development of a tree and therefore its wood as the result of unusual forces acting on it, such as an atypical gravitational pull.

readthrough. Transcription beyond a termination sequence due to failure of ribonucleic acid polymerase to recognize the termination codon.

receptacle. 1: In fungi, any hymenium supporting structure.

2: In fungi, an axis having one or more organs, as the stem in Phallales.

3: In algae, the swollen end of a branch bearing sporangia or gametangia.

4: In ferns, the cushion tissue bearing the sporangia.

5: The top of the stalk of a flower, bearing the perianth, stamens and pestel.

6: In flowering plants, the enlarged end of a peduncle bearing the flowers of a crowded inflorescence.

receptive body. A small branched or unbranched process from the stroma, capable of being spermatized by microconidia, e.g., in *Sclerotinia gladioli*, a member of Discomycetes, receptive hypha. Flexuous hypha, trichogyne, and other similar structures found in fungi used to receive male cells.

receptive papilla. In some oomycetes, a class of fungi, a small outgrowth from the Oogonium into the antheridium, to which the antheridium becomes attached.

receptive spot. The clear area in the female gamete of some algae, e.g., in *Oedogonium*, through which the male gamete enters.

receptiveness. The condition of a stigma when effective pollination, and fertilization is possible.

recessive. Alleles whose effects can only be seen in the homozygous condition. When in the heterozygous condition, it is the dominant allele and not the recessive allele which controls the phenotype; recessive trait; *cf.* dominant.

reciprocal cross. A cross where the sources of male and female gametes are reversed as in A♀B♂, A♂B♀.

recolonization. The re-establishment of vegetation on an area which has been stripped of plants.

recombination. 1: The occurrence of gene combinations in the progeny that differ from those of the parents as a result of independent assortment, linkage, and crossing-over.

2: The production of genetic information in which there are elements of one line of descent replaced by those of another line, or additional elements.

recombination mosaic. A mosaic produced as the result of somatic crossing-over.

recon. The smallest deoxyribonucleic acid unit capable of recombination.

rectinerved. Having straight or parallel veins.

rectipetaly. The tendency of plant members to grow in a straight line.

rectiseral. Arranged in straight rows.

recumbent. Of or pertaining to a plant or plant part that tends to rest on the surface of the soil.

recurrent. Said of the small veins of a leaf, when they bend back towards the midrib.

recurved. Of a pileus in mushroom, when it is bent or curved backwards.

red drop. Decrease in quantum yield of photosynthesis at wavelengths greater than 680 nm, an area of the spectrum occupied by the red absorption band of chlorophyll a.

red snow. Snow stained by the surface growth of a unicellular green alga, *Chlamydomonas nivalis*, which are rich in haematochrome.

red tide. A reddish discoloration of coastal surface waters due to concentrations of certain toxin-producing dinoflagellates; red water.

reducing sugar. A sugar which is able to reduce Benedict's solution or Fehling's solution, i.e, a sugar containing a free —H and —OH at C1. They exist in neither α nor β form when in free solution, but the molecules constantly alternate between two states. Monosaccharides, e.g., glucose and disaccharides, e.g., lactose and maltose are reducing sugars.

reduction division. A name sometimes given to meiosis, because the daughter cells receive a haploid set of chromosomes from the diploid parent cell.

reductive pentose pathway. The set of reactions in photosynthesis in which CO_2 is fixed by ribulose-diphosphate, a pentose sugar, to give PGA, which is used to produce a hexose sugar and more ribulose-diphosphate, which is used for further CO_2 fixation. This pathway is driven by energy from ATP produced in the light reaction. It also requires $NADPH_2$ from the light reaction for the reduction of PGA.

redundancy. 1: Repetition of a specified deoxyribonucleic acid sequence in a nucleus.

2: Multiplicity of codons for individual amino acids.

red water. *See* red tide.

redwood. *Sequoia sempervirens.* An evergreen tree of the pine family, it is the tallest tree in the Americas, attaining 350 feet (107 meters); its soft heartwood is a valuable building material.

reed. Any tall grass characterized by a slender jointed stem.

reflexed. Turned abruptly backward.

regeneration. 1: Renewal or restoration of structures or tissues. typically after loss or damage.

2: The growth of new plants from perennating organs, e.g., rhizomes, regenerate.

3: The growth of new vegetation in a place where the old vegetation has been removed or damaged.

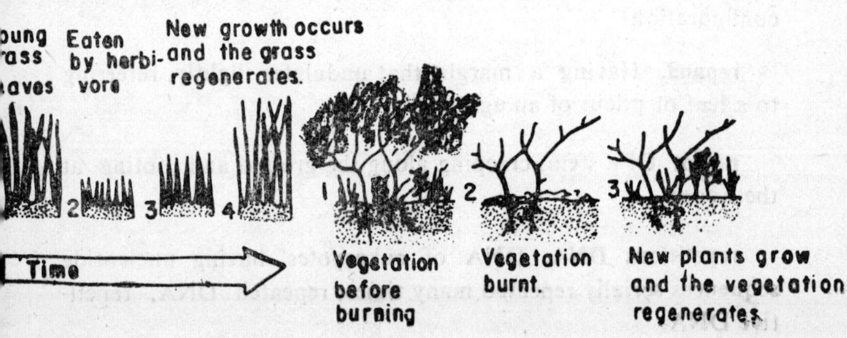

regular. Having radial symmetry, referring to a flower.

regulin. An antitumour antibiotic from *Aspergillus restrictus.*

reforestation. The process of re-establishing a forest on previously cleared land.

reindeer lichen. *Cladonia alpestris* and *C. rangiferina*, Species grazed by reindeer and caribou.

reindeer moss. *See* reindeer lichen.

relative growth rate (RGR). A measure of growth in terms of increase in weight (usually dry weignt) per unit time with respect to the initial weight (W) of the plant; RGR$=dW/dt\times 1/W$.

relative humidity. The ratio of the actual water vapour present to that of total saturation at the existing temperature, expressed as a percentage.

relative sexuality. The state or condition of a gamete that is able to act as a male or as a female when mated to different gametes.

remote. Pertaining to the lamellae of agarics which do not reach the stipe but leave a space around it.

renaturation. The return of a protein or nucleic acid from a denatured and nonfunctioning state to its' 'native", functioning configuration.

repand. Having a margin that undulates slightly, referring to a leaf or pileus of an agaric.

repent. Of a stem, creeping along the ground and rooting at the nodes.

repetitious DNA. DNA of eukaryotes having nucleotide sequences serially repeated many times; repeated DNA, repetitive DNA.

replacement. Substitution of inorganic matter for the original organic constituents of an organism during fossilization.

replica plating. A method for the isolation of nutritional mutants in microorganisms; colonies are grown from a microorganism suspension previously exposed to a mutagenic agent, on a complete medium in a petri dish; a velour surface is used to transfer the impression of all these colonies to a petri dish containing a minimal medium; colonies that do not grow on the minimal medium are the mutants.

replication. The process by which new DNA is made. The two strands in the DNA double helix separate, and a new strand of nucleotide polymer is synthesized on each one. Because each nitrogen-containing base in the nucleotide units of the polymer will only pair with one other base, the new DNA has exactly the same sequence of bases as the old. This self-copying process is the basis of heredity; replicate.

replum. A thin wall separating the two valves or chambers of cruciferous fruits, not a true part of the carpel but is an ingrowth of the placentas.

repressor. An end product of metabolism which represses the synthesis of enzymes in the metabolic pathway. The product of a regulator gene that acts to repress the transcription of another gene.

reproduction. The process in which an organism produces offspring like itself. Reproduction can be sexual or asexual, and is one of the most important characteristics of living organisms; reproduce; reproductive.

reproductive isolation A situation in which two individuals or populations cannot breed with each other.

reproductocentric. Pertaining to Chytridiales, having development of one or more reproductive structures at the centre of gravity of the thallus; genocentric *cf.* polycentric.

resin duct. A canal (intercellular space) lined with secretory cells that release resins into the canal; common in gymnosperms.

resinous. Containing or producing resin, said of bud scales when coated with a sticky exudate of resin, e.g., in *Aesculus* spp.

resistance. The ability of a plant to remain relatively unaffected by a disease because of its inherent genetic and physiological or structural characteristics.

respiration. 1: The process in which cells release energy stored in carbohydrates in order to drive the chemical reactions of metabolism. Aerobic respiration involves glycolysis, the Krebs cycle, and the electron transfer chain; this process uses oxygen, and produces carbon dioxide and ATP. The Krebs cycle and electron transfer chain take place in the mitochondria. Anaerobic respiration involves glycolysis and fermentation, in which oxygen is not used.
2: The exchange of gases between an organism and its surrounding medium; breathing.
3: In ecological energetics, that part of the assimilated energy of an individual, population or trophic unit that is dissipated as heat as the result of exothermic metabolic processes, per unit time.

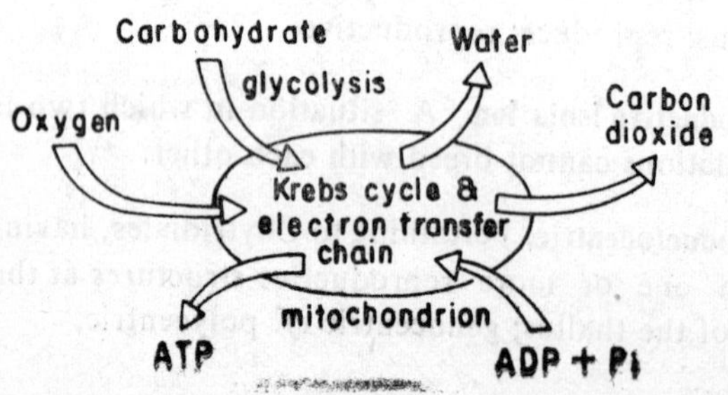

respiratory index. The amount of carbon dioxide produced per unit dry weight per hour.

respiratory quotient (RQ). The ratio of the volume of oxygen consumed to the volume of carbon dioxide released; respiratory ratio.

response. The change produced in an organism by a stimulus, usually adaptive.

resting spore. 1: A fungal spore, usually thick-walled, that can remain viable in a dormant condition for an extended period, e.g., chlamydospores, oospores, amphiospores, teliospores.

2: A spore which germinates after a resting period.

restriction enzymes. Ability to cleave double—stranded DNA at a defined nucleotide sequence.

resupinate. Inverted, usually through 180° so as to appear upside down or reversed, e.g., the hymenium of some fungi being on the outer side or the twisting of the flower as in orchids.

reticulate. 1: Of the veins of leaves, when their pattern is like a network.

2: Having or resembling a network of fibers, veins, or lines.

3: Of or relating to evolutionary change resulting from genetic recombination between strains in an interbreeding population.

reticulate walls. Spore walls having a pattern of superficial lines or ridges.

retroculture. The reisolation of a pathogen from a host into wihch it had been introduced experimentally.

retrogressive. Pertaining to a process of dedifferentiation or reversal to a simpler state or form; regressive; retrogression.

retrorse. Bent downward or backward.

retroversion. A turning back.

retuse. Having a rounded apex with a slight central notch.

reverse graft. A plant graft made by inserting the scion in an inverted position.

reverse mutation. Back mutation.

reverse transcriptase. An enzyme encoded by the RNA of certain viruses that is able to make complementary single-stranded DNA chains from RNA templates.

reverse transcription. Synthesis of DNA using RNA as template.

revolute. Rolled backward and downward.

R factor. A self-replicating, infection like agent that carries genetic information and transmits drug resistance from bacterium to bacterium by conjugation of cell; resistance factor.

Rf-value (relative front). A value used in chromatography such as paper chromatography. It represents the ratio of the distance travelled by the sample or by a reference substance to the distance travelled by the solvent (mobile phase). As the mobile phase always travels at least as far as the sample, the Rf-value is $\leqslant 1$. By comparing Rf-values of unknown substances with those of known substances a sample may be identified.

rhagadiose. Having deep cracks or fissures.

rhamnose. $C_6H_{12}O_5$. A deoxysugar occurring free in poison sumac, and in glycoside combination in many plants; isodulcitol.

rhamphoid. Beak-shaped.

rheophilous bog. A bog which draws its source of water from drainage.

rheoplankton. Plankton found in flowing water.

rheotaxis. A directed response of a motile organism to a water or air current, either into the current (positive rheotaxis) or with the current (negative rheotaxis); rheotactic.

rhinosporidiosis. Polyploid growths in the nose and other organs of man, horse, etc., caused by *Rhinosporidium seeberi.*

rhipidium. A fan-shaped inflorescence with cymose branching where branches lie in the same plane and are suppressed alternately on each side.

rhizanthous. Producing flowers directly from the root.

rhizina (*pl.* rhizinae). A root-like bunch of hyphae growing from the bottom of the thallus of foliose lichens. These may be in the form of a single robust hypha, or wooly-hirsute, or have brush-like tips or intertwined into strande; rhizine.

rhizocarpous. 1: Used of plants having perennial roots and annual stems and foliage.

2: Pertaining to perennial herbs that produce flowers and fruit below ground level as well as above ground.

rhizoid. Thread-like cells which grow from the lower surface or base of fungi, liver worts, mosses, lichens and ferns. This helps to hold the plant to substrate and perform the function of roots, i.e., act as a feeding organ.

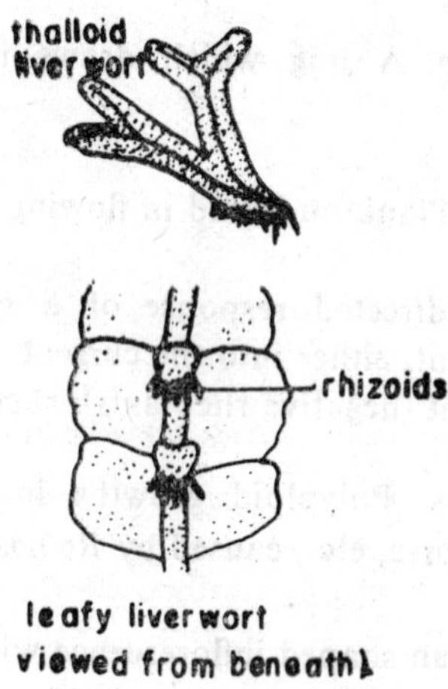

rhizomatous. Producing or possessing rhizome.

rhizome. An underground horizontal stem, often thickened and tuber-shaped, and possessing buds, nodes, and scalelike leaves. This is a way of vegetative reproduction and perennation, as the shoots grow into whole new plants; rhizomatous.

rhizomorph. A compact strand of hyphae formed by the longitudinal joining of hyphal strands into a bundle; it has a hard outer covering, grows from the apex, and serves as a survival organ and in the transport of food materials within the thallus, characteristics of many Basidiomycetes.

rhizophagous. 1: Feeding on roots; rhizophage, rhizophagy.

2: Used of a plant that obtains nourishment through its own roots.

rhizophilous. Growing on roots; rhizophile, rhizophily.

rhizoplane. The root-surface component of the rhizosphere.

rhizophore. A leafless, downward-growing dichotomous *Selaginella* shoot that has tufts of adventitious roots at the apex.

rhizoplast. A delicate fiber or thread running between the nucleus and the blepharoplast in cells bearing flagella.

rhizosphere. 1: The soil immediately surrounding plant roots that is influenced structurally or biologically by the presence of such roots; root zone; *cf.* phyllosphere.

2: The general name given by some ecologists to the parts of the biosphere in which roots grow.

L-rhodeose. *See* L-fucose.

rhodoplast. A reddish chromatophore occurring in red algae.

rhodosporous. Having light red spores.

rhodotorulic acid. A bacterial growth-promoting substance from *Rhodotorula* spp.

rhodoxanthin. $C_{40}H_{50}O_2$. A xanthophyll carotenoid pigment.

rhynchosporous. Having beaked spores.

rib. In a leaf or similar organ, the primary vein; also any prominent vein or nerne.

riboflavin. Riboflavin is required by cells as a coenzyme in many oxidation reactions, including those of photosynthesis. It is found in all plant cells vitamin B_2.

riboflavin 5'-phosphate. $C_{17}H_{21}N_4O_9P$. The phosphoric acid ester of riboflavin; flavin phosphate; flavin mononucleotide; FMN; isolloxazine mononucleotide; vitamin B_2 phosphate.

ribonuclease. $C_{587}H_{909}N_{171}O_{197}S_{12}$. An enzyme that catalyzes the depolymerization of ribonucleic acid.

ribonucleic acid. *See* RNA.

ribonucleoprotein. Any of a large group of conjugated proteins in which molecules of ribonucleic acid are closely associated with molecules of protein.

ribophorins. Integral membrane proteins of the rough endoplasmic reticulum that appears to be receptor and binding sites for ribosomes.

ribose. $C_5H_{10}O_5$. A pentose sugar occurring as a component of various nucleotides, including ribonucleic acid.

riboside. Any glycoside containing ribose as the sugar component.

ribosomal deoxy ribonucleic acid. (r DNA). The genes of the nucleolar organizing region that code for ribosomal RNA.

ribosomal proteins. A group of proteins that combine with rRNA and give the ribosome its three-dimensional structure.

ribosomal ribonucleic acid. (r RNA). Any of three large types of ribonucleic acid found in ribosomes: 5S RNA, with molecular weight 40,000; 14—16S RNA, with molecular weight 600,000; and 18—22S RNA with molecular weight 1,200,000. The molecule is irregularly twisted, but in places is doubled back to form short regions of base-pairing.

ribosome. One of the small, complex particles composed of various proteins and three molecules of ribonucleic acid which synthesize proteins within the living cell. Cells can contain many thousands of ribosomes, which are found either on the endoplasmic reticulum or as polysomes.

ribulose. $C_5H_{10}O_5$. A pentose sugar that exists only as a syrup synthesized from arabinose by isomerization with pyridine;important in carbohydrate metabolism; D-erythropentose; D-riboketose.

ribulose diphosphate (RuDP) $C_5H_{12}O_{11}P_2$. A compound consisting of a molecule of the pentose sugar ribulose and two phosphate groups. It is the main compound involved in CO_2 fixation during photosynthesis.

ribulose-diphosphate carboxylase. An enzyme that catalyzes CO_2 fixation by ribulose-diphosphate.

rickettsia. Plant disease agents belonging to schizomycetes group of bacteria, which may cause virus-like symptoms.

rigens. Latin meaning 'rigid'.

rimose. Having the surface marked by a network of intersecting cracks.

rimulose. Having small cracks.

rind. 1: The bark of a tree.

2: The thick outer covering of certain fruits.

3: The outer layers of a fruit-body or sclerotium of a fungus.

ring. 1: Of bacterial growth at the surface of the liquid culture which sticks to the container.

2: Annulus of mushroom.

3: In meiosis, chromosomes associated in ring usually by terminal chiasmata in twos, fours or sixes etc.

ringent. Having widely separated, gaping lips of a corolla.

ring porous. Said of a wood which contains more vessels, or larger vessels in the spring wood than elsewhere, so that it is marked in cross-section by rings or portions of rings of small holes.

ripe. Of fruit, fully developed, having mature seed and so usable as food.

rivulose. Marked by irregular, narrow lines like little rivers as shown on a map.

RNA. Ribonucleic acid. The nucleic acid directly involved in protein synthesis. RNA differs from DNA by having uracil instead of thymine and ribose instead of deoxyribose in its nucleotides. The RNA polymer is usually a single strand. There are three main kinds of RNA: *messenger RNA* (mRNA), which carries the genetic code from the nucleus to the cytoplasm; *transfer RNA* (tRNA), to which amino acids are attached before protein synthesis; and *ribosomal RNA* (rRNA), which is a structural part of the ribosomes.

RNA polymerase. Enzyme that catalyzes the formation of RNA from ribonucleoside triphosphates, using DNA as a template.

rock-hair. Pendant grey to black species of *Alectoria* (a lichen) which resemble human hair.

roestelium (*pl.* roestelia). In *Gymnosporangium* spp., an aecium which is long to tube like with a strongly developed peridium extending beyond the chains of aeciospores; *cf.* peridermium.

rogue. A variation from the standard varietal type; also, to remove such undesirable plants, especially those infected by viruses, from the growing crop.

root. The organ of a plant, that grows down into the soil. Roots anchor the plant in the ground and take up water and nutrients from the soil. In some plants the roots also store food. They differ from stems in not having nodes and leaves.

root cap. A thick, mass of parenchymal cells covering the meristematic tip of the root, which protects the root as it grows and lubricates its passage through the soil.

root hair. One of the hair-like outgrowths of the root epidermis. Root hairs increase the surface area of a root and help in the uptake of water and nutrients.

root nodule. 1: An abnormal nodular growth on a plant root system caused by the establishment of symbiotic nitrogen-fixing bacteria of the genus *Rhizobium* in the host tissue.

2: Members of Plasmodiophorales form nodules in *Alnus*, *Myrica*, and *Elaeagnus*.

root parasitism. A condition found in a semi-parasitic plant the roots of which penetrate the roots of the host plant and remove elaborated food material.

root pressure. The pressure which partly causes upward movement of xylem sap resulting from the active transport of solutes into the xylem which itself causes osmotic flow of water into the xylem.

rootstock. 1: A root or part of a root used as the stock for grafting.

2: Subterranean stem; rhizome.

roridous. Covered with drops of liquid like dew.

rose hip. The ripened false fruit of a rose plant.

rosette. 1: Any structure or marking resembling a rose.
2: Any of various plant diseases in which the leaves become clustered in the form of a rosette.
3: A structure in which leaves are arranged in a tight spiral on a short stem with very short internodes.

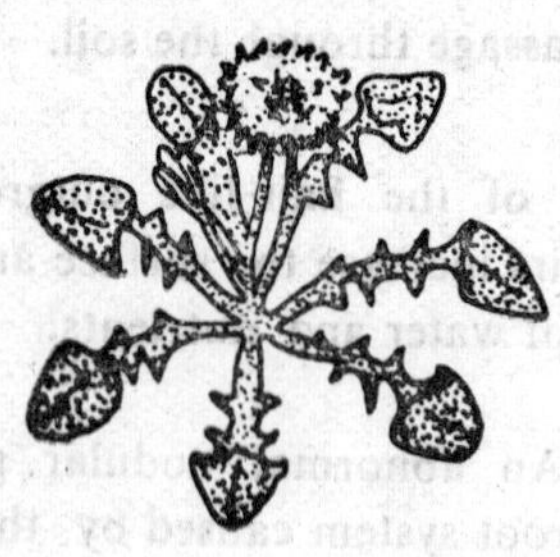

rostrum. Any beak-like process.

rosulate. Forming a small rosette.

rot. Decay; decomposition.

rotate. Of a gamopetalous corolla, having a short tube and petals radiating like the spokes of a wheel.

rotund. Nearly circular; orbicular inclining to be oblong.

rough colony. A flattened, irregular, and wrinkled colony of bacteria indicative of decreased capsule formation and virulence.

rough endoplasmic reticulum (RER). Portion of the endoplasmic reticulum bearing ribosomes.

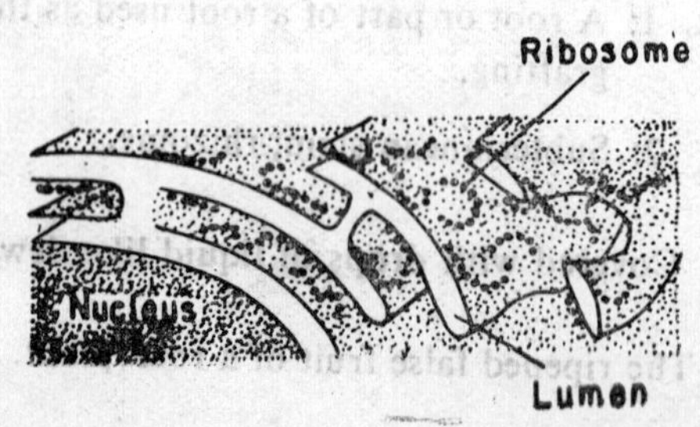

RQ. Respiratory quotient.

rRNA. Ribosomal RNA.

ruderal. Living in waste places or waste material.

rufous. Having a reddish-brown colour.

rugose. 1: Having a wrinkled surface, the venation seeming impressed into the surface.

2: Said of leaves roughened or crinkled by viral diseases, e.g., rugose mosaic of potatoes.

rugulose. Delicately wrinkled.

ruminate. Mottled in appearance; applied to a surface or tissue showing dark and light zones of irregular outline.

runcinate. Coarsely serrate to sharply incised with the teeth pointing toward the base as in *Taraxacum*.

runner. A horizontally growing, sympodial stem system; adventitious roots from near the apex, and a new runner emerges from the axil of a reduced leaf. This is away of vegetative reproduction. After the new plant has started to grow, the runner dies and decays stolon.

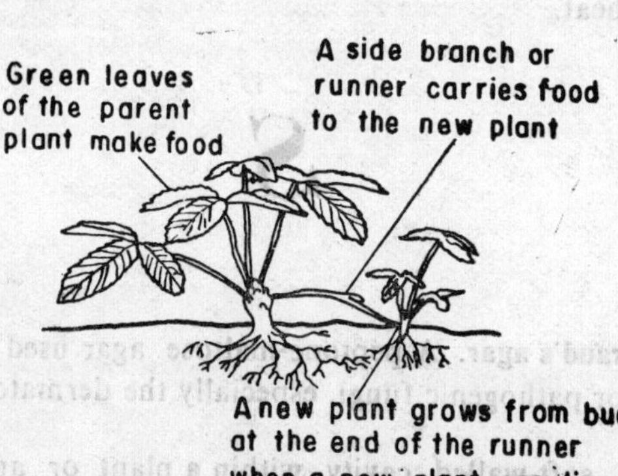

rupestral. Pertaining to or living on walls or rocks; mural; rupestrine.

rupicolous. Living on walls or rocks; rupestral; mural; rupicole.

ruralis. Latin meaning 'living in rustic places'.

russetting. A hyperplastic symptom of disease in which brownish, roughened areas form on the skin of fruit or tubers because of excessive cork-cell production.

rust. 1: A parasitic basidiomycete fungus of the order Uredinales. Rusts cause dark spots on the stems and leaves of plants. Some rusts, e.g., *Puccinia graminis* (cereal rust) are economically serious parasites.

2: The common name of the fungus itself.

3: A disease with rusty symptoms; black stem rust of cereals, *Puccinia graminis*, blister rust of *Pinus* and *Ribes*, *Cronartium ribicola*, red rust of tea, *Cephaleuros* (an alga); yellow rust of cereals, *P. striiformis*.

ryo. *Secale cereale.* A cereal plant of the family Gramineae cultivated for its grain, which contains the most desirable gluten, next to wheat.

S

Sabouraud's agar. A peptone-maltose agar used as a culture medium for pathogenic fungi, especially the dermatophytes.

sac. A soft-walled cavity within a plant or animal, often containing a special fluid and usually having a narrow opening or none at all.

saccate. Having a sac-like or pouch-like form.

saccharase. An enzyme that catalyzes the hydrolysis of disaccharide to monosaccharides, specifically of sucrose to dextrose and levulose; invertase; invertin; sucrase.

safflower. *Carthamnus tinctorius.* An annual thistlelike herb belonging to the family Compositae; the leaves are edible, flowers yield dye, and seeds yield a cooking oil.

saffron. *Crocus sativus*, family Iridaceae, the source of a yellow dye used for colouring food and medicine.

sagittate. Like an arrowhead in form; triangular, with the basal lobes pointing downward or concavely toward the stalk.

saltation. 1: A drastic and sudden mutational change; an abrupt evolutionary change; macrogenesis.

2: To move by leaping or bounding; saltatory.

3: A mutation occurring in the asexual state of fungal growth, especially one occurring in culture.

saltmarsh. A coastal habitat found in temperate regions, but typically associated with tropical and subtropical mangrove swamps, in which excess sodium chloride is the predominant environmental feature due to flooding by very high tides. The vegetation of salt marshes is mostly herbaceous.

saltmarsh plain. A salt marsh that has been raised above the level of the highest tide and has become dry land.

salt-spray climax. A climax community along exposed Atlantic and Gulf seacoasts composed of plants able to tolerate the harmful effects of salt picked up and carried by onshore winds from seawater.

salver form. Said of a gamopetalous corolla with the slender tube and an abruptly expanded flat limb, as that of the phlox; hypocrateriform.

samara. A dry, indehiscent, winged fruit usually containing a single seed, such as of sugar maple. (*Acer saccharum*) and ash (*Fraxinus*). Wing-like outgrowths assist in dispersal by wind.

sampling. The process of taking a sample.

sandalwood. 1: Any species of the genus *Santalum* of the family Santalaceae characterized by a fragrant wood.

2: *S. album*. A parastic tree with hard, close-grained, aromatic heartwood used in ornamental carving and cabinetwork.

sandblow. A patch of coarse, sandy soil denuded of vegetation by wind action.

sap. 1: The water and nutrients contained and transported in the xylem or phloem.

2: Sap is also a general name for any liquid exuded from the Plant when it is cut.

sapling. A young tree with a trunk less than 4 inches (10 cm) in diameter at a point approximately 4 feet (1.2 m) above the ground.

saprobiontic. Saprophagous; saprobiotic; saprobiont.

saprogenesis. Survival; that phase of the life cycle of a pathogen during which it is not actively causing disease in a living suscept.

saprophage. An organism that lives on decaying organic matter.

saprophagous. Feeding on dead or decaying organic matter; saprobiontic; saprophagic; saprophage, saprophagy; *cf.* biophagous.

saprophilous. Thriving in humus-rich substrates, or on decaying organic matter; saprophile, saprophily.

saprophytic. Used of a plant obtaining nutrient from dead or decaying organic matter. It usually lacks chlorophyll e.g., *Monotropa* and several fungi; humicular; hysterophytic; saprophyte.

sapwood. The younger, softer outer layers of the xylem of a stem, containing some living cells. The sapwood lies outside the heartwood, and its main function is translocation; alburnum.

sarcochore. A plant dispersing minute, light disseminules.

sarcody. A hyperplastic symptom in which swellings occur above and below portions of organs that are tightly encircled, as a stem might be chocked by a twining vine.

sarmentose. Producing slender, flexuous prostrate stems or runners.

saturation deficit. A measure of humidity, derived by subtracting the actual water vapour pressure from the maximum possible vapour pressure at a given temperature, expressed as a percentage of total saturation in mm of mercury; water vapour deficit; *cf.* relative humidity.

savannah. The tropical and subtropical grassland biome., transitional in character between grassland or desert and rain forest, typically having drought resistant vegetation dominated by grasses with scattered tall trees of the families Leguminosae, Bombacaceae, Bignoniaceae or Dilleniaceae; Tropical grassland biome; savanna.

savanna-woodland. *See* tropical woodland.

saxicolous. Living or growing on or among rocks or stones; lapidicolous; saxatile; saxigenous; saxicole, saxicoline.

saxifragous. Living in rock crevices; saxifrage.

scab. A hyperplastic symptom characterized by rough, crusty lesions formed by excessive cork production.

scabrous. Having a rough surface covered with stiff hairs or scales.

scalariform. 1: Resembling a ladder; having transverse markings or bars.

2: Of tracheids and vessels whose cell walls have ladder-like ridges of thickening.

scale. 1: A name given to many kinds of small, mostly dry, and appressed leaves or bracts, often only vestigial.

2: A small outgrowth, e.g., on the petiole of a fern frond.

3: The bract of a catkin.

scale scar. A mark left on a stem after bud scales have fallen off.

scandent. Climbing without the aid of tendrils.

scanning electron microscope (SEM). Electron microscope that permits observation of a specimen's surface structure; the beam is not transmitted through the sample but causes the release of secondary electrons from the metal-coated surface of the specimen. The results give a three-dimensional effect.

scansorial. Adapted for climbing.

scape. A flower stalk growing from the level of the ground, as in herbaceous plants; it may bear scales or bracts but no foliage leaves and may be one or many flowered.

scapose. Bearing flowers or inflorescence on a scape.

scarious. 1: Having a thin, membranous texture.

2: Applied to leaflike parts or bracts that are not green, but thin, dry, and membranaceous, often more or less translucent.

scavenger. An organism that feeds on carrion, refuse, and similar matter.

schizocarp. A dry dehiscent fruit developed from a syncarpous ovary. A schizocarp splits into two halves when ripe. Each half is a single carpel known as mericarp, as in most umbellifers.

schizophyte. A plant that reproduces by fission.

sciad. A plant growing in shady situations; a shade plant; sciophyte; skiophyte; skiaphyte.

scion. That part usually a stem or a bud or structure transplanted from one plant to another during a graft.

sciophilous. Thriving in shaded situations. or in habitats of low light intensity; heliophobous; skiophilous; umbrophilic; sciophile, sciophilic, sciophily.

sciophyte. A plant that thrives at lowered light intensity.

sciophyllous. Having leaves that can tolerate shaded situations; skiophyllous.

sclereid. A kind of cell found in the sclerenchyma of some plants, with heavily lignified walls. Sclereids are usually found in groups.

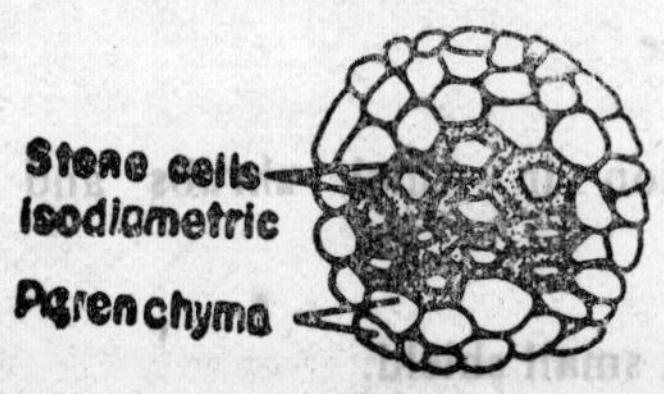

sclerenchyma. A hard, lignified tissue consisting of fibres and sclereids. It is found in the stems, roots, leaves or fruits of many plants, and its function is support.

sclerocaulous. Having a hard, dry stem because of exceptional development of sclerenchyma.

sclerochore. A plant that disperses disseminules without apparent morphological adaptations.

sclerophyllous. Of plants whose leaves contain sclerenchyma. Such leaves are usually thick and leathery.

sclerotium (*pl.* sclerotia) 1: A hard, compact mass of fungal tissue consisting of an outer sclerotized rind and an inner parenchymatous medulla; capable of surviving long periods of adverse environmental conditions.

2: The hardened, resting or encysted condition of the plasmodium of Myxo mycetes.

scobiform. Resembling sawdust.

scorpioid. Said of a circinnately coiled determinate inflorescence in which the flowers are two-ranked and borne alternately at the right and the left.

scorpioid cyme. A cyme with a curved axis and flowers arising two-ranked on alternate sides of the axis and sometimes appearing racemose.

scrub. Vegetation in which shrubs and small trees are dominant.

scutate. Like a small shield.

scutellum. 1: A rounded apothecium with an elevated rim found in certain lichens.
2: The flattened cotyledon of a monocotyledonous plant embryo, such as a grass.

seaweed. The general name for any large, parenchymatous alga in the sea.

sacalose. A polysaccharide consisting of fructose units; occurs in green rye and oats, and in rye flour.

secondary cambium. One of the tissue layers formed after the initial cambial layers in certain plant roots, and that produce a ring of tissue.

secondary constriction. Any constriction site along a chromosome other than the primary constriction at the centromere.

secondary cycle. A cycle initiated by inoculum from a primary from another secondary cycle without an interposed resting or dormant period for the pathogen; *cf.* primary cycle.

secondary periderm. Any layer of the periderm except the first and outermost layer.

secondary phloem. Phloem produced by the cambium, consisting of two interpenetrating systems, the vertical of axial and the horizontal or ray.

secondary structure. Structure of a polypeptide chain describing the location, extent, and types of helices (as well as nonhelical regions).

secondary thickening. The thickening of a stem or root because of the activity of the cambium to give xylem and phloem. It provides the plant with extra support and vascular tissue.

secondary vegctation. Vegetation growing in a place which has been disturbed by man, e.g. roadsides, old farmland, etc.,

secondary wall. The portion of a plant cell wall produced internal to and following deposition of the primary wall; usually consists of several anisotropic layers, and often has prominent internal rings, spirals, bars, or reticulations; a highly ordered wall, containing parallel microfibril.

secondary xylem. Xylem produced by cambium, composed of two interpenetrating systems, the horizontal (ray) and vertical (axial).

secretory structure. Plant cells or organizations of plant cells which produce a variety of secretions.

secund. Having lateral members arranged on one side only; said of inflorescences when the flowers appear as if borne from only one side.

sedge. A monocotyledon of the genus *Carex* in the family Cyperaceae.

sedimentation coefficient. A quantitative measure of the rate of sedimentation of a given substance through water at 20°C in a unit centrifugal field; expressed in *Svedberg units* (S).

sedoheptulose. A monosaccharide with seven carbon atoms. It is a ketose sugar found in phosphorylated form probably in all green plants (produced in Calvin cycle), but it is accumulated in the stonecrops e.g., in *Sedum* species.

seed. A fertilized ripe ovule of an angiosperm or gymnosperm. The seed is the product of sexual reproduction, and the means by which the progeny of a plant can be spread. The seed is covered by a testa, and contains an embryo and endosperm. The seeds of angiosperms are produced in fruits, and those of gymnosperms are produced in cones or strobili.

seed coat. The envelope which encloses the seed except for a tiny pore, the micropyle.

seed fern. The common name for the extinct plants classified as Pteridospermae, characterized by naked seeds borne on large, fern-like fronds.

seed leaf. A cotyledon.

seedling. 1: A young plant growing from its seed. It is usually called a seedling until it loses its cotyledons.
2: A tree younger and smaller than a sapling.
3: A tree grown from a seed.

seed plant. A plant which reproduces by seed, i.e., a spermatophyte or phanerogam.

segregation. The separation of each of a pair of alleles into different gametes, as a result of meiosis. This is the mechanism behind Mendel's first law, which states that alleles brought together in the F_1 generation can be segregated in the F_2 generation.

seismotaxis. A directed response of a motile organism to mechanical vibration or a shock stimulus; vibrotaxis; seismotactic.

seismotropism. An orientation response to mechanical vibration or a shock stimulus; seismotropic.

self compatible. Of an individual plant which can fertilize its female gametes with its own male gametes.

self-fertilization. The union of male and female gametes produced by the same individual; selfing; idiogamy; mychogamy; *cf.* cross fertilization.

Self fertilization

self-incompatible. Of an individual plant which cannot fertilize its female gametes with its own male gametes; self sterile.

self-pollination. Transfer of pollen from anther to stigma of the same flower or to another flower on the same plant; *cf.* cross pollination.

self-sterility. The condition of hermaphroditic organisms that cannot produce viable offspring by self-fertilization; self-incompatibility.

SEM. Scanning electron microscope.

semiconservative replication. Replication of deoxyribonuleic acid by longitudinal separation of the two complementary strands of the molecule, each being conserved and acting as a template for synthesis of a new complementary strand.

semidormancy. Decrease in plant growth rate; may be seasonal or associated with unfavourable environmental conditions.

semiparasitic. Pertaining to an organism (partial parasite) that derives only part of its nourishment from its host, or that lives for only part of its life cycle as a parasite; semiparasite,

semipermeable. Of membranes which let some substances pass through them, but not others. The membranes in plant cells are mostly permeable to small molecules, e g., water (H_2O), monosaccharides and amino acids, but not to large molecules, e.g., polypeptides.

semisaprophyte. A partially saprophytic plant.

semispecies. 1: The species that compose a superspecies.

2: Populations thht have acquired some attributes of species rank.

3: Organisms that are borderline between species and subspecies.

senescence. The study of the biological process of ageing; senescent.

sepal. A usually green, leaf-like organ. A whorl of sepals forms the calyx of a flower. The sepals are the outer layer of the flower bud before it opens.

sepaloid. Said of an involucre or corolla that simulates a calyx in appearance.

separation disk. A layer of gelatinous material between two-adjacent negative cells in some blue-green algae; associated with hormogonium formation.

separation layer. A structurally distinct layer of the abscission zone of a plant containing abundant starch and dense cytoplasm.

septate. Partitioned; divided into compartments by cross-walls.

septicidal. Dehiscence along or into the partitions, not opening directly into the locule.

septum (*pl.* septa). A partition or crosswall, as of a hypha or a spore. Hyphal septa are either complete or are breached by a number of pores or by a single central pore which in some fungi may have a complicated structure as in dolipore septum),

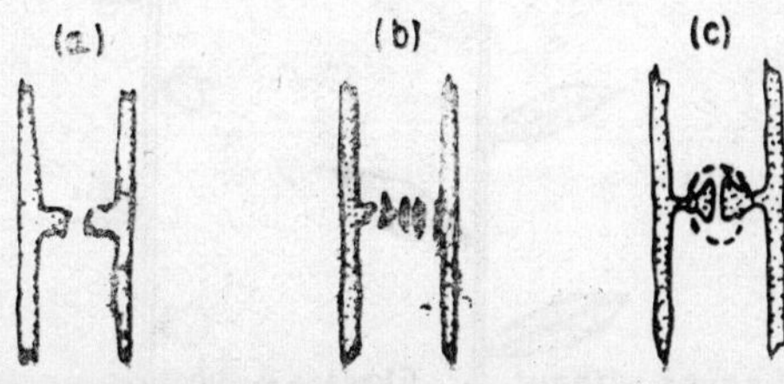

Sequoia. A genus of conifers having overlapping, scale-like evergreen leaves and vertical grooves in the trunk; the giant sequoia (*Sequoia gigantea*) is the largest and oldest of all living trees.

sere. A succession of plant communities in a given habitat leading to a particular climax association; a stage in a community succession; seral; variously classified according to havitat or major influencing factors as: aquatosere, clisere, geosere, hydrosere, lithosere, microsere, oxysere, plagiosere, preclisere, prisere, psammosere, subsere, xerosere.

seriate. In series, usually in whorls or apparent whorls.

sericeous. Of, pertinaing to, or consisting of silk; silky.

serology. The study of antigens and antibodies; serological.

serotaxonomy. Taxonomy based on serological characters.

serotinal. Pertaining to the late summer.

serotinous. Coming late; used particularly with reference to late flowering plants, and for behaviour that occurs late in the day, or late in the season.

serrate. Possessing a notched or toothed edge.

serrulate. Finely serrate.

sessile. 1: Of organs without a stalk, e.g., a leaf which has no petiole and is attached directly to the stem; or flowers without pedicels.

2: Non motile or sitting.

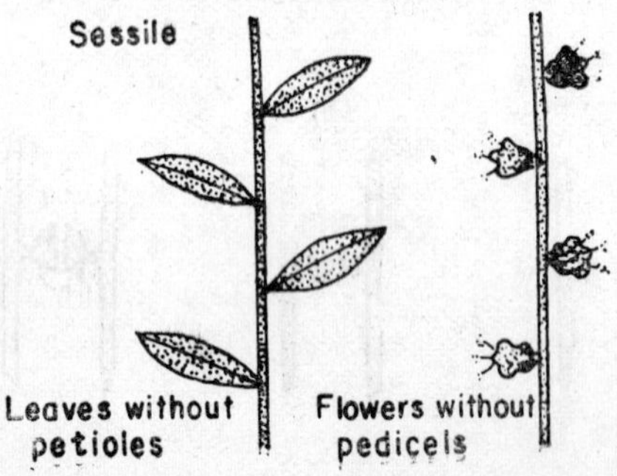

seta (*pl.* setae). 1: A slender, usually rigid bristle or hair.

2: Of fungi, bristle-like or hair-like structure occurring in certain fruiting bodies in acervuli, and on certain spores.

3: The stalk of the sporophyte of a bryophyte.

setaceous. Bearing bristles.

setiform. Bristle-shaped.

setose. Covered with bristles.

sex cell. A gamete.

sex factor. *See* fertility factor.

sex-influenced inheritance. That part of the inheritance pattern on which sex differences operate to promote character differences.

sex limited inheritance. Expression of a phenotype in only one sex; may be due to either a sex-linked or autosomal gene.

sex-linked inheritance. The transmission to successive generations of differences that are due to genes located in the sex chromosomes.

sexual. Of reproduction which involves the fusion of two cells and their nuclei from two parent individuals, so that the offspring receives genetic material from both parents. Sexual reproduction occurs in all the divisions of the plant kingdom; sex.

shake culture. 1: A method for isolating anaerobic bacteria by shaking a deep liquid culture of an agar or gelatin to distribute the inoculum before solidification of the medium.

2: A liquid medium in a flask that has been inoculated with an aerobic microoganism and placed on a shaking machine; action of the machine continually aerates the culture.

sheath. Any long or more or less tubular structure surrounding an organ or part, e.g., the lower part of the leaf of a grass, which is rolled around the stem.

shield cell. Peripheral cells constituting the boundary of the globules in Charales.

shikimic acid. $C_7H_{10}O_5$. A crystalline acid that is a plant constitutent, and an intermediate in the biochemical pathway from phosphoenolpyruvic acid to tyrosine.

shoot. 1: The aerial portion of a plant, including stem, branches and leaves.

2: A new, immature growth on a plant.

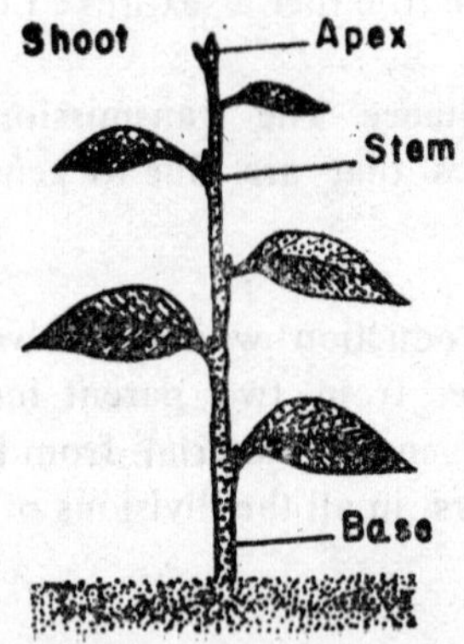

short-day plant. (SDP). A plant which has a flowering period initiated and accelerated by exposure to a short photoperiod of less than 12 h duration. A short day plant may have an obligate requirement of a favourable day length for flowering, and may remain permanently vegetative otherwise (absolute SDP.) e.g., *Xanthium pennsylvanicum*; other species may show hastened flowering under SD (quantitative SDP) e.g., *Cannabis sativa.*

shot hole. A symptom characterized by the dropping out of roundish fragments of leaf tissue attacked by certain leaf-spotting pathogens.

shrub. A low woody plant with several shoots and trunks arising from the base, not tree-like nor with a single bole.

sib pollination. Pollination of one plant by another plant derived from the same parents; adelphogamy.

sieve area. An area in the wall of a sievetube element, sieve cell, or parenchyma cell characterized by clusters of pores through which strands of cytoplasm pass to adjoining cells; sieve plate.

sieve cell. A long, tapering cell that is charocteristic of phloem in gymnosperms and lower vascular plants, in which all the sieve areas are of equal specialization.

sieve element. A cell in the sieve-tube of the phloem. Sieve elements are long, thin, living cells with thin cell walls and sieve plates at their ends. Translocation of substances takes place in the sieve elements.

sieve tissue. *See* pholem.

sieve tube. A phloem element consisting of a series of thin-walled cells arranged end to end, in which some sieve areas are more specialized than others.

sigmoid. Said of a leaflet or segment that is curved side-wise in opposite directions; S-shaped.

signs. Of pathogens, visible structures produced in or on diseased tissues.

silage. Green grass and vegetable crops stored damp and anaerobically in a pit. Primary fermentation of starch and cellulose occurs, giving a product rich in fatty acids.

silent mutation. A mutation that does not result in amino acid sequence change.

siliceous. Containing minute particles of silica.

siliceous skeleton. The silicon-containing cell wall of a diatom.

silicle. A short many-seeded capsule formed from two united carpels, usually not more than twice as long as wide and divided on the inside by a replum; e.g., in family in Cruciferae.

silicolous. 1: Living in soil rich in silica or silicates; silicicolous; silicole.

2: Growing on feints.

silique. A silicle-like capsule, but usually at least four times as long as it is wide, which opens by sutures at either margin and has parietal placentation; e.g., in certain members of family Cruciferae.

silvery. With a whitish, metallic, more or less shining lust.

silvicolous. Inhabiting woodland; sylvicolous; silvicole.

silviculture. The management aud exploitation of forests.

simple. 1: Said of a leaf when not compounded into leaflets.

2: Of an inflorseence when not branched.

sinuate. Having a wavy margin with strong indentations.

sinus. The space or recess between two lobes or divisions of a leaf or other expanded organs.

siphoneous. Of algae in which the plant body is not divided into cells, i.e., it is multinucleate.

siphonostele. A type of stele consisting of pith surrounded by xylem and phloem.

sirenin. A reproductive hormone secreted by the female gametes of *Allomyces*, a member of class Chytridiomycetes which attracts the male gametes.

skiophyte. A plant growing in shady situations or inhabitats of low light intensity; sciad; sciophyte; skiaphyte; skiarophyte; *cf.* heliophyte.

slop culture. A method of growing plants in which surplus nutrient fluid is allowed to run through the sand or other medium in which the plants are growing.

smathophyte. Psamathophyte; a plant of sandy soils.

smooth. Said of surfaces that have no hairiness, roughness, or pubescence, particularly of those not rough or scabrous.

smooth endoplasmic reticulum (SER). Endoplasmic reticulum that does not have attached ribosomes, the membraues usually forming a network of branching and fusing tubes-Smooth ER appears to be the site of lipid synthesis destined to be included in lipoproteins.

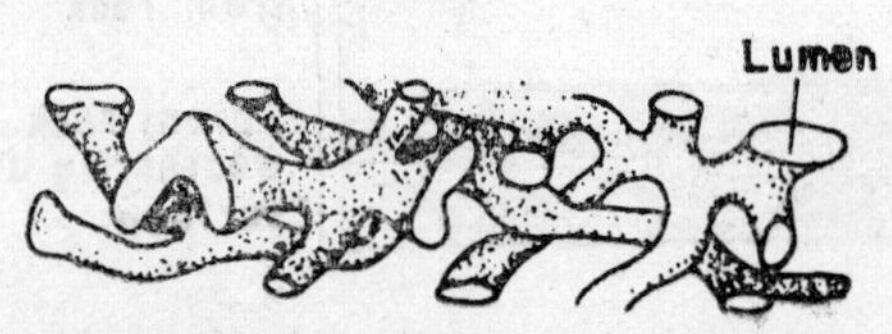

smut. 1: A disease caused by smut fungi (order Ustilaginales).

2: The fungus itself,

smut spore. A dark, thick-walled resting spore of a smut fungus; may germinate to produce a promycelium, the organ of meiois; often improperly termed chlamy dospore; *cf.* teliospore.

sobole. An underground creeping stem.

soboliferous. Bearing or producing shoots from the ground, clump-forming; usually applied to shrubs or small trees as spp. of *Syringa*, *Rhus*, and some palms.

soil profile. The sequence of different layers of material in the soil. The layers, or horizons differ in chemical composition and thickness. The top layers are usually organic derived from the litter, and the layers underneath are inorganic, derived from the rock beneath them. Soils in different places have their own characteristic soil profiles, depending on the climate and the type of rock on which they occur.

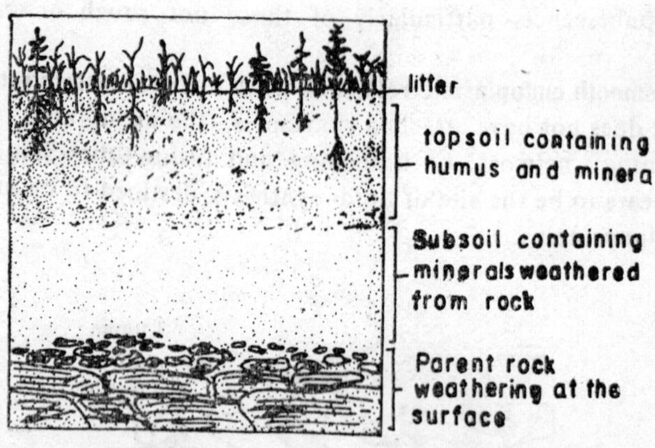

solar propagation. A method of rooting plant cuttings involving the use of a modified hotbed; bottom heat is provided by radiation of stored solar heat from bricks of stones in the bottom of the hotbed frame.

solarization. The inhibitory effect of extremely high light intensities on photosynthesis, resulting largely from photo-oxidation; heliosis.

solitary. Of organs which are borne singly, on their own, e.g., a flower in a one-flowered inflorescence.

somatic. Of any process, or part of an organism, that is not connected with sexual reproduction, e.g., mitosis is somatic cell division.

somatic copulation. A form of reproduction in Ascomycetes and Basidiomycetes involving sexual fusion of undifferentiated vegetative cells.

somatic crossing-over. Crossing-over during mitosis in somatic or vegetative cells.

sorbin. *See* sorbose.

sorbitol. This is the sugar alcohol of D-glucose. Where the terminal—CHO group is reduced to—CH_2OH. Sorbitol is found in ripe fruits of family Rosaceae, especially *Sorbus*.

sorbose. $C_6H_{12}O_6$. A carbohydrate prepared by fermentation; produced as water-soluble crystals that melt at 165°C; used in the production of vitamin C; sorbin; L-sorbose.

sorus (*pl.* sori). 1: A heap or cluster. An organ on the dorsal surface of the leaf of a fern, in which sporangia are produced. Sori have different shapes and are found on different parts of the leaf in different species. Their function is to protect the sporangia.

2: Of smut and rust fungi, fruiting bodies in which masses of spores are produced.

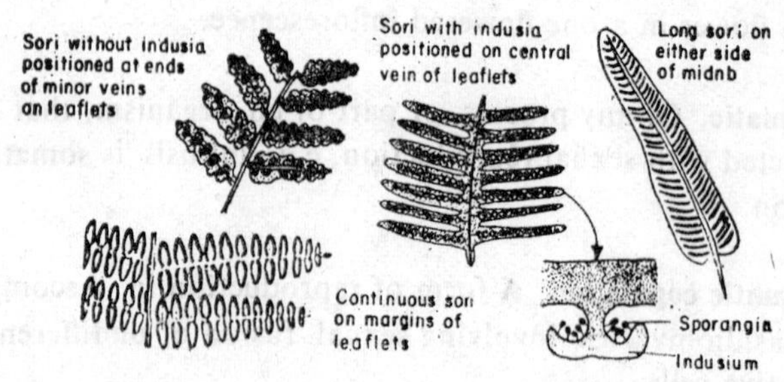

space biology. A term ferm for the various biological sciences and disciplines that are concerned with the study of living things in the space environment.

spadix. (pl spadices). An inflorescence consisting of a thick, or fleshy axis with many small sessile flowers that is enclosed in a leaflike spathe as in the monocotyledon family Araceae.

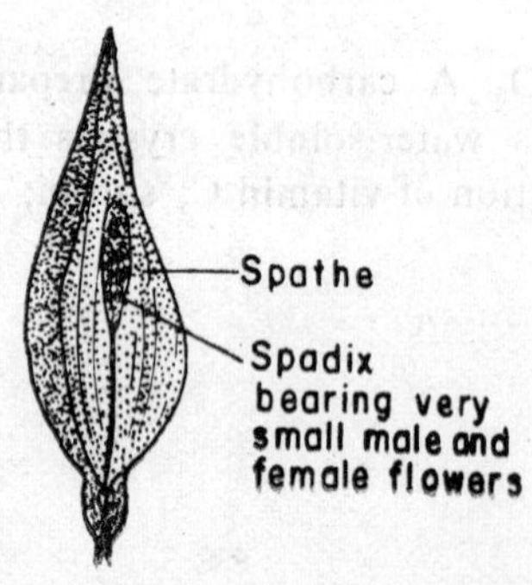

spathaceous. Spathe-like.

spathe. The single large bract or a leaf which encloses a young spadix; it is sometimes coloured and flowerlike.

spatulate. Shaped like a spoon.

spathe-valves. One or more herbaceous or scarious bracts that subtend an inflorescence or flower and generally envelop the substended unit when in bud.

spawn. 1: The eggs of certain aquatic organisms; also the act of producing such eggs or egg masses.

2: The mycelium that is used to start mushroom culture; to put inoculum (spawn) into a mushroom bed or other substratum.

speciation. The evolutionary process in which new species are produced; speciate.

species. A taxonomic category ranking immediately below a genus and including closely related, morphologically similar individuals which actually or potentially interbreed. Species are given Latin binomial names. They are sometimes divided into subspecies and varieties on the basis of small differences between populations; specific.

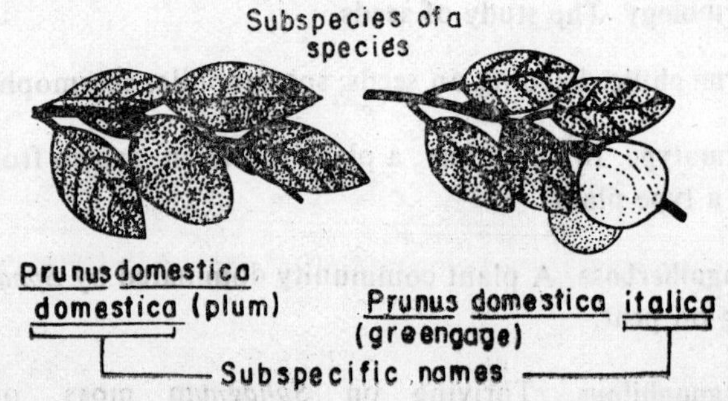

species-abundance curve. A graph of species numbers against the number of individuals per species.

species-area curve. A graph of the relationship between number of species found and the size of the area surveyed; can be used to determine the minimal area.

species concept. The idea that the diversity of nature is divisible into a finite number of definable species.

species population. A group of similar organisms residing in a defined space at a certain time.

S period. The stage in the cell cycle during which DNA replication occurs.

spermagonium (*pl.* spermagonia). A pycnium; spermogonium; spermagone.

spermatophyte. A member of the division Spermatophyta, or seed plants. This division includes all angiosperms and gymnosperms.

spermatozoid. A motile male gamete, or antherozoid in bryophytes, ferns and many algae.

sperm nucleus. One of the two nuclei in a pollen grain that function in double fertilization in seed plants.

spermology. The study of seeds.

spermophilic. Feeding on seeds; spermophile, spermophily.

spermotype. In taxonomy, a plant specimen grown from the seed of a type plant.

sphagniherbosa. A plant community dominated by *Sphagnum* growing on peat.

sphagnophilous. Thriving on *Sphagnum* moss, or in *Sphagnum*-rich habitats; sphagnophile, sphagnophily.

sphenopsid. Any of a group of *Equisetum* or horsetail-like plants; usually used for prehistoric ancestors or relatives of modern *Equisetum* spp.

spheroplast. 1: A plant cell which possesses only a partialor modified cell wall.

2: A bacterial cell that assumes a spherical shape due to partial or complete absence of the wall.

spicate. Spike-like.

spicute. An empty diatom shell.

spike. A usually unbranched, elongated, simple indeterminate inflorescence whose flowers are sessile, the flowers either congested or remote; a seemingly simple inflorescence whose flowers may actually be composite heads, or other inflorescence types.

Spike

spikelet. A secondary spike; a part of a compound inflorescence which itself is spicate; the floral unit of a grass inflorescence composed of flowers and their subtending bracts.

spindle. The minute curved threads of protein which appear during cell division, spreading across the cell from either end. The movement of chromosomes during cell division is organized on the spindle. The protein threads of the spindle are microtubules, which are formed during metaphase.

spine. A long, thin, sharp, stiff organ mostly arising from the wood of stems, and also sometimes of the leaves, of some plants, which is a defence against attack by herbivores.

spinescent. Terminated by a sharp spine or tip.

spinney. A small grove of trees or a thicket with undergrowth.

spinulose. With small spines over the surface.

spiral. 1: Of nodes and leaves which are arranged on the stem in a helical way.

2: Of the thickening of the walls of xylem cells.

split gene. A gene that is not continuous; interrupted gene.

splitter. A taxonomist who divides taxa very finely.

spongy mesophyll. A tissue in the leaves of many plants, e.g., dicotyledons, lying underneath the palisade parenchyma. It is composed of large cells with many intercellular spaces between them.

sporangioles. Walled vessels containing myxospores; occur in the fruiting bodies formed by certain myxobacters; cysts.

sporangiophore. A stalk or filament on which sporangia are borne, e.g., in *Equisetum*, ferns and in fungi.

sporangiospore. A spore that forms in a sporangium,

sporangium (*pl.* sporangia). 1: A small round organ in which spores are produced, by meiosis, from spore mother cells; in most eusporangiate ferns it is composed of a stalk (sporangiophore), an annulus, and a capsule.

2: Pertaining to algae and fungi more or less spherical body in which asexual spores are produced: Sporangia in *Ectocarpus* can be of two types unilocular or pluriloocular.

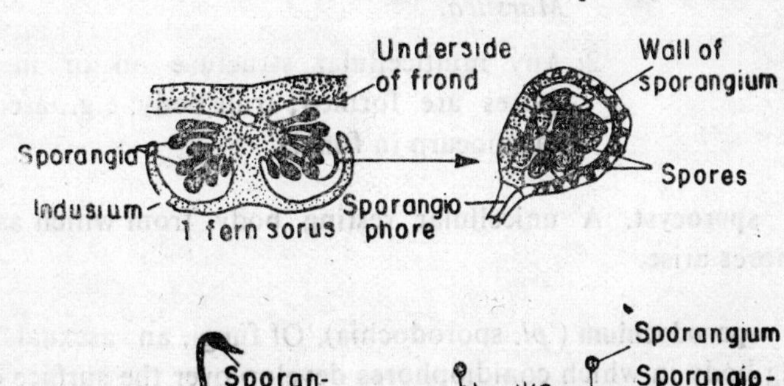

spore. A small round cell with a thick wall from which a whole new plant is produced. In bryophytes, pteridophytes and spermatophytes, spores are haploid and are produced by the sporophyte. In bryophytes and pteridophytes, dispersal is achieved by spores. In angiosperms, the spores develop into small gametophytes in the ovules and pollen grains. In all these plants, spores are produced as a result of meiosis. (meiospores). Fungi also produce spores, but these are of many kinds and are different from those of green plants i.e., they are either one celled or multicelled (e.g., phragmospore, spore ball, etc.) and every cell is capable of germination;

spore ball. A unit of dispersal comprised of a more or less firmly aggregated group of spores in smuts e.g., *Sorosporium, Tolyposprium* or spores and sterile cells as in *Urocystis.*

spore mother cell. One of the cells of the archespore of a sporebearing plant from which a spore, but usually a tetrad of spores, is produced.

sporidium (*pl.* sporidia): A small spore produced on a promycelium, as in the smuts and rusts.

sporocarp. 1: A receptale containing sporangia as in *Salvinia, Marsilea.*

2: Any multicellular structure on or in which spores are formed; fruit body; e.g., ascocarp, basidiocarp in fungi.

sporocyst. A unicellular resting body from which asexual spores arise.

sporodochium (*pl.* sporodochia). Of fungi, an asexual fruiting body in which conidiophores develop over the surface of an erumpent, cushion-like fungal structure.

sporogenesis. 1: The formation of spores.

2: Reproduction by means of spores; sporogenic, spory.

sporogenous. Of tissues in which reproductive parts or organs are produced; sporiparous; sporigenous.

sporogonium. The sporophyte of a moss or a liverwort consisting of a foot, seta and capsule.

sporophore. A structure on the thallus of fungi which produces spores, especially a conidiophore of microfungi; ascocarp, basidiocarp of macrofungi.

sporophyll. A modified leaf, whose function is to produce sporangia and spores: Sporophylls may be similar to vegetative leaves, as in many pteridophytes, or organized into cones, as in gymnosperms. The sporophylls of angiosperms are the stamens and carpels.

sporophyte. 1: An individual of the sporebearing generation in plants exhibiting alternation of generation.

2: The spore-producing generation.

3: The diplophase in a plant life cycle. In angiosperms, gymnosperms and pteridophytes, the sporophyte is the main vegetative stage. In bryophytes, the sporophyte grows directly from the archegonium of the gametophyte, and depends on the gametophyte for its nutrition.

sporopollenin. A substance related to suberin and cutin but more resistant to decay that is found in the exine of pollen grains, and in the zygospores of *Mucor mucedo*; a member of class Zygomycetes, under the right conditions the exine may last for many thousands of years, although its contents die and the zygospores of *Mucor* remain dormant but undamaged in the soil for long periods. Sporopollenin is formed by oxidative polymerization of β.carotene.

sporulation. The act and process of spore formation; sporulate.

spot. A symptom of disease characterized by a limited necrotic area, as on leaves, flowers, and stems.

spreading. Standing outward or horizontally.

springwood. The portion of an annual ring that is formed principally during the growing season; it is softer, more porous, and lighter than summerwood because of its higher proportion of large. thin-walled cells.

spur. A tubular or sac-like projection from a blossom, as of a petal or sepal; it usually contains a nectar-secreting gland.

squalene. Isoprenoid hydrocarbon, $C_{30}H_{50}$, found in many natural oils. Biosynthetic precursor of sterols.

squamate. With small scale-like leaves or bracts; scaly.

squamous. Covered with or composed of small scales, more coarsely so than when lepidate.

squamulose. Covered with or composed of minute scales.

squarrose Having stiff divergent bracts, or other processes.

squarrulose. Mildly squarrose.

stab culture. A culture of anaerobic bacteria made by piercing a solid agar medium in a test tube with an inoculating needle covered with the bacterial inoculum.

stain Any one of many dye substances used in microscopy to show up particular parts of cells or tissues.

stalk. 1: The "stem" of any organ, as the petiole, peduncle, pedicel, filament, stipe.

2: A non-living, ribbon-like or tubular appendage excreted by a bacterial cell.

stamen The male reproductive organ of a flo ver, consisting of a filament bearing an anther sometimes reduced to only an anther; the pollen bearing organ of a seed plant. The stamen is attached to the receptacle between the petals and the pistil. The number, shape, and position of the stamens in a flower are important characters in the classification of angiosperms; staminal.

staminate. Of flowers which have stamens but no pistil i.e., male flowers.

staminode. A sterile stamen, which does not produce pollen and is borne in the staminal part of the flower; in some flowers as in *Canna* or family Aizoaceae the staminodes are petal-like and showy; staminodium.

stand. A group of plants, distinguishable from adjacent vegetation, which is generally uniform in species composition, age, and condition.

standard. 1: The upper and broad, more or less erect petal of a papilionaceous flower.

2: The narrow, usually erect or ascending unit of the inner series of perianth of an *Iris* flower as opposed to the broader, often drooping falls.

standing crop. The number of individuals or total biomass present in a community at one particular time.

starch. A nutrient polysaccharide $(C_6H_{10}O_5)n$, found in plant cells, protists and certain bacteria; occurs in cells in the form of granules several μms in diameter; contains a mixture of two different polysaccharides, amylose a linear polymer of glucose and amylopectin, a glucose polymer with interlinked chains. The relative amounts of these two polysaccharides vary according to the source of the starch. Iodine solution turns blue black on contact with starch; this reaction is used as a test for either starch or iodine.

statenchyma. A tissue composed of statocyte cells.

stationary phase. The period following termination of exponential growth in a bacterial culture when the number of viable micro-organisms remains relatively constant for a time.

statocyst. A gravity sensing cell containing statoliths in a fluid medium; statocyte.

statolith. Very small grains of starch, surrounded by a membrane or other solid inclusions; which moves readily in the fluid contents of a statocyst, comes to rest on the lower surface of the cell, and is believed to function in gravity perception.

stele. The part of a plant stem or root including all tissues and regions of plants from the cortex inward, including the pericycle, phloem, cambium, xylem, and pith if this is present.

stellate. Star-like; stellate hairs have radiating branches, or, when falsely stellate, are separate hairs aggregated into star-like cluster; hairs once or twice forked are often treated as stellate.

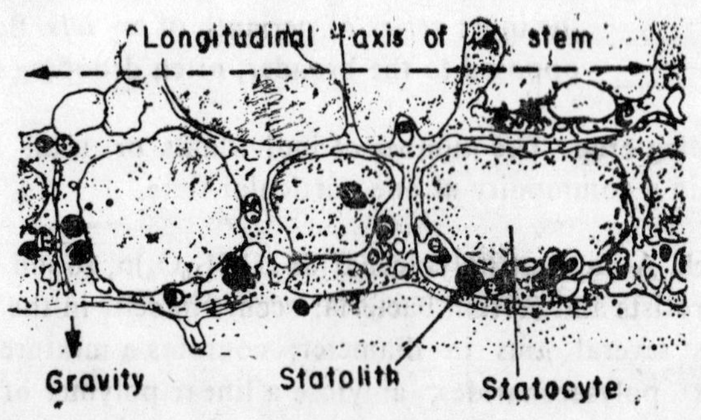

Effect of gravity on distribution of statoliths in statocyte cells.

stem. The main axis of a plant with nodes, buds, leaves and flowers. Most stems are above the ground, but some, e.g., rhizomes, are underground.

steppe (*pl.* steppes). Semi-arid areas of treeless grassland found in the mid-latitudes of Europe and Asia.

sterigma (*pl.* sterigmata). Of fungi, a minute stalk that bears spores; basidiospores are borne on sterigmata that protrude as parts of walls of basidia.

sterile. 1: Of organisms which cannot produce offspring.

2: Of reproductive organs which do not produce gametes e.g., staminodes.

3: A culture free from living microorganisms.

sterilized. Free from living microorganisms.

stigma. The rough or sticky apical surface of the pistil for reception of the pollen.

stilt root. A root which grows out from near the bottom of the trunk, in some trees, into the ground. Its function is support. Many palms have stilt roots. They are sometimes known as prop roots.

prop root. stilt root.

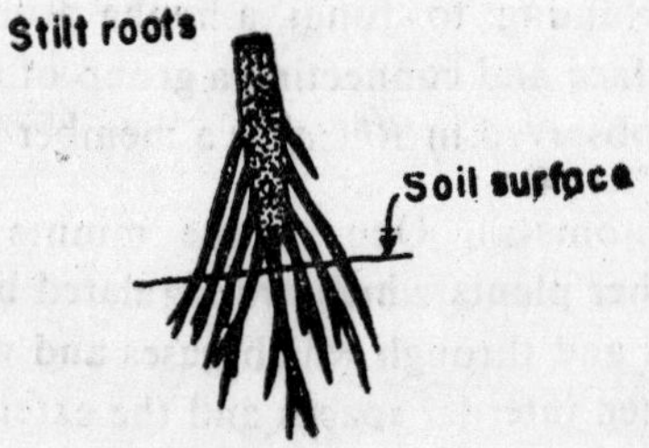

stimulus (*pl.* stimuli) Any internal or external agent that produces a reaction or change in an organism.

stipe. 1: A stalk, especially of a mushroom or toadstool.

2: The petiole of a fern frond.

3: The stem-like portion of the thallus in certain algae.

stipel. Stipule of a leaflet.

stipitate. Borne on a stipe or short stalk.

stipule. A small, leaf-like, appendage, found in many plants, which grows at the base of a petiole, sometimes protecting an axillary bud; the 3 parts of a complete leaf are blade, petiole and stipules (usually 2).

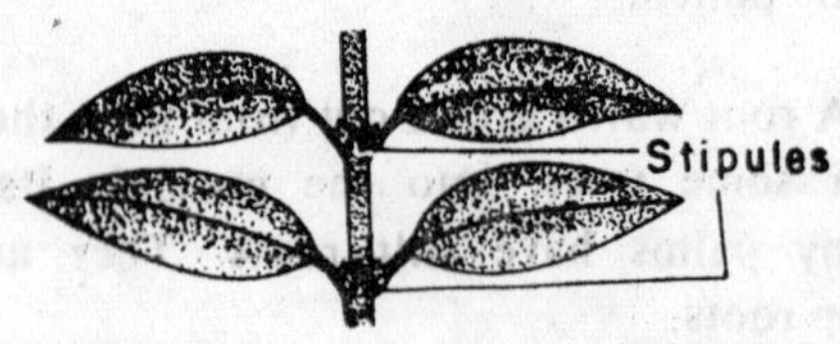

stolon. 1: A stem which grows along the ground, producing at its nodes new plants with roots and upright stems; stoloniferous.

2: Pertaining to fungi, a hypha produced above the surface and connecting a group of sporangiophores as observed in *Rhizopus* a member of Zygomycetes,

stoma (*pl.* stomata). One of the minute openings in the epidermis of higher plants which are regulated by two sausage-like guard cells and through which gases and water vapour are exchanged between internal spaces and the external atmosphere.

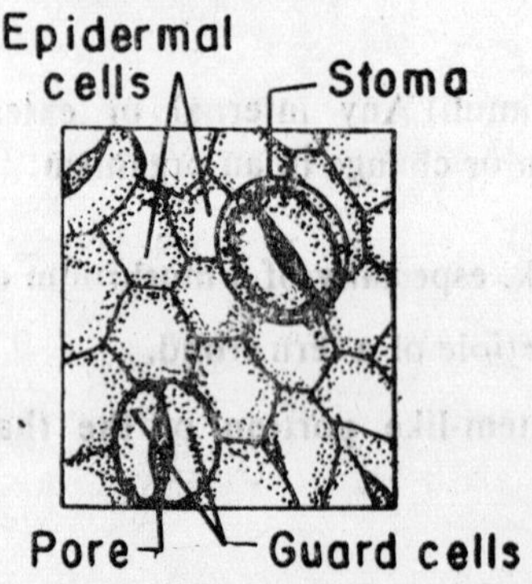

stomatal resistance. The property of the stomata in restricting the free exchange of carbon dioxide (CO_2) by a plant leaf; the major constraint on CO_2 uptake into the plant leaf governed largely by the diameter of the stomatal pores.

stomatal transpiration. The loss of water vapour through the open stomata of leaves; *cf.* cuticular transpiration.

stone. The hard endocarp of a drupe containing the seed.

stone cell. An isodiametric, sclereid. See brachysclereid.

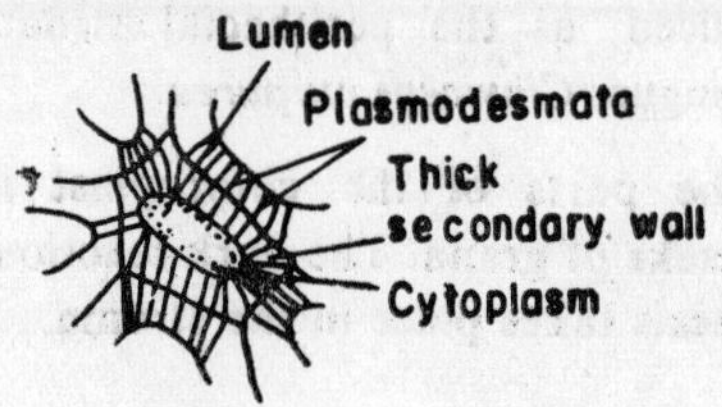

stone fruit. *See* drupe.

stonewort. The common name for algae comprising the class Charophyceae, so named because most species are lime-encrusted.

strain. A reproductively isolated population whose individuals have identical genotypes over many generations and show phenotypic differences from other populations.

straw. A stem of grain, such as wheat or oats.

striate. With five longitudinal lines, channels or ridges.

strict. Straight and upright, little if at all branched, often rigid.

strigose. With sharp, appressed straight hairs, stiff and often basally swollen.

strobilation. Reproduction by successive budding.

strobile. Cone.

strobilus. 1: A reproductive organ consisting of overlapping scales, as in some pteridophytes, like *Equisetum*, *Lycopodium*.

2: The cones of gymnosperms like *Cycas* and the conifers.

stroma. 1: (*pl.* stromata). A mass of fungal tissue, often sclerotized, in which fruiting bodies are produced, as the perithecial stromata of the ergot fungus, *Claviceps purpurea*.

2: The parts of the chloroplast in between the stacks of grana. The dark reaction of photosynthesis takes place in the stroma.

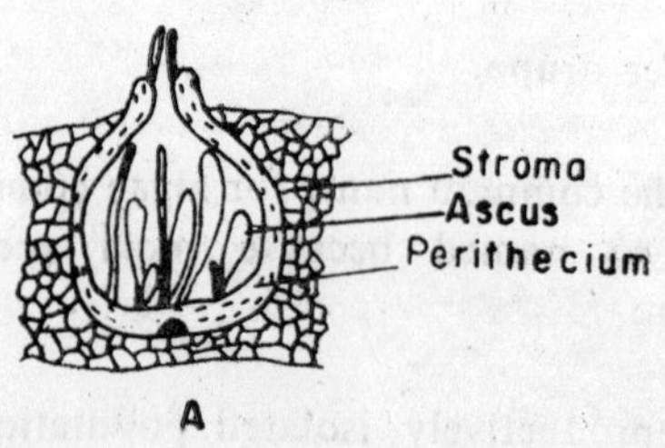

A

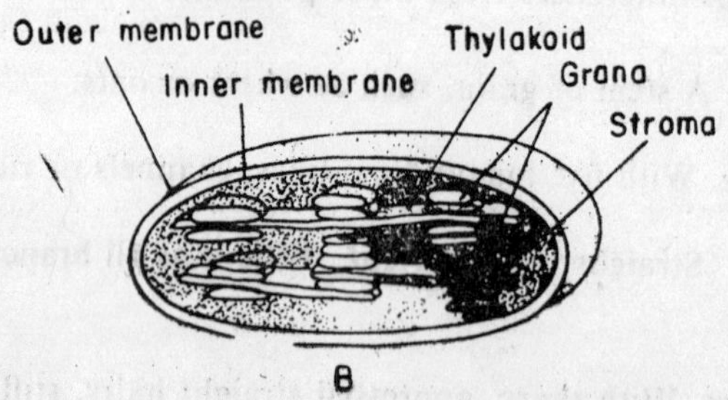

B

strophiole. A crestlike excrescence around the hilum in certain seeds.

stylar. Within the style or stigma; pertaining to the style.

style. A long upgrowth at the top of a carpel, bearing the stigma at its lip. The style positions the stigma so that it is likely to receive pollen. After pollen has reached the stigma, tubes grow down through the style to the ovary.

stylopodium. A conical or disk-shaped enlargement at the base of the style in plants of the family Umbelliferae.

subaudroecious. Incompletely androecious; used of a plant having mostly male flowers but with a few female or hermaphrodite flowers.

subboreal. A biogeographic zone whose climatic condition approaches that of the boreal.

subdivision. A taxon within a division.

subdominant. A species which may appear more abundant at particular times of the year than the true dominant in a climax; etc., in a savannah trees and shrubs are more conspicuous than the grasses, which are the true dominants.

suberin. A fatty substance found in many plant cell walls, especially cork.

suberin. A mixture of substances formed from fatty acids, which is found in cork cell walls. Suberin prevents water from passing through the cork.

suberization. Infiltration of plant cell walls by suberin resulting in the formation of corky tissue that is impervious to water; suberized.

suberose. Having a texture like cork due to or resembling that due to suberization.

subgynoecious. Incompletely gynoecious; used of a plant having mostly female flowers but with a few male or hermaphrodite flowers.

subherbaceous. Herbaceous, but becoming woody as the season progresses.

sublittoral. 1: The deeper zone of a lake below the limit of rooted vegetation.

2: The marine zone extending from the lower margin of the intertidal (littoral) to the outer edge of the continental shelf at a depth of about 200 m; sometimes used for the zone between low tide and the greatest depth to which photosynthetic plants can grow; sublitoral.

submersed. Pertaining to a plant or plant structure growing entirely under water; immersed; submerged.

subpetiolar. Under the petiole and usually enveloped by it.

subsere. A secondary community that succeeds an interrupted climax.

subshrub. A suffrutescent perennial, or a very low shrub often loosely treated as a perennial.

subsoil. The general-term for the lower, inorganic horizons in a soil profile.

subspecies. 1: A group of interbreeding natural populations differing taxonomically and with respect to gene pool characteristics, and often isolated geographically, from other such groups within a biological species, and interbreeding successfully with these groups where their ranges overlap.In the naming of subspecies, a third Latin name is put after the binomial.

2: A taxon at the rank of subspecies.

subspontaneous. Used of a plant that has been introduced into a new area and has successfully established itself in the new habitat.

substrate. 1: The general name for the substance on which an enzyme acts.
2: General term for the soil or the surface on woich an organism is growing.

substratohygrophilous. Thriving on moist substrates; substratohygropnile, substratohygrophily.

subtend. To stand below and close to, as a bract underneath a flower, particularly when the bract is prominent or persistent. The flower is in axil of the bract.

subtropical. Of the regions of the world between the tropical and temperate regions.

subvalvate. Incompletely or somewhat valvate.

subulate. Awl shaped, linear, delicate, and tapering to a sharp point.

succession 1: The process of development of vegetation, involving changes of species and communities with time. Succession occurs because the growth of plants alters the biotic factors and edaphic factors of a habitat making possible the colonization successional of other species;
2: The chronological distribution of organisms within an area; the geological, ecological or seasonal sequence of species within a habitat or community.

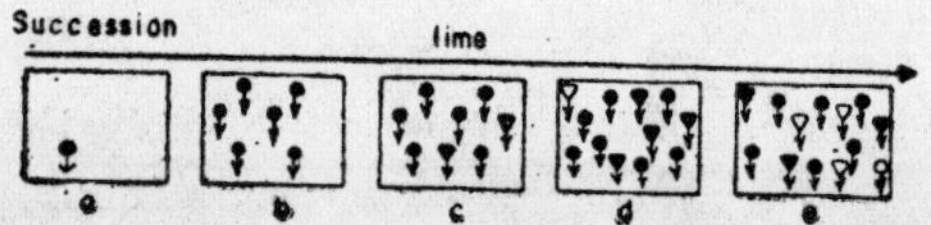

succinic acid dehydrogenase. An enzyme that catalyzes the dehydrogenation of succinic acid to fumaric acid in the presence of a hydrogen acceptor; succinic dehydrogenase.

succulency. 1: The condition of having specialized juicy, soft and fleshy tissue in a plant root or stem for the conservation of water; succulent.

2: Describing a plant having juicy fleshy tissue e.g., a cactus, family Cactaceae.

sucker. A new shoot which develops from the base of a plant or from its roots. This is a way of vegetative reproduction.

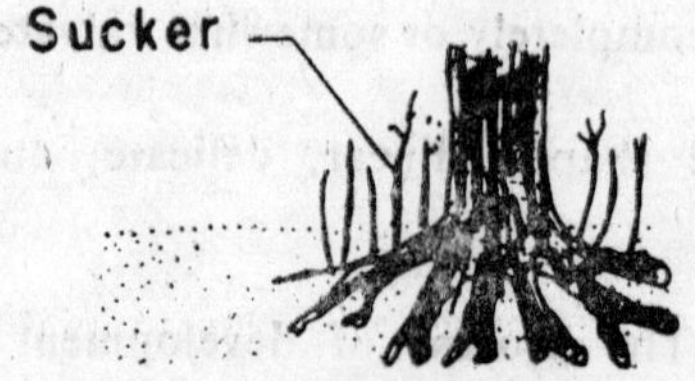

sucrose. $C_{12}H_{22}O_{11}$. A disaccharide, formed from a molecule of glucose and a molecule of fructose, found only in plants. It is the sugar which is obtained from sugar cane and sugar beet.

suffruticose. Low stems which are woody, grading into herbaceous at the top; suffrutiscent.

sugar. A carbohydrate which is soluble in water and sweet to the taste, e.g., sucrose, glucose. Sugars are produced in photosynthesis, and the energy obtained from them during respiration is used to drive the reactions of metabolism.

sulcate. Grooved or furrowed lengthwise.

sulfur bacteria. Any of various bacteria having the ability to oxidize sulfur compounds, e.g., *Thiobacillus*, *Desulfovibrio desulfuricans* etc.

superior ovary. One which is attached to the receptacle above the stamens and the perianth

summer spore. A fungal spore that germinates without resting and that is usually the inoculum for secondary infections during the growing season of the suscept.

superior. 1: Positioned above another organ or structure.

2: Referring to a calyx that is attached to the ovary.

3: Referring to an ovary that is above the insertion of the floral parts.

superparasitism. 1: Parasitism of a host that is itself a parasite; hyperparasitism; superparasite.

2: The multiple infection of a host by several parasites, usually of a similar kind.

superposed. 1: Growing vertically over another part.

2: Of or pertaining to floral parts that are opposite each other.

supine. Lying flat and with face upward.

suppressed. Vestigial to the degree of not being evident superficially or macroscopically, but whose presence in ancestral forms may be indicated by other features, as anatomy.

suppression. A hypoplastic symptom characterized by the failure of plant organs or substances to develop.

suprafoliar. Above the leaves.

suscept. Any plant or species of plant that is susceptible to disease; an abbreviated term denoting susceptible plant or susceptible species.

susceptibility. The condition of being susceptible; the inability of a plant to resist or avoid disease.

suspensor. 1: A group or chain of cells developed from the fertilized ovum in seed plants, which attaches the embryo to the wall of the embryo sac.

2: A hypha which bears an apical gametangium in fungi of the Mucorales.

suture. 1: A line or mark of splitting open.

2: A groove marking a natural division or union.

3: The lengthwise groove of a plum or a similar fruit.

sward. An area of vegetation consisting mainly of grasses.

swamp. A waterlogged land supporting a natural vegetation predominantly of shrubs and trees.

swarm spore. Any independently motile spore.

syconium. The fruit of fig (*Ficus*).

sycophagous. Feeding on figs; sycophage, sycophagy.

symbiont. An organism living in a symbiotic relationship with another organism.

symbiosis. The living together of two organisms; the relationship between two interacting organisms or populations, commonly used to describe all relationships between members of two different species, and also to include intraspecific associations; sometimes restricted to those associations that are mutually beneficial; symbioses; symbiont, symbiote, symbiotic.

symbiotrophic. Used of an organism obtaining nourishment through a symbiotic relationship; symbiotrophy.

symmetrical. Of structures whose parts are arranged equally and regularly on either side of a line or plane (bilaterally symmetrical), e.g., in a zygomorphic flower, or around a central point (radially symmetrical), e.g., in an actinomorphic flower; symmetry.

sympatric. Used of populations, species or taxa occurring together in the same geographical area; the populations may occupy the same habitat (biotic sympatry) or different habitats (neighbouring sympatry) within the same geographical area; sympatry; *cf.* allopatric.

sympatric hybridization. The occasional production of hybrids between two well-defined sympatric species

sympatric speciation. The differentiation and attainment of reproductive isolation of populations that are not geographically separated and which overlap in their distributions.

sympetalous. The petals united at least at the base; gamopetalous.

symplast. The living parts of a plant, i.e., the cells containing cytoplasm.

sympodial. Of a kind of growth in which the main axis of the plant is formed by the growth of lateral buds near the apex of the shoot, instead of by continuous growth from the apex.

Growth of lateral branches causes height increase

sympodial inflorescence. A determinate inflorescence that simulates an indeterminate inflorescence as a scorpioid cyme.

symptom. A visible expression by a suscept of a pathologic condition.

symptomatology. The study of symptoms of disease and signs of pathogens for the purpose of diagnosis.

synandrium. An androecium coherent by the anthers as in some aroids; when anthers are connate they are termed syngenesious.

synangium. A compound sorus made up of united sporangia.

synapsis. Pairing of homologous chromosomes during the zygotene of prophase I in meiosis.

synaptinemal complex. A complex ribbon-like structure that extend the length of synapsing chromosomes and are believed to function in exchange pairing; synaptonemal complex.

synaptospermy. The clumping of seeds by means of cohering or interlocking surface structures.

synaptospory. The clumping of spores by means of cohering or interlocking surface structures.

syncarp. A compound fleshy fruit.

syncarpous. 1: Of ovaries formed by two or more carpels joined together. This is an important character in the classification of angiosperms.

2: Sometimes used when separate pislils within one flower are partially united.

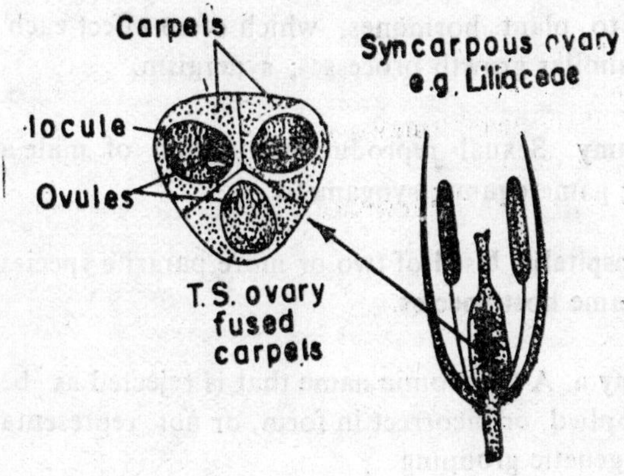

synchorology. Study of the occurrence, distribution and classification of plant communities into the following regional units; region, province, sector, subsector, district and sub-district.

synchronous growth. A population of bacteria in which all cells divide at approximately the same time.

syncytium. A multinucleate protoplasmic mass not differentiated into cells.

syndynamics. The study of the causes of and trends in successional changes within a plant community.

synecology. Study of the ecology of organisms, populations, communities, or systems; ecological sociology; *cf.* autecology.

synergid. Either of two small cells lying in the embryo sac in seed plants adjacent to the egg cell toward the micropylar end.

synergism. An ecological association in which the physiological processes or behaviour of an individual are enhanced by the nearby presence of another organism.

synergistic. Of processes in which one substance reinforces the action of another substance or substances. This is usually applied to plant hormones, which often affect each other and control similar growth processes; synergism.

syngamy. Sexual reproduction; fusion of male and female gametes; gametogamy; syngametic.

synhospitalic. Used of two or more parasite species occurring on the same host species.

synonym. A taxonomic name that is rejected as being incorrectly applied, or incorrect in form, or not representative of a natural genetic grouping.

synopsis. In taxonomy, a brief summary of current knowledge about a taxon.

synsepalous. Gamosepalous; the sepals marginally connate, at least basally.

syntaxonomy. The classification and nomenclature of plant communities; *cf.* idiotaxonomy.

synxenic. Used of a culture comprising two or more organisms under controlled conditions.

synzoochorous. Dispersed by the agency of animal species; synzoochore.

systemic. Of disease, one in which there is general spread of the pathogen throughout the plant.

systematics. The classification of living organisms into hierarchical series of group emphasizing their phylogenetic interrelationships; often used as equivalent to taxonomy.

T

2, 4, 5-T. 2, 4, 5-trichlorophenoxyacetic acid, a translocated hormone weedkiller used to control scrub and woody vegetation.

tachysporous. Used of a plant that disperses its seeds quickly.

tactic movements. A movement of an organism or freely motile part of it, in response to an external directional stimulus. If the movement is towards the stimulus the taxis is positive, whereas a movement away form the stimulus is a negative taxis.

tactile. Pertaining to the sense of touch; tactual.

tailed. Said of authers having caudal appendages.

tangelo. A tree that is hybrid between a tangerine or other mandarin and a grapefruit or shaddock; produces an edible fruit.

tangerine. Any of several trees of the species *Citrus reticulata*; the fruit is a loose-skinned mandarin with a deep-orange or scarlet rind.

tang line. The highest continuous line on the shore along which a particular seaweed grows; has variously been applied to wracks, laminarians and other algal forms; *cf.* algal line.

tannase. An enzyme that catalyzes the hydrolysis of tannic acid to gallic acid; found in cultures of *Aspergillus* and *Penicillium*.

tannins. A group of complex aromatic compounds which may contain saccharides combined with various phenols; common in the outer tissues of many plants, which are bitter to the taste and are a defence against herbivores Tannins are used in the tanning of leather and in making dyes and inks.

tapetum. A layer of nutritive cells surrounding the spore mother cells in the sporangium in higher plants; it is broken down to provide nourishment for developing spores.

taphrophyte. A ditch plant; taphrad.

taproot. A root system in which the primary root forms a dominant central axis that penetrates vertically and rather deeply into the soil; it is generally larger in diameter than its branches.

tartareous. Of a surface of a lichen thallus when it is thick rough and crumbling.

tartaric acid. $HOOC(CHOH)_2COOH$. A white, crystalline, soluble, organic acid whose normal salts are called tartrates and whose acid salts are called hydrogen-tartrates. It is found in many plants and fruits, especially grapes. Potassium sodium tartrate is a double salt called Rochelle salt. It is used as an ingredient of Fehling's solution. Tartaric acid is used in dyeing and in effervescent powders.

tassel. The staminate inflorescence of maize.

tautology. Unnecessary repetition of a word or statement; tautological, tautomer, tautomeric.

tautonymy. In taxonomy, the identical spelling of the generic name and the epithet within a binomen or trinomen; tautonym; tautonymous name.

taxis (*pl.* taxes). *See* tactic movements.

taxon (*pl.* taxa). Any taxonomic group, e.g., a species, a family. All the members of a taxon share similar characteristics, which are different from those of other groups.

taxonomy. The theory and practice of describing, naming and classifying organisms; systematics; biosystematics; genonomy; taxonomic.

tea. 1: The dried, crushed leaves of the tea plant, *Camellia sinensis*; family Theaceae, which is an evergreen shrub grown in tropical and sub-tropical regions. Tea is the most popular of the alkaloid beverages and is used by over half of the world's population.

2: *See* cannabis.

tea fungus. A symbiotic association of yeasts (*Schizosaccharomycodes ludwigii*) and bacteria (esp. *Acetobacter xylinum*).

tegmen. 1: An integument or covering.

2: The inner layer of a seed coat.

teichoic acid. A polymer of ribitol or glycerol phosphate with additional compounds such as glucose linked to the backbone of the peptidoglycan found in the cell walls of some Gram+bacteria.

teleblem. Universal veil; teleoblema.

Teliomycetes. A class of Basidiomycotina; include the rusts and smuts, parasites of many economically important crops. Basidiocarp lacking and replaced by teliospores or chlamydospores grouped in sori or scattered within the host tissue.

teliospore. A teleutospore: a spore of the rust fungi, usually a resting spore, that germinates to produce a promycelium in which meiosis occurs; it eventually produces basidiospores; teleutospore.

telium (*pl.* telia). A sorus of a rust fungus in which teliospores are produced; teleutosorus.

telmatad. A wet meadow plant; telmatophyte.

telmatium. A wet meadow or marsh community; telmathium.

telocentric. Pertaining to a chromosome with a terminal centromere.

telomere. A centromere in the terminal position on a chromosome.

telome theory. A theory of the evolution of the organs of plants, especially the branches and leaves, starting with very simple leafless primitive pteridophytes with green stems.

telophase. The phase of meiosis or mitosis at which the chromosomes, having reached the poles reorganize into interphase nuclei with the disappearance of the spindle and the reappearance of the nuclear membrane; in many organisms telophase does not occur at the end of the first meiotic division.

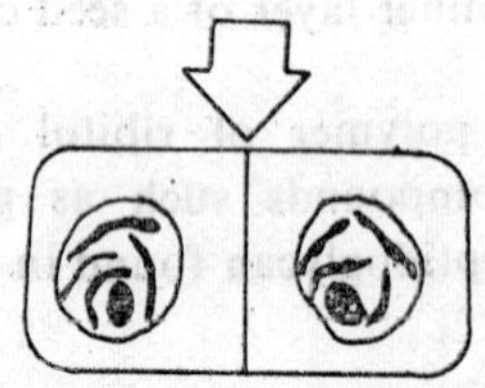

TEM. Transmission electron microscope.

tempeh. An oriental food composed of soybeans fermented with *Rhizopus oligosporus*.

temperate. 1: Moderate; not excessive or extreme; used of climates with alternating long warm summers with long days and short cold winters with long nights.

2: Pertaining to latitudes between the tropics and the polar circles in each hemisphere.

temperate and cold savanpah. A regional vegetation zone, very extensively represented in North America and in Eurasia at high altitudes; consists of scattered or clumped trees (very often conifers and mostly needle-leaved evergreens) and a shrub layer of varying coverage; mosses and even more abundantly, lichens form an almost continuous carpet.

temperate and cold scrub. Regional vegetation zone whose density and periodicity vary a good deal; requires a considerable amount of moisture in the soil, whether from mist, seasonal downpour, or snowmelt; shrubs may be evergreen or deciduous; and undergrowth of ferns and other large-leaved herbs are quite frequent, especially at subalpine level; wind shearing and very cold winters prevent tree growth; bosque; fourre; heath.

temperate mixed forest. A forest of the North Temperate Zone containing a high proportion of conifers with a few broad-leaved species.

temperate phage. A deoxyribonucleic acid phage, the genome (DNA) of which can under certain circumstances become integrated with the genome of the host, being thus transmitted through cell divisions without causing host lysis, e.g., lambda (λ) phage.

temperate rainforest. A vegetation class in temperate areas of high and evenly distributed rainfall characterized by comparatively few species with large populations of each species; evergreens are somewhat short with small leaves, and there is an abundance of large tree ferns; cloud forest; laurel forest; laurisilva; moss forest; subtropical forest.

temperate woodland. A vegetation class similar to tropical woodland in spacing, height, and stratification, but it can be either deciduous or evergreen, broad-leaved or needle-leaved; parkland; woodland.

temperature coefficient Q_{10}. A measure of the increase in rate of a reaction or process produced by an increase in temperature of 10°C.

temperature-efficiency index. For a given location, a measure of the long-range effectiveness of temperature (thermal efficiency) in promoting plant growth. Abbreviated T-E index; thermal-efficiency index.

temperature-efficiency ratio. For a given location and month, a measure of thermal efficiency; it is equal to the departure, in degrees Fahrenheit, of the normal monthly temperature above 32°F (0°C) divided by 4: $(T-32)/4$. Abbreviated T-E ratio; thermal-efficiency ratio.

temperature-sensitive mutant. A mutant gene that is functional at high (low) temperature but is inactivated by lowering (elevating) the temperature.

template. The macromolecular model for the synthesis of another macromolecule.

temporary parasite. A parasite that makes contact with its host only for feeding.

tendril. A rotating or twisting thread like process or extension by which a plant grasps an object and clings to it for support; morphologically it may be stem or leaf.

tentacle. A sticky structure of insectivorous plants such as the sundew.

tepal. An organ of a perianth in which there is no difference between the calyx and corolla, e.g., in tulips, onion etc.

teratology. The study of malformations and monstrosities; teratologic.

terebrate. Having scattered perforations.

terebrator. A trichogyne of lichens.

terete. Of a stem, cylindrical in section, but tapering at one or both ends.

terminal bud. A bud that develops at the apex of a stem; apical bud.

terminator codon. A codon that acts as a stopping point in the code sequence for protein synthesis.

termiticole. An organism that lives in a termites nest.

ternate. Composed of three subdivisions, as a leaf with three leaflets.

Tertiary. One of two geological sub-eras into which the Cenozoic is divided. It began about 65 million years ago and ended about 2 million years ago; comprising the Palaeocene Eocene, Oligocene, Miocene, Pliocene epochs. The general geography of the Earth had a modern look-and flowering plant dominated.

tertiary parasite. An organism parasitic on a hyperparasite.

tertiary structure. The characteristic three-dimensional folding of the polypeptide chains in a protein molecule brought about by interactions among side chains of amino acids.

terrestrial. Of the ground; a land plant as opposed to aquatics, epiphytes or saprophytes.

terricolous. Growing on the ground; *cf.* terrestrial.

tessellate. Marked by a pattern of small squares resembling a tiled pavement.

testa. The hard, outer coat of a seed which protects the embryo and prevents water from entering the seed until it is ready to germinate; episperm.

tetracycline. 1: Any of a group of broad-spectrum antibiotics produced biosynthetically by fermentation with a strain of *Streptcmyces aureofaciens* and certain other species or chemically by hydrogenolysis of chlortetracycline.

2: ($C_{22}H_{24}O_8N_2$) A broad-spectrum antibiotic belonging to the tetracycline group of antibiotics; useful because of broad antimicrobial action, with low toxicity, in the therapy of infections caused by gram-+and Gram—bacteria as well as rickettsiae and large viruses such as psittacosis-lymphogranuloma viruses.

3: An antibiotic that acts by inhibiting protein synthesis.

tetracytes. The spores resulting from meiosis.

tetrad. 1: A group of four chromatids lying parallel to each other as a result of the longitudinal division of each of a pair of homologous chromosomes during the pachytene and later stages of the prophase of meiosis.

2: A group of four haploid spores which are the product of meiosis of the spore mother cell.

3: A tetrahedron.

tetrad analysis. A method of genetic analysis possible in fungi, algae, bryophytes, and orchids in which the four products of an individual cell which has gone through meiosis are recovered as a group; it provides more direct and complete information regarding segregation and recombination mechanisms than is possible to obtain from meiotic products collected at random.

tetradynamous. An androecium of 6 stamens, 4 longer than the outer 2, as in most Cruciferae.

tetrahedral. Four sided, as a 3-sided pyramid and its base.

tetramerous. Characterized by or having four parts; fourmerous.

tetrandrous. With 4 stamens.

tetraploid. Of cells which have four sets of homologous chromosomes in their nuclei.

tetraploidy. The occurrence of related forms possessing in the somatic cells chromosome numbers four times the haploid number.

tetrasomic. Pertaining to an aneuploid having one or more homologous chromosomes represented four times.

tetraspore. One of the haploid asexual spores of the red algae formed in groups of four.

tetratomic. Fourtimes furcate at one node.

tetrose. Any of a group of monosaccharides that have a four-carbon chain; an example is erythrose, $CH_2OH\text{-}(CHOH)_2\text{-}CHO$, an important plant metabolite in the hexose monophosphate shunts and in Calvin cycle.

thalamus. The receptacle of a flower belonging to Thalamiflorae, a taxon whose floral parts are hypogynous, distinct, and separate from one another on the receptacle.

thalloid liverwort. A liverwort in which the gametophyte is a flat, more or less undifferentiated thallus. About 20% of liverwort species are thalloid.

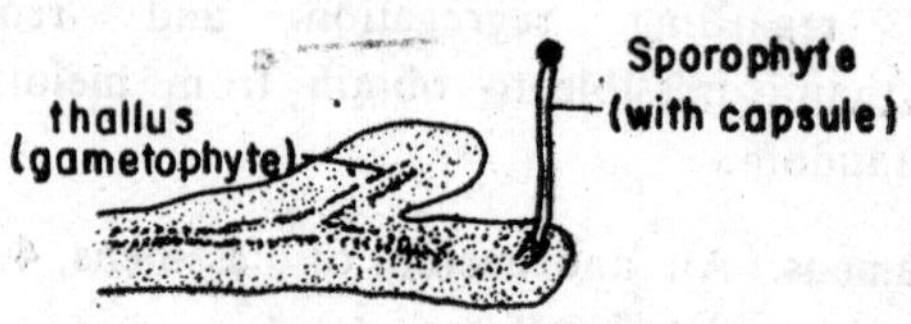

Thallophyta. A major division of the plant kingdom which includes algae and fungi. These 'plants' have plant bodies which are not differentiated into stem, leaves and roots.

thallospore. An asexual spore having no conidiophore, or one which is not separated from the hypha or conidiophore producing it, i.e., an arthrospore, blastospore, or chlamydospore.

thallus (*pl.* thalli). A more or less undifferentiated plant body, without distinct roots, stems and leaves, e.g., the gametophyte of thalloid liverworts or the plant body of an alga.

thamnisophagous. Of endotrophic mycorrhiza, forming haustorial arbuscles which are finally digested by the host.

thamnophilic. Thriving on, or living in, bushes or shrubs; thamnophilous; thamnophile, thamnophily.

theca. The pollensac of an anther.

thecium. The part of an apothecium containing the asci between the epithecium and hypothecium; sometimes used for the whole sporocarp or as equivalent to hymenium.

theine. *See* caffeine.

thelytoky. Obligatory parthenogenesis in which a female gives rise only to female offspring; thelyotoky; thelytokous, thelytocous, thelyotocous.

therapy. Principle of plant-disease control marked by the cure of diseases, as with heat or systemic chemicals.

thermal belt. Any one of several possible horizontal belts of a vegetation type found in mountainous terrain, resulting primarily from vertical temperature variation; thermal zone.

thermal death point (T.D.P.). That high temperature at which death of an organism occurs after a specified length of time, usually 10 minutes.

thermal-efficiency index. *See* temperature-efficiency index.

thermal-efficiency ratio. *See* temperature-efficiency ratio.

Thermoactinomyces. A genus of bacteria in the family Micromonosporaceae; single spores form on aerial and substrate mycelia; optimum growth is at temperatures between 45 and 60°C.

thermoduric bacteria. Bacteria which survive pasteurization, i.e., they are heat resistant, but do not grow at temperatures used in a pasteurizing process.

thermoperiodicity. The totality of responses of a plant to appropriately fluctuating temperatures.

thermophilic. Thriving in warm environmental conditions; used of microorganisms having an optimum for growth above 45°C; thermophilous; thermophilus; thermophil, thermophile, thermophily.

thermophobic. Intolerant of high temperatures.

thermophyte. 1: A plant tolerant of, or thriving at, high temperatures.
2: A hot-spring plant; thermad.

thermotaxis. 1: The directed response of a motile organism towards (positive) or away from (negative) temperature stimulus; thermotactic.
2: Regulation of body temperature.

therophyte. An annual plant; a plant that completes its life cycle in a single season, passing unfavourable seasons as a seed; *cf.* Raunkiaerian life forms.

thiamine. $C_{12}H_{18}ON_4Cl_2S$. A vitamin B_1, found in yeast, husks of cereal grains, egg white and liver. A deficiency may lead to beriberi and several other disorders.

thigmomorphosis. A change in form due to contact; thigmomorphic.

thigmotaxis. A directed response of a motile organism to a continuous contact with a solid surface; stereotaxis; thigmotactic.

thigmotropism. Curving growth due to contact with an object, e.g., the coiling of tendrils of a climbing plant around a stake. This is sometimes known as haptotropism.

thistle. Any of the various prickly plants comprising the family Compositae.

thorn. A sharp short pointed outgrowth on the surface of a plant, especially on the stem. Thorns can be simple outgrowths of the epidermis or can be modifications of other organs e.g., the stipules.

thornbush. A vegetation class that is dominated by tall succulents and profusely branching smooth-barked deciduous hardwoods which vary in density from mesquite bush in the Caribbean to the open spurge thicket in Central Africa; the climate is that of a warm desert, except for a rather short intense rainy season; Dorngeholz; Dorngestrauch; dornveld; savane armee; savane epineuse; thorn scrub.

thorn forest. A type of forest formation, mostly tropical and subtropical, intermediate between desert and steppe; dominated by small trees and shrubs, many armed with thorns and spines; leaves are absent, succulent, or deciduous during long dry periods, which may also be cool; an example is the caatinga of northeastern Brazil.

thorn scrub. *See* thornbush.

throat. 1: The opening or orifice into a gamopetalous corolla, or perianth.

2: The place where the limb joins the tube.

thrush. A throat disease of man, especially children, caused by *Candida albicans.*

thryptogen. An organism increasing the sensitivity of a suscept to outside factors, e.g., to cold.

thylakoid. An internal membrane system which occupies the main body of a plastid; particularly well developed in chloroplasts. In cyano-bacteria they occur within the cytoplasm. A thylakoid may be disk-shaped (as in grana) or elongated; it is the site of the light-requiring reactions of photosynthesis.

thyme. A perennial mint plant of the genus *Thymus*; pungent aromatic herb is made from the leaves.

thymidine. $C_{10}H_{14}N_2O_5$. A nucleoside derived from deoxyribonucleic acid; essential growth factor for certain microorganisms in mediums lacking vitamin B_{12} and folic acid.

thymidylic acid. $C_{10}H_{15}N_2O_8P$. A mononucleotide component of deoxyribonucleic acid which yields thymine, D-ribose, and phosphoric acid on complete hydrolysis.

thymine. $C_5H_6N_2O_2$. A pyrimidine component of nucleic acid, first isolated from the thymus. It pairs with adenine in DNA.

thyrse. An inflorescence with a racemose primary axis and cymose secondary and later axes, or panicle like cluster with main axis indeterminate and lateral axes determinate as in most lilacs.

thyrsus (*pl.* thyrsi). Densely branched; used for the densely branched apices of some lichens. e.g., *Cladonia alpestris.*

tigers's milk. *Polyporus sacer*; used as a medicine in Malaysia.

tiller. A single new plant growing from the base of an old plant, especially in grasses.

timberline. The elevation or latitudinal limits for arboreal growth. tree line.

tinea. Ringworm or other skin diseases in man or animals caused by dermatophytes.

tinea barbae. Beard ringworm.

tinea capitis. Head ringworm.

tinea cruris. Groin ringworm.

tinea pedis. 'Athlete's foot', foot ringworm.

tissue. A group of cells, of similar shape and size, and of same function. Plant organs usually have several different kinds of tissues, e.g., leaves have epidermal, mesophyll and vascular tissue.

tissue culture. A process in which cells from an organism are grown on a medium in isolation from the organism from which they were taken. Plant tissue cultures, which usually consist of calluses of undifferentiated cells, are sometimes used for the production of drugs.

T-method. A budding method in which a T-shaped cut is made through the bark at the internode of the stock, the bark of the scion is separated from the xylem along the cambium and removed, and the scion is forced into the incision on the stock.

toadstool Any mushroom which is poisonous. Toad was a symbol of evil, associated with the devil and black magic. As well as being venomous toads were regarded as low, creeping animals, similar to serpents. The association of poisonous fungi with toads obviously reinforced the idea that they were vegetable vermin, to be avoided. Some toadstools are *Amanita muscaria*, *Hydnum repandum*, *Lycoperdon perlatum*, *Boletus edulis* etc.

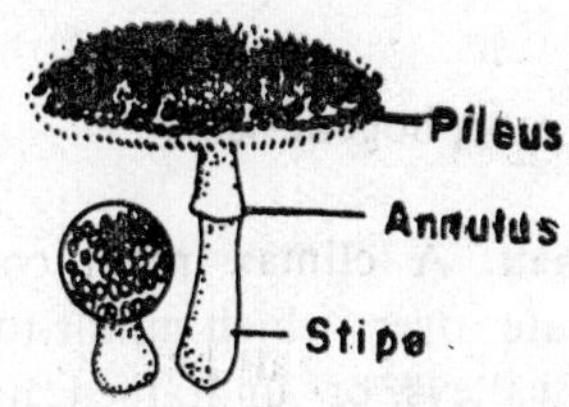

tolerance. Of diseased plants, the ability to endure disease without serious injury or crop loss.

tomentose. Covered with densely matted hairs; densely wooly or pubescent.

tomentulose. Somewhat or delicately tomentose.

tonicity. A measure of osmotic tension; defined by the osmotic response of cells immersed in a solution; *cf.* hypertonic, hypotonic, isotonic.

tonoplast. The membrane surrounding a plant-cell vacuole.

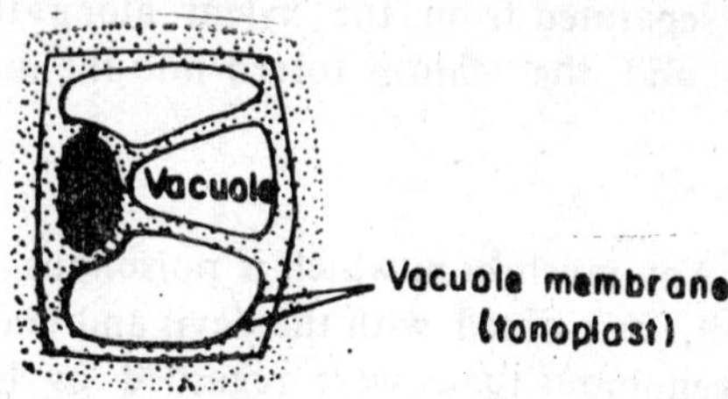

top grafting. Grafting a scion of one variety of tree onto the main branch of another.

topocline. A graded series of characters exhibited by a species or other closely related organisms along a geographical axis.

topogenesis. *See* morphogenesis.

topographic climax. A climax plant community under a uniform macroclimate over which minor topographic features such as hills, rivers, valleys, or undrained depressions exert a controlling influence.

topotype. A specimen of a species not of the original type series collected at the type locality.

topsoil. The general term for the upper, organic horizons in a soil profile.

topwork. A procedure employed to propagate seedless varieties of fruit and hybrids, to change the variety of fruit, and to correct pollination problems, using any of three methods; root grafting, crown grafting, and top grafting.

torulose. Twisted or knobby; irregularly swollen at close intervals; moniliform.

torus. The name sometimes given to the receptacle of a flower.

totipotency. The potentiality of a cell isolated from a plant, to give rise to all of the various cell types that comprise the tissues of an organism.

totipotent. 1: Bisexual.

2: Having the potential for various specializations during development, in response to external or internal factors.

toxicology. The study of poisonous substances and their effects.

toxin. 1: A substance which is poisonous. Plants produce toxins, e.g., alkaloids to protect themselves from attack by herbivores.

2: Fungal toxins, e.g., alternaric acid, baccatin, cochliobolin. diaporthin, fusaric acid, lycomarasmin, skyrin, victorin.

T-phage. Any of a series (T1—T7) of deoxyribonucleic acid phages which lyse strains of the Gram—bacterium, *Escherichia coli* and its relatives.

TPN. *See* triphosphopyridine nucleotide.

trabecula. A transverse partition, complete or incomplete, as in aspo rangium of *Isoetes*.

trace-element. An element required by a plant in very small amounts, e.g., boron, molybdenum, iodine, zinc, copper etc. Trace elements are often important constituents of enzymes, hormones and vitamins.

tracheid A long spindle shaped, dead cell in the xylem with closed ends and lignified secondary walls laid in various thicknesses and patterns over the primary wall. Xylem sap passes from one tracheid to the next through pits in the cell walls.

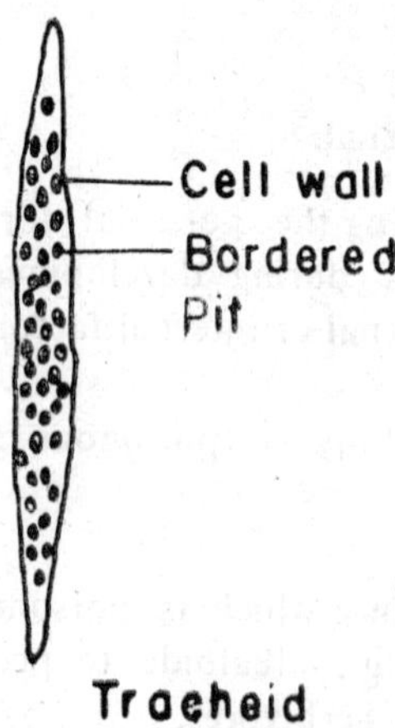

Tracheid

trama. The loosely woven hyphal tissue between adjacent hymenia in basidiomycetes.

transaminase, One of a group of enzymes that catalyze the transfer of the amino group of an amino acid to a keto acid to form another amino acid; aminotransferase.

transcriptase. *See* transcription.

transcription. A process which takes place in the nucleus of a cell. The genetic code in chromosomal DNA is transferred to messenger RNA (mRNA) by the production of a molecule of mRNA from a length of DNA. The mRNA then leaves the nucleus and aims for the ribosomes, where it acts as a template for protein synthesis. The process of transcription is catalyzed by the enzyme transcriptase.

transduction. The transfer of genetic material from donor to recipient microorganism, as from one bacterial cell to another by a bacteriophage; *cf.* transformation,

transect. A long rectangular area or a set of quadrats in a line marked out by an ecologist for counting and sampling organisms in one or several habitats.

transfection. Infection of a cell with viral deoxyribonucleic acid or ribonucleic acid.

transferase. Any of various enzymes that catalyze the transfer of a chemical group from one molecule to another.

transfer ribonucleic acid (tRNA). The smallest ribonucleic acid molecule found in cells; its structure is complementary to messenger ribonucleic acid and it functions by transferring amino acids from the free state to the growing polypeptide chains of messenger RNA at the site of protein synthesis in a ribosome during translation. There are different molecules of tRNA for each of the fundamental amino acids.

transformation. 1: A change of form; metamorphosis.

2: A change resulting from the acquisition of foreign genetic material.

3: In statistics, a change of variable; used to simplify calculations or to transform the distributions of random variables to a normal distribution.

transforming principle. Deoxyribonucleic acid which effects transformation in bacterial cells.

transition. A mutation resulting from the substitution in deoxyribonucleic acid or ribonucleic acid of one purine or pyrimidine for another.

translation. That part of protein synthesis in which molecules of tRNA carrying amino acids are matched with the genetic code carried by mRNA, so that the amino acids are joined together in the correct sequence to make a polypeptide. This takes place on the ribosomes.

translocation. 1: Movement of water, mineral salts, and organic substances from one part of a plant to another.

2: The transfer of a chromosome segment from its usual position to a new position in the same or in a different chromosome.

transmethylase. A transferase enzyme involved in catalyzing chemical reactions in which methyl groups are transferred from a substrate to a new compound.

transmethylation. A metabolic reaction in which a methyl group is transferred from one compound to another; methionine and choline are important donors of methyl groups.

transmission. The dissemination of pathogens and the inoculation of suscepts.

transpiration. The loss of water vapour from a plant, mainly through the stomata. Transpiration results from the need for the stomata to be open for gaseous exchange to take place. Transpiration cools the leaves and the continuous movement of water through the plant also carries mineral salts from the roots to the leaves. Factors affecting transpiration are humidity, temperature, atmospheric pressure, air movements (wind), light and plant's water supply. See also guttation and potometer.

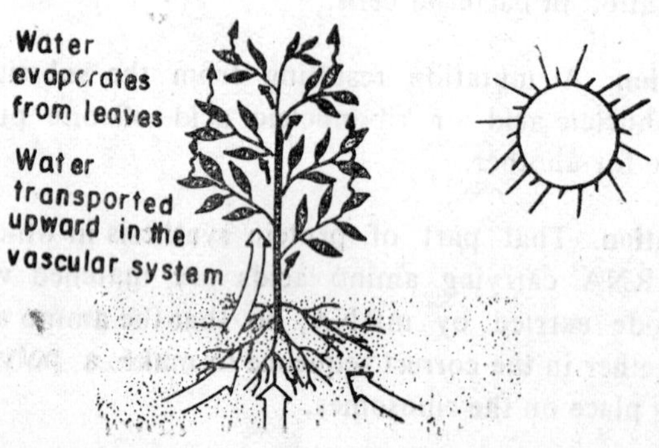

transpirational flux. It is the driving force for the movement of water vapour out of a leaf and the total resistance to the movement in the pathway. The driving force is the difference in concentration of water vapour between a leaf and the external atmosphere. The total resistance is the sum of the stomatal resistance and the boundary layer resistance.

$$\text{Transpirational flux} = \frac{Cl - Ca}{Rs + Ra}$$

where Cl and Ca are the water vapour concentrations in substomatal cavities in the leaf and in the external atmosphere; Rs is the stomatal resistance; and Ra is the resistance of the boundary layer external to the leaf.

transpiration stream. The continuous flow of water through a plant from the roots, through the stem and out through the stomata of the leaves. During transpiration, most evaporation occurs from the walls of spongy mesophyll cells. Loss of water from these cells results in their cell sap becoming more concentrated than that of neighbouring cells. Thus the osmotic pressure of the surface cells is greater than that of the cells deeper in the leaf and water is drawn into them. Cells which lose water to the surface cells will then take water from the neighbouring cells and so on. Thus a continuous stream of water moves through the plant as long as evaporation continues.

transplantation. 1: The artificial removal of part of an organism and its replacement in the body of the same or of a different individual.

2: To remove a plant from one location and replant it in another place.

transversion. A mutation resulting from the substitution in deoxyribonucleic acid or ribonucleic acid of a purine for pyrimidine or a pyrimidine for a purine.

tree. A perennial woody plant at least 20 feet (6 m) in height at maturity, having an erect stem or trunk and a well-developed crown or leaf canopy.

tree fern. A fern of the family Cyatheaceae, in which the fronds grow from the top of a trunk. The trunk in tree ferns consists partly of the woody bases of dead fronds. Most tree ferns are tropical.

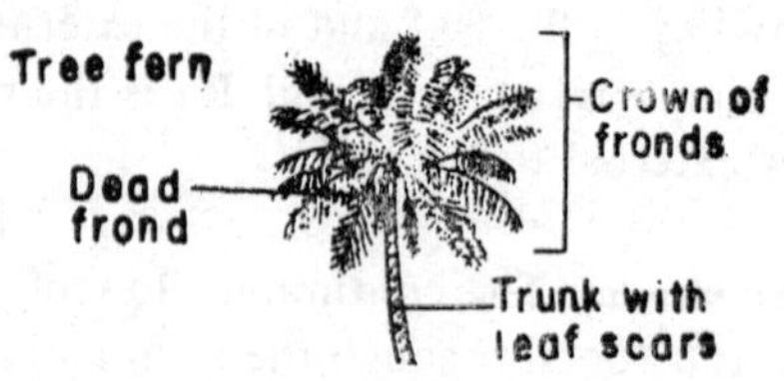

tree hair. *See* rock hair.

tree line. The level on the side of a mountain above which trees do not grow. Vegetation at higher levels is herbaceous or absent altogether; timber line.

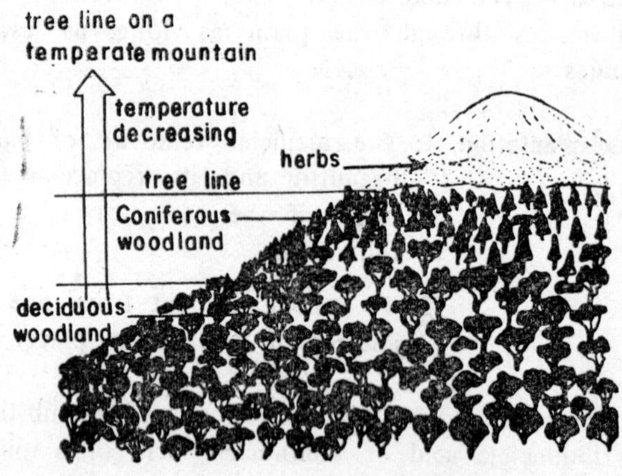

tremelloid. Like jelly or wet gelatin; gelatinous.

tribe. A group of related genera within a family.

tricarboxylic acid cycle. TCA cycle;=Krebs cycle.

trichogyne. In some algae, lichens, and fungi, a projection from procarp or archicarp the female sex organ that receives the male gamete or nuclei before fertilization (karyogamy).

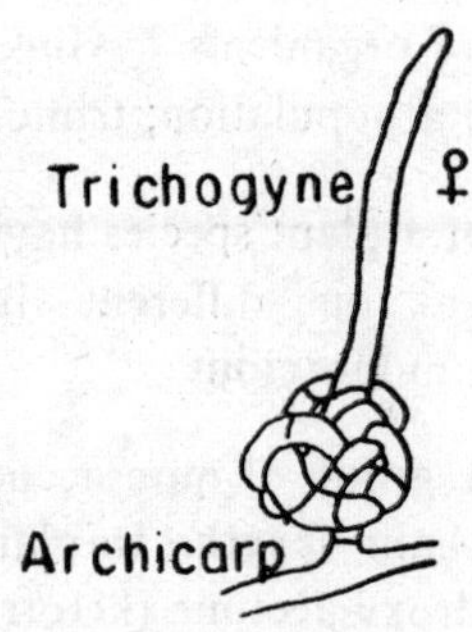

trichome. A hair on the epidermis of a plant.

tricolpate. Three-grooved.

trifoliate. Having three leaves or leaflets.

trifoliolate. Having a leaf or leaves of 3 leaflets, as most clovers.

trigenetic Pertaining to a symbiont requiring three different hosts during the life cycle.

trigenic. Used of characters or traits controlled by the integrated action of three genes; *cf.* monogenic, oligogenic, polygenic.

trigonal. Three-angled.

trihybrid. An individual heterozygous in respect of three pairs of alleles.

trilocular. Having three cavities or cells.

trimerous. Having parts in sets of three.

trimonoecious. Used of an individual plant bearing male, female and hermaphrodite flowers: coenomonoecious.

trimorphic. Used of organisms having three distinct forms in the life cycle, or in a population; trimorphism.

trioecious. Used of a plant species having male, female and hermaphrodite flowers on different individuals; trioikous; trioecy; *cf.* dioecious, monoecious.

triose. $C_3H_6O_3$. A group of monsaccharide compounds that have a three-carbon chain length. In-plants, D-glyceraldehyde (aldotriose) and dihydroxy-acetone (ketotriose) are found.

triphosphopyridine nucleotide (TPN). $C_{12}H_{28}N_7O_{17}P_3$. A greyish-white powder, soluble in methanol and in water; a coenzyme and an important component of enzymatic systems concerned with biological oxidation-reduction systems; codehydrogenase II; coenzyme II; triphosphopyridine dinucleotide.

tripinnate. Being bipinnate and having each division pinnate.

triplet code. A name for the genetic code so called because the code consists of nitrogen-containing bases in groups of three.

triploid. A polyploid having three sets of homologous chromosomes; triploidy.

triquetrous. Three-angled in cross section.

trisaccate pollen. A three-pored pollen grain, often having a triangular outline in cross section.

trisomic. Used of an aneuploid having one or more homologous chromosomes represented three times; trisomy.

trisporic acids. These are oxidised unsaturated derivatives of trimethyl cyclohexane, and act as sex hormones in members of Mucorales. Each mating type (+ or —) possess the ability to manufacture low levels of precursors. Cross diffusion of these precursors the prohormones (p^+ and p^-) lead to the formation of trisporic acids, resulting in the induction of aerial zygophores.

triternate. Three times 3; the leaflets or segments of a twice ternate leaf again in 3 parts.

trivalent. An association of three chromosomes synapsed during the first meiotic division.

tRNA. *See* transfer ribonucleic acid.

trophic. Pertaining to nutrition; trophism.

trophic levels. 1: The sequence of steps in a food chain or food pyramid, from producer to primary, secondary or tertiary consumer.
2: The nutrient status of a body of fresh water; categories include eutrophic, mesotrophic, oligotrophic, dystrophic.

trophotropism. An orientation response to food; trophotropic.

tropic. Pertaining to a tropism.

tropical. Of the regions of the world where the average monthly temperature changes little during the year, and the length of the day changes only slightly at different times of year.

tropical rainforest. A vegetation class consisting of tall, close-growing trees, their columnar trunks more or less unbranched in the lower two-thirds, and forming a spreading and frequently flat crown; occurs in areas of high temperature and high rainfall; hylaea; selva.

tropical savanna. *See* tropical woodland.

tropical scrub. A class of vegetation composed of low woody plants (shrubs), sometimes growing quite close together, but more often separated by large patches of bare ground, with clumps of herbs scattered throughout; an example is the Ghanaian evergreen coastal thicket; brush; bush; fourre; mallee; thicket.

tropical woodland. A vegetation class similar to a forest but with wider spacing between trees and sparse lower strata characterized by evergreen shrubs and seasonal graminoids; the climate is warm and moist; parkland; savanna-woodland; tropical savanna.

tropic movement. A directional plant movement in response to a stimulus. Such movements are the result of growth curvatures. A movement towards the stimulus is a positive tropism; a movement away from the stimulus is a negative tropism. There are various kinds of tropic responses like chemotropism, geotropism, hydrotropism, phototropism and thigmotropism.

tropism. Curving growth of a plant organ due to a stimulus coming from a particular direction, e.g., light or gravity.

tropophytia. Plants that thrive in a climate that undergoes marked periodic changes.

tropophyte. 1: A tropical plant.

2: A plant that thrives under mesic conditions for part of the year and xeric conditions at other times.

truffle. The edible underground fruiting body of various European fungi in the family Tuberaceae of class Discomycetes especially the genus *Tuber*.

trunk. The main woody stem of a tree, consisting of heartwood, sapwood and bark.

tryptophan. $C_{11}H_{12}O_2N_2$. An amino acid obtained from casein, fibrin, and certain other proteins; it is a precursor of indoleacetic acid, serotonin, and nicotinic acid.

tuber. A thick underground stem in which food is stored. Tubers have buds in modified leaf axils, from which new plants can grow, e.g., potato. Tubers are organs of perennation and vegetative reproduction.

tubercle. A small tuber, or rounded protruding body.

tuberous. Bearing or producing tubers.

tubule. Minute tubular structures peculiar to pore fungi like *Boletus* and *Polyporus* sp. The tubules are gathered together in a mass, parallel and vertical and form the lower part of the cap of the above-mentioned fungi.

tumefaction. Localized swellings like, galls, knots, clubbed roots—a pathological symptom in plants.

tumid. Swollen, inflated.

tunic. A loose membranous outer skin not the epidermis; the loose membrane about a corm or bulb.

tunica. The outer layer or layers of cells in the apical meristem of angiosperm shoots. Tunica cells divide periclinally, producing the surface tssues of the shoot.

tunicated. Provided with a tunic; having concentric or enwrapping coats or layers, as bulb of onion.

turbidostat. A device in which a bacterial culture is maintained at a constant volume and cell density (turbidity) by adjusting the flow rate of fresh medium into the growth tube by means of a photocell and appropriate electrical connections.

turbinate. Shaped like an inverted cone; top-shaped.

turbnlence. Irregular motion or agitation of liquids or gases; *cf.* laminar flow.

turfaceous. Pertaining to bogs; torfaceous.

turfophilour. Thriving in bogs; turfophile, turfophily.

turgid. Of the state of a cell which can absorb no more water by osmosis, because the cell wall prevents an increase in size. Turgid cells are the main means of support in herbaceous plants.

turgor. Distension of a plant cell wall and membrane by the fluid contents, inside the cell thus excerting a pressure (turgor pressure) on the cell wall.

turgor movement. A reversible change in the position of plant parts due to a change in turgor pressure in certain specialized cells; movement of *Mimosa* leaves when touched is an example.

turgor pressure (TP). The pressure caused by water in the vacuole which pushes against the cytoplasm and the cell wall. The turgor pressure is balanced by the osmotic pressure when the cell is well supplied with water when the cell loses water, the turgor pressure decreases and more water can enter the cell by osmosis.

turlon. A young shoot, or sucker developed from an underground bud, as observed in asparagus.

turnover. The number of substrate molecules transformed by a single molecule of enzyme per minute, when the enzyme is operating at maximum rate.

turnover number. The number of molecules of a substrate acted upon in a period of 1 minute by a single enzyme molecule, with the enzyme working at a maximum rate.

twig A young woody stem; more precisely the shoot of a woody plant representing the growth of the current season and terminated basally by circumferential terminal bud scar.

twiner. A climbing stem that winds about its support, as pole beans or many tropical lianas.

tylosis (*pl.* tyloses). A balloon-like outgrowth of a membrane of a vessel or a tracheid pit that protrudes into and blocks the vessel or tracheid cavity.

type concept. In taxonomy a method of nomenclature that attaches a name permanently to a type specimen, a name which is retained by the type throughout all subsequent taxonomic acts; type method.

type host. The host species from which the type of a parasite species or subspecies was collected.

type specimen. The original collection of an organism on which the description of a new species is based.

U

ubiquitous. Having a world-wide distribution, effect or influence; widespread; cosmopolitan; ecumenical; pandemic; ubiquist, ubiquitist.

ulmification The process of peat formation.

ultimate factor. The environmental factor to which the timing of a biological event is ultimately associated; ultimate causation.

ultracentrifuge. Centrifuge capable of producing rotor speeds up to 100,000 rpm and able to rapidly sediment tiny particles and macromolecules.

ultraplankton. Planktonic organisms less than 5 μm in length or diameter.

ultrastructure. Cellular structure studied with the aid of a transmission electron microscope; fine structure.

ultraviolet light. Electromagnetic radiation having a wavelength shorter than that of visible light (390-200 nm). Causes DNA base pair mutations and chromosome breaks.

ultrahaline. Used of sea water having a salinity greater than 30 parts per thousand.

umbel. An indeterminate inflorescence with the pedicels all arising at the top of the peduncle and radiating like umbrella ribs; there are two types, simple and compound.

umbellate. Umbelled; with umbels; pertaining to umbels.

umbelliform. Umbel-like; resembling an umbel.

umbilicate. 1: Having a small hollow.

2: A pileus of an agaric having a hollow on the top opposite the stipe.

umbilicus. Central holdfast occuring in some foliose lichen genera, e.g., *Umbilicaria*; navel, umbo.

umbo, 1: A conical projection arising from the surface.

2: A central swelling at the centre on top of the pileus above the stipe in agarics.

umbraticolous. Living in shaded habitats; skiophilous; umbraticole.

umbrophilic. Thriving in shaded habitats; heliophobous; skiophilic; umbrophile, umbrophily.

uncinate. Hooked obtusely at the tip.

uncoupling agent. A substance (e.g., 2, 4-dinitrophenol) that can uncouple phosphorylation of ADP from electron transport.

underground stem. Any of the stems that grow underground and are often mistaken for roots; principal kinds are rhizomes, tubers, corms, bulbs, and rhizomorphic droppers.

understorey. The vegetation layer between the overstorey or canopy and the ground-storey of a forest community, formed by shade tolerant trees of moderate height, shrubs, saplings and herbs; three component layers may be recognized, the upper-understorey, middlestorey and the lowerunderstorey; understorey stratum.

undifferentiated. Of cells in an embryo or young part of a plant, e.g., a meristem, which are all the same and have not developed into different tissues. In many simple plants, e.g., the prothalli of ferns, most of the cells are undifferentiated.

undulate. Wavy (up and down, not in and out), as some leaf or petal margins.

unguiculate. Narrowed into a petiole-like base; clawed.

ungulate. Shaped like a horse's hoof.

unialgal culture. A culture of one algal species or only of one strain.

uniaxial. Having only one axis.

unicellular. Composed of a single cell.

unifoliolate. Said of a compound leaf reduced to a single and usually the terminal, leaflet.

union. A ranked category in the classification of vegetation; a stratal plant community comprising one or more species of similar physiognomy and life form.

unigeneric. Said of a family composed of a single genus; monogeneric.

unijugate. Applied to a compound leaf composed of one pair of leaflets.

unilateral. One sided.

unilocular. Containing a single chamber or cell.

unilocular sporangium. A sporangium having zoospores only in one locule, e.g., in *Ectocarpus* an alga of class Phaeophyceae.

uniseriate. Arranged in a single row.

unisexual. 1: Used of a population or generation composed of individuals of one sex only.

2: Used of an individual having either male or female reproductive organs and producing only male or female gametes; gonochoristic; dioecious; diclinous.

3: Used of a flower possessing only male or female reproductive organs; imperfect flower; *cf.* bisexual.

unit character. 1: In taxonomy, a single character which cannot logically or conveniently be subdivided.

2: A character transmitted as a unit in heredity, controlled by one pair of alleles.

unit membrane. Membrane showing a three-layer or dark-light-dark pattern of electron density in the electron microscope; a former incorrect model of membrane structure proposing that phospholipid bilayer is coated on its outer and inner surfaces by proteins.

unitegmic Referring to an ovule having a single integument.

universal veil. The membrane which, in many fungi, covers the whole body of the young fungus. As the fungus develops, it tears to form the volva e.g., in the death cap, *Amanita phalloides.*

univorous. Feeding on only one type of food, or parasitizing a single host species; monophagous; univore, univory.

unoriented. Not arranged in any particular direction.

unstratified. Pertaining to lichen thallus, when the photobiont cells are dispersed throughout the thickness of the thallus; homoiomerous.

upperunderstorey. *See* understorey.

uptake. The process of taking water and nutrients from the soil into the roots of a plant, or of taking substances into a cell or organelle.

upwelling. An upward movement of cold nutrient-rich water from ocean depths.

uracil. $C_4H_4N_2O_2$. A pyrimidine base important as a component of ribonucleic acid, which pairs with adenine during transcription and translation.

urceolate. Shaped like an urn i.e., pitcher-like in form.

uredinium (*pl.* uredinia). In a rust fungus, a group of spore-bearing filaments crowded together beneath the cuticle or epidermis of a host plant, on which masses of red summer spores (urediniospores) are formed; Uredium, uredosorus.

urediospore. A thin-walled dikaryotic spore produced by rust fungi; gives rise to a vegetative mycelium which may produce more urediospores, considered as asexual stage of rust fungi; urediniospore, uredospore.

uredium (*pl.* uredia). *See* uredinium; uredosorus.

uredosorus (*pl.* uredosori). Uredinium, uredium.

uridine. $C_9H_{12}N_2O_6$. A crystalline nucleoside composed of one molecule of uracil and one molecule of D-ribose; a component of ribonucleic acid.

uridine diphosphate (UDP). The chief transferring coenzyme in carbohydrate metabolism.

uridine diphosphoglucose (UDPG). A compound in which α-glucopyranose is esterified, at carbon atoms 1, with the terminal phosphate group of uridine-5′-pyrophosphate (that is, uridine diphosphate); occurs in animal, plant, and microbial cells; functions as a key in the transformation of glucose to other sugars.

uridine monophosphate. *See* uridylic acid.

uridine phosphoric acid. *See* uridylic acid.

uridylic acid. $C_9H_{13}N_2O_9P$. Water-and alcohol-soluble crystals, melting at 202°C; used in biochemical research; uridine monophosphate (UMP); uridine phosphoric acid.

urkaryote. One of the three primary kingdoms of living organisms, comprising all eukaryote organisms.

urkingdoms. The three primary kingdoms, namely eubacteria, archaebacteria and urkaryote proposed recently as a basic tripartite scheme for grouping all living organisms, as an alternative to the widely used five-kingdom, scheme of Whittaker. Whereas eubacteria comprise all typical bacteria, the archaebacteria comprise all methanogenic bacteria and urkaryote all living organisms.

urn. The theca of a moss.

urophilic. Thriving in habitats rich in ammonia; urophile, urophily.

ursolic acid. $C_{30}H_{48}O_3$. A pentacyclic terpene that crystallizes from absolute alcohol solution, found in leaves and berries of plants; used in pharmaceutical and food industries as an emulsifying agent.

usnic acid. $C_{18}H_{16}O_7$. Yellow, dibenzofuran derivative crystalline, insoluble in water, slightly soluble in alcohol and ether, melts at about 198°C; found in lichens e.g , spp. of *Cladonia*, *Usnea*; used as an antibiotic against Gram+bacteria and fungi. Usually occurs in lichen cortices; usninic acid.

usninic acid. *See* usnic acid.

ustic acid. 1: A hydroxyquinol from *Penicillium ustus.*

2: An antimycobacterial product from *Ustilago maydis.*

utricle. 1: A small bladder.

2: A bladdery, one-seeded, usually indehiscent fruit, as in some amaranths.

3: The bladder-like covering of certain fungi. e.g., *Dendrogaster.*

utriform. Bag-like.

uvarious. Latin meaning 'like a bunch of grapes'.

ultrasonic waves. Sound waves of high intensity (beyond the audible range), used for the destruction of microbes or the cleaning of materials.

ultrafilteration. A method for the removal of all but the smallest particles, e.g., viruses, from a fluid medium.

urease. An enzyme that catalyzes the hydrolysis of urea.

$$NH_2-\overset{\overset{\displaystyle O}{\|}}{C}-NH_2 \xrightarrow{\text{urease}} 2NH_3+CO_2.$$

This was the first enzyme to be isolated by Sumner (1926), where he also discovered that enzymes are proteins.

vacuolar membrane. The protoplasmic membrane known as tonoplast which bounds a vacuole. separating; it from the surrounding protoplasm.

vacuolar sap. The liquid inside a vacuole.

vacuole. A cavity in a cell, surrounded by a membrane and filled with molecules and particles in a watery medium. Many plant cells, especially in leaves, have a single large vacuole and a thin layer of cytoplasm between it and the cell membrane; vacuolar.

vaginate. Sheathed.

vaginule. A minute sheath, surrounding the base of the seta in the bryophytes.

valence. The relative ability of a biological substance to react or combine.

valid. In taxonomy, used of a name or nomenclatural act that is correct according to the provisions of the Code.

valid name. The correct name for a particular taxon.

vallecula. Latin for 'grooves in flowers.'

vallecular canal. A large intercelluler space running in the cortex of *Equisetum*, outside the carinal canal.

volsoid. 1: Having the perithecia in a circle in the stroma.

2: Having groups of perithecia with their beaks pointing inwards, or even parallel to the surface.

valvate. Having valve-like parts, as those which meet edge to edge or which open as if by valves.

valvatc aestivation. Floral leaves in a whorl may touch one another at the margins without overlapping as observed in the family Annonaceae or family Mimosaeeae.

valvate dehiscence. Spontaneous release of the pollen from anthers or contents of a ripe fruit through a valve, flap or other large aperture.

valves. 1: A part of a fruit-wall which separates at dehiscence.

2: One of the two halves of the cell-wall of a diatom.

valve view. View from the top surface of epitheca or hypotheca of diatoms.

van der Waals force. A weak, attractive force between atoms; particularly important in hydrophobic bonding of amino acids in proteins.

variability. The property of being variable in form or quality.

variation. 1: Differences in the characteristics of individuals of a species or a population.

2: The occurrence of differences in the permanent structure of cells.

varicose. Dilated.

variegated. Of leaves with patches of different colour.

variegated position effect. A phenomenon observed in some cases when a chromosome aberration causes a wild-type gene from the euchromatin to be relocated adjacent to heterochromatin; the phenotypic expression of the wildtype allele will be unstable, producing patches of phenotypically mutant tissue that differ from the surrounding wide-type tissue.

variegation. The occurrence within an individual of a mosaic phenotype, typically with respect to colouration; partial albinism in plants.

variety. 1: A rank in the hierarchy of botanical classfication; the principal category between species and form.

2: An infrasubspecific taxon, such as varietas or cultivar; an ambiguous term often used for any variant group within a species; a group that differs from other varieties of the same subspecies including non-genetic variants and microgeographic races.

variolarioid. Having powdery or granular tubercules.

varnish tree. *Rhus vernicifera.* A member of the sumac family (Anacardiaceae) cultivated in Japan; the cut bark exudes a juicy milk which darkens and thickens on exposure and is applied as a thin film to become a varnish of extreme hardness; lacquer tree.

vascular. Of plants with a vascular system. Vascular plants include all pteridophytes and spermatophytes.

vascular bundle. A thread of vascular tissue in the vein of a leaf, or in a stem.

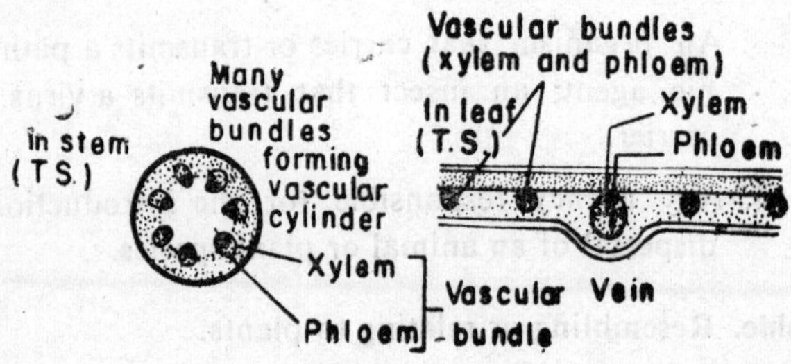

vascular cambium. The lateral meristem which produces secondary xylem and phloem.

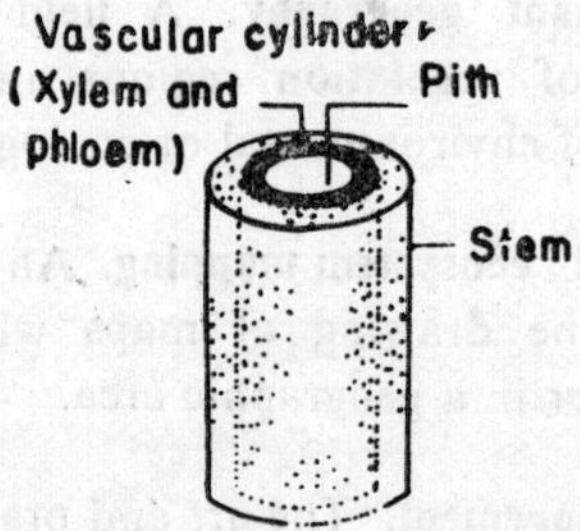

vascular cylinder. A tube of vascular tissue consising of xylem and phloem in a root or stem; stele.

vascular ray. A ray derived from cambium and found in the stele of some vascular plants often separating vascular bundles; medullary ray.

vascular system. The tissues consisting of xylem and phloem cells, which translocate substances from one part of a plant to another. The development of vascular systems has made it possible for plants to evolve on land.

vasculiform. Shaped like a little pot.

vector 1: In statistics, a quantity whose complete description requires two or more numbers.

2: A matrix with only one row or column; its elements are referred to as components.

3: An organism that carries or transmits a pathogenic agent; an insect that transmits a virus. *cf.* carrier.

4: Any agency responsible for the introduction or dispersal of an animal or plant species.

vegetable. Resembling or relating to plants.

vegetation. The total plant life or cover in an area; also used as a general term for plant life; the assemblage of plant species in a given area.

vegetational plant geography. A field of study concerned with the mapping of vegetation regions and the interpretation of these in terms of environmental or ecological influences.

vegetation and ecosystem mapping. An art and a science concerned with the drawing of maps which locate different kinds of plant cover in a geographic area.

vegetation management. The art and practice of manipulating vegetation such as timber, forage crops, or wild life, so as to produce a desired part or aspect of that material in higher quantity or quality.

vegetation zone. 1: An extensive, even transcontinental, band of physiognomically similar vegetation on the earth's surface.

2: Plant communities assembled into regional patterns by the area's physiography, geological parent material, and history.

vegetative. Pertaining to assimilative, somatic and trophic growth processes rather than to reproduction. Stems, leaves and roots are vegetative organs.

vegetative eunuch. A plant which completes its growth and dies without producing seeds, thereby consuming resources but leaving no descendants.

vegetative growth. Growth of the tissues and organs not involved in sexual reproduction. Vegetative growth occurs by mitosis and the lengthening and enlargement of cells.

vegetative reproduction. 1: A type of asexual reproduction in which a whole new plant is produced from an organ, e.g., a rhizome, bulb, or tuber, which is not involved in sexual reproduction.

2: In lower plants, reproduction by asexual processes such as budding or fragmentation; vegetative division.

veins. 1: Lines which can be seen on the surface of a leaf, marking the position of the vascular bundle.

2: Strands of tissue on the lower surface of foliose lichens esp. *Peltigera* where they may replace a lower cortex.

vein banding. A symptom of virus-infected leaves in which tissues along the veins are darker green than other laminar tissue.

vein clearing. A symptom of virus-infected leaves in which veinal tissue is lighter green than that of healthy plants.

velamen. The corky epidermis covering the aerial roots of an epiphytic orchid; consisting of several layers of dead cells often with spirally thickened and perforated walls, which soak up water running over it.

velar. Pertaining to a veil.

veldt. Grasslands of eastern and southern Africa that are usually level and mixed with trees and shrubs. Also spelled veld.

velum. 1: A veil-or curtain-like membrane round the stipe of basidiocarp left after the expansion of the pileus *i.e.*, partial veil; a sheath of hyphae forming a complete membrane over the fruit body of an agaric is complete veil or universal veil.

2: A membranous outgrowth from the base of a leaf of *Isoetes.* It covers or partly covers the sporangium.

velutinate. Thickly covered with delicate hairs; like velvet; velutinous.

venation. The pattern of veins on the surface of a leaf. In most dicotyledons the venation is reticulate, and in most monocotyledons it is parallel.

venenatous. Latin meaning 'poisonous'

venter. The swollen basal region of an archegonium, containing the egg-cell.

ventilation tissue. The sum total of the intercellular spaces in a plant, by means of which gases can circulate through the plant body.

ventral. The lower surface of a dorsiventral plant structure, such as thallus.

ventral canal cell. An unwalled cell which lies in the venter of an archegonium, above the egg.

ventral suture. The presumed line of junction of the edges of the unfolded carpel.

ventral trace. One of the two laterally placed vascular strands often present in the wall of a carpel.

ventricose. 1: Swollen in the middle.

2: Having an inflated bulge at one side.

verbascose. It is one of the major non-reducing carbohydrates transported in higher plants in phloem. They are oligosaccharides consisting of galactose, glucose and fructose.

vermicular. Shaped like a worm.

vernalin. It is a hypothetical vernalization stimulus postulated by Melchers, which has not been isolated or characterised till date.

vernalization. A process of thermal induction in plants, in which growth and flowering are promoted by exposure to low temperature; that is, at temperatures below the optimal temperature for growth; jarovization, yarovization.

vernation. The characteristic arrangement of young leaves within the bud; prefoliation.

verruca. A wart-like elevation on the surface of a plant or animal.

verruciform. Resembling a wart.

verrucose. Having the surface covered with wart-like protuberances; verrucous.

versatile anther. An anther whose attachment is near its middle, thus enabling it to swing freely.

versicoloured. Exhibiting colour change, or different colours; variegated.

versiform. 1: Said of organs of the same kind which are not all of the same shape.
2: Changing form with age.

verticel. Whorl.

verticillaster. A kind of inflorescence which looks like a a dense whorl of flowers, but is really a combination of two crowded dichasial cymes, one at each side of the stem.

verticillate. Arranged in whorls, resembling the spokes of a wheel.

vertical resistance. Almost absolute resistance, but only to some races of a pathogen; an epidemiological term. Vertically resistant plants are usually hypersensitive.

vesicle. 1: Any small body in a cell or organelle which is surrounded by a membrane and contains products of metabolism. Vesicles in the cytoplasm are produced mainly by the Golgi body.
2: A thin-walled globular swelling usually at the end of a hypha.
3: A bladder-like sac.
4: Pertaining to *Aspergillus*, the swollen apex of the conidiophore.
5: In *Pythium*, the evanescent extra sporangial structure in which zoospores are differentiated.

vesicular. 1: Like, or pertaining to, a vesicle.

2: Like a bladder.

vesicular bodies: Thin walled vesicles in the sub-hymenium of certain Hymenomycetes.

vesicular-arbuscular mycorrhiza (VAM). It is a kind of endotrophic mycorrhiza formed by species of *Glomus* and related Zygomycotina in plants in natural and agricultural environments; so named because the fungi produce large swollen vesicles and intricately branched arbuscules within the plant cells. The fungus is a biotroph and has a symbiotic relationship with host.

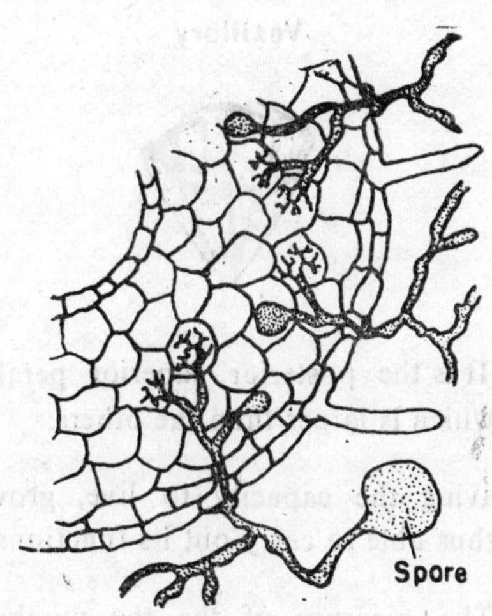

vesiculose. Made from or full of vesicles.

vessel. Conducting tissue in the xylem consisting of vessel elements, found mainly in angiosperms.

vessel element. A long, often thin, dead cell in the vessels of the xylem. Vessel elements are arranged end to end with large holes in the end walls, through which the xylem sap can pass. The cell walls of vessel elements are thickened with lignin.

vestiture. A covering, especially of hairs.

veterinary mycology. Study of the fungal diseases of higher animals. Among the common fungi parasitising animals are *Aspergillus*, *Blastcmyces*, *Candida*. *Cryptococcus*, *Histoplasma*, *Microsporum* etc.

vexillary. This is the typical aestivation of the papilionaceous corolla. The posterior vexillum overlaps the two lateral petals which again overlap the paired anterior petals.

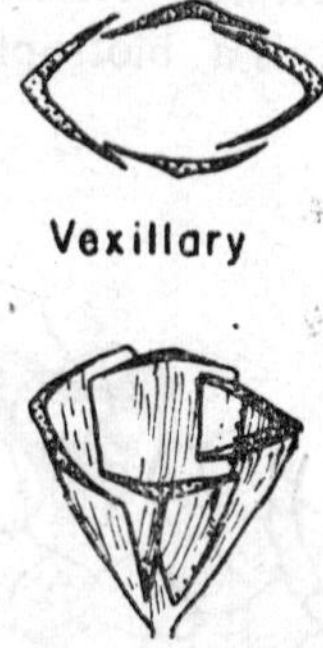

Vexillary

vexillum. It is the posterior superior petal of a papilionaceous flower which is larger than the others.

viable. Having the capacity to live, grow, germinate or develop and thus able to carry out its functions; viability.

viability. The measure of the the number of individual surviving in one class, relative to another standard class.

vibrio. 1: A bacterium that is curved with a twist; has less than one complete turn or twist (in contrast to a helical bacterium).

2: *Vibrio* is a genus of Gram negative bacteria.

vicid. Sticky, slimy, glutinous, lubricous, mucilaginous, viscous.

vigour. The intensity of growth or general metabolic activity of an organism, population or community.

villose. Covered with long, weak, soft hairs, which are not matted; villous. *cf.* tomentose.

vinecsent. Turning wine-red.

violaxanthin. It is a naturally occurring carotenoid pigment exist which in the chloroplast envelope, giving it a yellowish colour. It is also considered to be a precursor of abscissic acid —-anatural growth inhibitor.

virens. Latin meaning 'green'.

virescence. A symptom in which green pigmentation occurs in plant tissues not normally green.

virgate. 1: Banded.

2: Long, slender and stiff, much branched.

virion. The complete mature virus particle.

viroid. Agents of infectious disease, characterized by the absence of encapsidated proteins; they consist only of low molecular weight RNA molecules, that can replicate themselves and can cause disease.

virology. The study of viruses; virusology.

virose. 1: Poisonous.

2: Having a strong unpleasant smell.

virotoxin. One of a group of toxins present in the mushroom *Amanita virosa*.

virtual tautonymy. In nomenclature, the nearly identical spelling, or same origin or meaning, of a generic name and of the epithet of one of the species or subspecies originally included in the genus,

virucide. An agent that kills viruses.

virulence. 1: The capacity of a pathogen to invade host tissue and reproduce.

2: The degree of pathogenicity.

viruliferous. Containing a virus; of an insect vector containing virus and being capable of introducing it into a suscept.

virus. 1: A group of very simple organisms, which consist of a strand of nucleic acid surrounded by a coat of protein. Viruses do not metabolize, and can only reproduce inside the cells of other organisms, where the nucleic acid of the virus directs the synthesis of more viruses using the protein synthesis machinery already in the cell. Viruses can destroy cells in this way, and cause many diseases in other organisms. They are sometimes classified in a kingdom of their own. Viruses are very small, ranging from 10-250 nm in diameter.

2: The ending of a name of a genus in virological nomenclature.

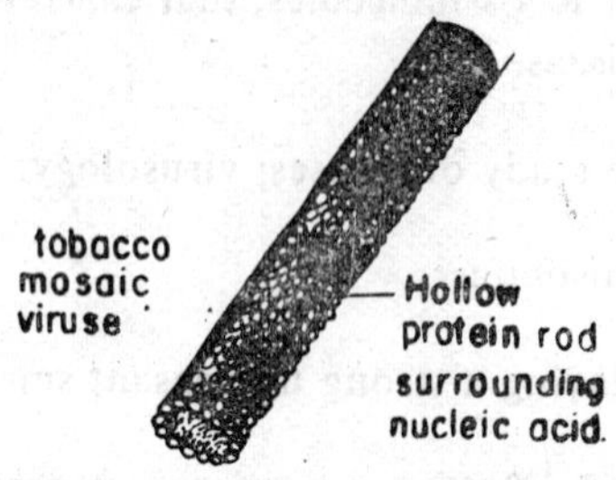

virus interference. A phenomenon which may be defined as protection of host cells against one virus, conferred as a result of prior infection with a different virus.

virusology. The study of viruses; virology.

viscid. Having a sticky surface, as certain leaves.

viscid disseminule. A seed or a spore which has a sticky surface, or sticky hairs, and is dispersed by being attached to an animal.

viscosity. 1: Internal friction in fluids.

2: The relatively slow rate of dispersal of individuals in a population, and consequently of gene flow; population viscosity.

vital staining. The staining of cells or their parts while they are living, and without killing them.

vitamin. An organic substance required as a coenzyme in many of the chemical reactions of metabolism. There are many different kinds of vitamins, and they are required by organisms in very small amounts.

vitamin E. $C_{29}H_{50}O_2$. Any of a series of eight related compounds called tocopherols, α-tocopherol having the highest biological activity; occurs in wheat germ and other oils and is believed to be needed in certain human physiological processes.

viteline. Yellow like the yolk of an egg.

viticolous. Growing on vines; viticole.

vittate. 1: Having longitudinal stripes, bands, or ridges.

2: Bearing specialized oil tubes (vittae), in the pericarps of some fruits.

viviparous. 1: Producing bulbils or young plants, instead of and in place of flowers.

2: Said of seed which germinates within the fruit before it is detached from the parent plant.

volunteer. A crop growing from unplanted seeds, e.g., seed shed by the previous crop.

volutin. A basophilic substance, thought to be a nucleic acid, occurring as granules in the cytoplasm and vacuoles of algae and fungi especially yeasts. These granules stain a deep violet when the cells are stained with dilute methylene blue.

volva. A sheath of hyphae enclosing the whole of the fruit-body of agarica and Gasteromycets (i.e., universal veil), becom-ingruptured as thefru it-body enlarges and sometimes remaining as a cup or pouch around the base of the stipe or receptacle.

voucher specimen. Any specimen identified by a recognized authority for the purposes of forming a reference collection.

vulgaris. Latin meaning 'common'.

wall pressure (WP). *See* turgor pressure.

wandering lichens. Lichens with an epigeic habit; e.g., *Parmelia revoluta.*

wanderplasm. The cytoplasmic contents along with many nuclei which accumulate close to the base of the antheridial branch of *Vaucheria* at the time of the development of oogonium.

wart. A small blunt-tipped rounded outgrowth.

warted. Bearing warts.

water balance. The difference between the rate of water intake (absorption) by a plant and water loss (transpiration).

water bloom. *See* bloom.

water capacity. The amount of water that a soil can retain against gravity; usually expressed as a percentage per volume of soil.

water conservation. The protection, development, and efficient management of water resources for beneficial purposes.

water cultures. Aqueous solutions containing different concentrations of various mineral salts. They are used in experiments to show the relative importance of individual elements to plant growth.

water cycle. The global biogeochemical cycle of water involving exchange between the hydrosphere, atmosphere, lithosphere and living organisms.

water microbiology. An aspect of microbiology that deals with the normal and adventitious microflora of natural and artificial water bodies.

water parasite. A parasite that derives only water from its host.

water pollution. Contamination of water by materials such as sewage effluent, chemicals, detergents, and fertilizer runoff.

water potential. The thermodynamic state of the water within a plant cell expressed in units of pressure equal to the difference in free energy per unit volume between matricaily bound, pressurized, or osmotically constrained, water and that of pure water. Water diffuses from a solution of high water potential to a solution of low water potential if they are separated by a semi-permeable membrane. In turgid plant cells, the water potential is equal to the sum of the osmotic potential, and matric potentials.

water-use efficiency of photosynthesis. The ratio of photosynthesis to transpiration; calculated as W_f=Photosynthesis/Amount of water transpired.

water-use efficiency of productivity. The ratio of dry matter produced to amount of water consumed; calculated as W_p=Dry matter production/Water consumption.

wavelength. The length of a wave of light. Different wavelengths have different colours and different levels of energy.

wax. The general term for a group of insoluble, organic esters of fatty acids and a variety of higher monohydric alcohols. Waxes are prodnced by certain plants, e.g., beewax and carnauba wax from a Brazilian palm. They form a waterproof layer on many stems, leaves, fruits etc. Waxes are used in the impregnation of paper and wood, in lacquers, polishes, paints, cosmetics, etc.

weathering. The process in which rocks are broken down to smaller particles by the action of external agencies such as wind, rain, temperature changes, plants, etc. Weathering may take place when water expands on freezing in pores or cracks in rocks, when rocks expand strongly heated by the Sun or when plant roots grow into cracks in rocks. Weathering may also take place when oxygen in the air oxidises the rock or when dissolved carbon dioxide in rain water dissolves calcium carbonate present in rocks.

weed. Any plant growing where it is not wanted. A plant that is useless or of low economic value, especially one growing on cultivated land to the detriment of the crop.

wet and dry bulb hygrometer. This consist of two thermometers placed side by side. The bulb of one is in air and the bulb of other is wrapped in muslin which dips into the water. The wet bulb is cooled due to the cooling effect of evaporation, the amount of which dopends on the humidity of the air. The relative humidity of the air can be obtained from tables of dry bulb temperatures and wet bulb depressions.

westwood. Wood having a water soaked appearance because of a high water content; may be a caused by bacteria or by physiological factors.

wheat germ. The embryo of a wheat grain.

whip grafting. A method of grafting by fitting a small tongue and notch cut in the base of the scion into corresponding cuts in the stock,

white blister. A disease of plants caused by spp. of *Albugo*, class Oomycetes; on members of the families Cruciferae, Amaranthaceae, Convolvulaceae etc.; white rust.

whorl. A group of three or more organs of the same kind, arising at the same level on a stem and arranged in a circle, e.g., the petals of a flower or the branches of a horsetail or leaves on the stem.

wild type. 1: The natural or typical form of an organism, strain or gene, arbitrarily designated as standard or normal for comparison with mutant or aberrant individuals or alleles.

2: The phenotype characteristic of the majority of individuals of a species under natural conditions.

willow. A deciduous tree and shrub of the genus *Salix*, order Salicales; twigs are often yellow-green and bear alternate leaves which are characteristically long, narrow, and pointed, usually with fine teeth along the margins.

wilting. 1: Loss of turgidity in plants caused by an excess of water loss by transpiration over water uptake by absorption from the soil.

2: Of plant disease, a plesionecrotic symptom characterized by loss of turgor, which results in drooping of leaves, stems, and flowers. Wilting can be due to fungal or bacterial infections.

wilting coefficient. The amount of water remaining in a soil when a plant is in a state of permanent wilting.

wilting point. The time at which the water content of a soi is reduced to a level at which permanent wilting occurs.

windburn. Injury to plant foliage, caused by strong, hot, dry winds.

windchill. The effect of the wind in causing excessive cold penetration into plants.

wind dispersal. The dispersal of sporee, seeds and fruits by wlnd.

windkill. Plant death resulting from the effect of wind, often through windchill, or physiological drought.

wind pollination. The transport of pollen from anthers to stigmas by wind. Anmophily.

wing. 1: One of the lateral petals in the Papilionaceae.

2: A flattened outgrowth from a fruit or seed increasing the, area witnout increasing the weight, serving in wind-dispersal.

3: The downwaldly continuing base of a decurrent leaf.

winnowing. Grading and separation of the fine fractions of particulate matter by wind or water eurrents.

winter annual. A plant which lives for a short time, grows, and sets its seeds in colder parts of the year, dies, and survives the rest of the year as seed.

witches'-broom. A hyperplastic symptom due to proleptic development of many weak shoots from adventitious buds in a stem.

witches'-broom disease. An abnormal cluster of small branches or twigs that grow on a tree or shrub formed as a result ofrepeated branching due to attack by fungi, especially rusts e.g., *Pucciniastrum*, *Gymnosporangium* etc., or due to viruses, dwarf mistletoes, or insect injury; hexenbesen; staghead.

wobble. Capability of the third base in the tRNA anticodon (5′ end) to form a hydrogen bond with any two or three bases at 3′ end of the messenger RNA's codon. Thus, a single tRNA species can recognize several different codons.

wolf's moss. A lichen, *Letharia vulpina* from the Scandinavian belief that it was poisonous to wolves.

wound hormone. A substance produced in wounded tissue, which stimulate cell division; differentiated cells of the stem or root become dedifferentiated when they resume meristmatic activity. Whether such substances are involved in all cases of cell division following wounding is not clear.

wood. 1: Hard tissue made of the remains of dead xylem cells in the stems of perennial plants. Wood contains lignin, and its function is to support the plant and to conduct water; woody.

2: A dense growth of trees, more extensive than a grove and smaller than a forest.

woodland. Vegetation in which trees are dominant. The trees in woodland are smaller and more spaced apart than those in forest.

wood's light. It is UV light filtered through nickel oxide-containing soda glass. *Microsporum* but not *Trichophyton*-infected hairs show a bright greenish fluorescence in wood's light, an effect made use of in diagnosis.

wood sugar. *See* xylose.

woody plant. A perennial plant having a secondarily thickened lignified stem.

WP. *See* wall pressure and turgor pressure.

Woronin bodies. Rounded or elongated-oval highly refractive bodies in the cells of certain Discomycetes, particularly in association with septa.

Woronin's hypha. 1: In Ascomycetes, a coiled hypha probably homologous with an archicarp.

2: A loosely coiled hypha of large diameter, at the center of a young perithecium, which later develops ascogenous hyphae; scolecite.

wortmannin. An antibiotic from *Pencillium wortmanii*; antifungal especially against *Botrycis*, *Cladosporium* and *Rhizopus*.

xanthochroic. Of a basidiocarp of Hymenomycetes, having a reddish or yellowish-brown content which darkens on treatment with KOH.

xanthomegnin. A pigment from *Trichophyton megninii.*

xanthophyll. $C_{40}H_{56}O_2$. Any of a group of yellow, alcohol-soluble carotenoid pigments that are oxygen derivatives of the carotenes, and are found in certain flowers, fruits, and leaves. Major xanthophylls found in green leaves are cryptoxanthin, lutein, zeaxanthin, violaxanthin, neoxanthin. They protect plants against the photooxidation of chlorophyll and absorb and transfer light energy to chlorophyll-a during photosynthesis.

X-body. An inclusion in the plant cell suffering from a virus disease.

X-chromosome. A heterochromosome associated with the determination of sex, and sometimes occuring alone.

xenia. Phenotypic changes in the maternal tissue of a fruit (endosperm) resulting from the direct influence of the pollen genotype.

xenogamy. Cross fertilization; fertilization between flowers on different plants.

xenogeneic. Originating in a member of another species.

xenogenesis. 1: Spontaneous generation; abiogenesis.
2: Any unusual method of reproduction.

xenogenous. Originating from outside the organism or system; exogenous; allochthonous.

xenograft. Tissue transplanted from one organism to another of a different species; heterograft.

xenology. The study of host-parasite relationships.

xenoparasite. A parasite infesting an organism that is not its normal host; xenoparasitism.

xenospore. A spore dispersed from its place of origin.

xerad. Xerophyte.

xerantic. Becoming parched or dried up; withering.

xerarch succession. An ecological succession beginning in a dry habitat; xerosere; *cf.* mesarch succession.

xerasium. An ecological succession following drought or drainage.

xeric. 1: Of or pertaining to a habitat having a low or inadequate supply of moisture.

2: Of or pertaining to an organism tolerating or adapted to dry condition.

xerochastic. Used of a fruit in which dehiscence is induced by desiccation; xerochasis, xerochasy.

xerochore. That region of the Earth's surface covered by dry desert.

xerocleistogamy. Self-pollination within flowers that remain unopened because of inadequate moisture; xerocleistogamic.

xerodrymium. A dry thicket community.

xerogeophyte. A plant which enters a resting stage during periods of drought.

xerohylad. A dry-forest plant; xerohylophyte.

xerohylium. A dry-forest community.

xerohylophilous. Thriving in dry forests; xerohylophile, xerohylophily.

xerophobous. Intolerant of dry conditions; xerophobic; xerophobe, xerophoby.

xerophyte. A plant living in a dry habitat, typically showing xeromorphic or succulent adaptations; a plant able to tolerate long periods of drought; zerophyte; xerophil; xerad; xerophytic; *cf.* hydrophyte, hygrophyte, mesophyte.

xeropoophyte. A heath plant; xeropoad; xeropoophyta.

xerosere. 1: A xerarch succession.

2: A temporary community in an ecological succession on dry, sterile ground such as rock, sand, or clay.

xerosium. An ecological succession on drained and dried soils.

xerostatic. Used of an ecological succession completed under xeric conditions.

xerotaxis. The condition of a plant succession that remains unaffected by drought.

xerothamnium. A spiny-shrub community.

xerotherm. A plant adapted to hot dry conditions.

xerothermic. Used of organisms tolerating or thriving in hot and dry environments; xerotherm.

xerotherous. Used of organisms adapted to dry summer conditions.

xerotropism. An orientation response of plants or plant structures to desiccation; xerotropic.

X-ray crystallography. The use of X-ray scattering by crystals to determine the three dimensional structure of molecules, especially proteins and nucleic acids.

xylan. A hemicellulose (a structural polysaccharide) formed from the sugar D-xylose; occur in the primary cell wall along with cellulose, usually acting as an additional strengthening agent. Jute hemicellulose is a branching xylan with xylopyranose as the main constituent.

xylic gap. A gap in the xylem opposite a leaf base.

xylochrome. A mixture of substances to which the colour of the heart wood is due. It contains tannin, gums and resins.

xylogenous. Living on wood.

xyloma. A sclerotium-like body of order Dothideales, producing sporogenous structures inside.

xylophagous. Referring to an organism feeding on wood; dendrophagous; hylophagous; lignivorous; xylophage, xylophagy.

xylophilous. Thriving on or in wood; xylophile, xylophily.

xylophyte. 1: A woody plant; xylophyta.

2: A plant living in or on wood.

xylem. Tissue in the vascular system of a plant, consisting of tracheids, vessels, parenchyma and sclerenchyma. The vessels, tracheids and sclerenchyma have lignified cell walls. Most xylem cells are dead, and contain no cytoplasm. The function of the xylem is to translocate water and, nutrients from the roots to the stems and leaves; xylem is also known to transport cytokinins, a plant growth regulator. It is the chief supporting tissue of higher plants. There are two types of xylem: primary xylem, which develops from the procambium, and secondary xylem, which is additional to primary xylem and is produced by the activity of cambium.

xylose. $C_5H_{10}O_5$ An aldopentose sugar found in the pentosans of straw, bran and wood. It is a white crystalline soluble solid, sweet in taste, melts at about 148°; used as a non-nutritive sweetner and in dyeing and tanning; also occurs as a regular constituent of glycoprotein, covalently bonded with the amino acid serine in the cell; wood sugar.

D-xylulose. A ketopentose, found in free or phosphary lated state in metabolic reactions in the cell.

Y

yam. 1: A plant of the genus *Dioscorea*; family Dioscoreaceae, grown for its edible fleshy root; yield a cortiones a steroid hormone, used in rheumatic diseases and ophthalmic disorders.

2: An erroneous name for the Puerto Rico variety of sweet potato; the edible, starchy tuberous root of the plant.

yarovization. See vernalization.

Y-chromosome. With diploid sex-differentiation, the sex chromosome which is present and pairs with the X-chromosome in sex heterozygote. With haploid sex differentiation, the sex chromosome of the male.

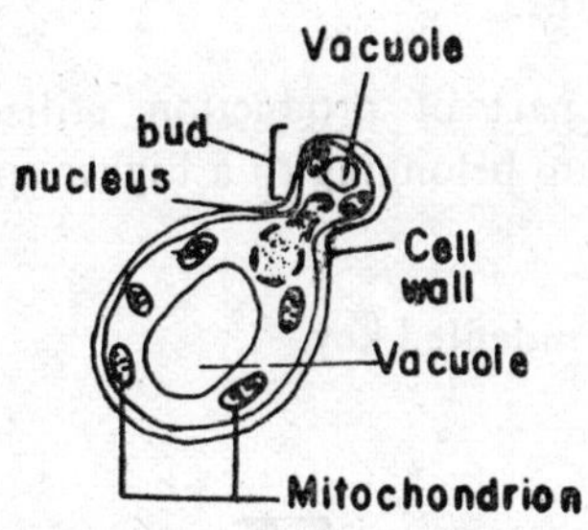

yeast. One of a number of unicellular fungi which usually reproduce asexually by budding or fission (asporogenous yeasts). However yeasts can also reproduce sexually by conjugation and can form ascospore (sporogenous yeast) e.g., *Saccharomyces cerevisiae*. Yeasts can respire both aerobically and anaerobically. They are important substances in brewing and baking where they produce ethanol and carbon dioxide. There are various kinds of yeasts e.g., apiculate yeasts, (e,g., *Saccharomycodes*, having minute polar projections); baker's yeasts (*Saccharomyces cerevisiae*); black yeast, yeast-like state of *Aureobasidium*, *Cladosporium*, etc.; Chinese yeast, *Mucor rouxii*; food yeast, dry *Candida utilis*.

yellows. A general term applied to plant diseases characterized by yellowing of the leaves and stunting. It may be caused by virus, fungus, insect or toxin.

yellow earths. A group of soils in tropcs, sub-tropics and warm-temperate regions. The yellow colour is probably connected with a high degree of hydration of the ferric oxide.

yellow rice. Rice discoloured by *Penicillium islandicum* rendered carcinogenic for rodents and possibly man.

yellow rust. It is a fungal disease on wheat caused by *Puccinia striiformis* a member of the class Teliomycetes; stripe rust.

yellow snow. Snow coloured yellow by the growth on it of an alga *Chlamydomonas yellowstonensis* sometimes seen in Alps and Antarctic regions.

yield (**Y**) That part of production utilized by a consumer species or by a group belonging to a higher trophic level, or by man.

yoked key. See indented key.

Z

Z-DNA. Double helical DNA in which the helices are left—handed and the sugar phosphate backbones of the two polynucleotides trace a zigzag course around the axis of the helices.

zearalenone. This is one of the two groups of mycotoxins (the other being trichothecine) produced by several spp. of *Fusarium* in molded corn. It is most toxic to swine in which it causes abnormalities and degeneration of genital system.

zeatin. (6-(4-hydroxy-3-methyl but-2-lenyl) amino purine). It is the first naturally occurring cytokinin isolated from milk stage kernels of corn (*Zea mays*); present in roots, stems and leaves with the greatest amount in kernels. The biological properties of zeatin are similar to those of kinetin or may be more active in some instances.

zein. This is one of the prolamines occurring in maize. Prolamines are insoluble in water but are soluble in 70-80 percent ethanol. On hydrolysis these proteins yield large quantities of proline and ammonia.

zenotropism. *See* negative tropism.

zerophyte. *See* xerophyte.

zeugite. The organ, *e.g.*, an ascus or a basidium, in which fertilization is completed and the dikaryophase ends.

zincophyte. A plant adapted to, or tolerating, high zinc levels in the soil; frequently used as an indicator of this particular soil type.

zineb. This is one of the organic fungicides sold as Dithane Z-78, Parzate etc., it is an excellent safe, multipurpose foliar and soil fungicide for the control of leaf spots, blights, fruit rots etc. of vegetables, flowers, fruit trees and shrubs. Its struc ture is:

$$\begin{matrix} CH_2NH.CS.S \\ CH_2NH.CS.S \end{matrix} \Big> Zn$$

zoidiogamy. Fertilization by a motile male gamete; zoidiogamic.

zoidiophilous. Pollinated by animals; zoophilous; zodiophilous; zoidiophily.

zonal centrifuge. A centrifuge that uses a rotating chamber of large capacity in which to separate cell organelles by density-gradient centrifugation.

zonate. Describes the caps of many fungi, especially of the genus *Lactarius*, and the thalli of crustose lichens, when they have a sequence of alternate concentric zones of varying colours.

zonation. 1: Regular concentric variation of texture, pigmentation or sporulation, observable in cultures, frequently associated with diurnal fluctuations in light, temperature or other factors.

2: Arrangement of organisms in biogeographic zones.

zone lines. Narrow, dark brown to black lines in hard wood part of decaying wood generally caused by fungi.

zoocecidium. A plant gall usually caused by an insect.

zoochlorellae. Unicellular green algae which live as symbionts in the cytoplasm of certain protozoans, sponges, and other invertebrates.

zoochorous. Dispersed by the agency of animals; zoochoric; zoochore, zoochory.

zoodomatia. Plant structures acting as shelters for animals.

zoogamete. A motile gamete; planogamete.

zoogloea. A bacterial colony embedded in slime.

zoogonidium. Zoospore.

zoophilous. Pollinated by animals; having an affinity for animals; zoidiophilous; zodiophilous; zoophile, zoophily.

zoophyte. Any animal that resembles a plant in morphology or mode of life; zoophytic.

zoosporangium. A spore case bearing zoospores.

zoospore. An independently motile spore in protistans, some fungi and algae.

Z-scheme. The Z-scheme, so named because of its shape, is the pathway of electron transport and the production of NADPH and ATP in chloroplasts.

zygomorphic. Of flowers which are symmetrical in one direction only (bilaterally symmetrical), *e.g.*, the flowers of orchids, often due to differences in sizes and shapes of petals and/or sepals.

zygophore. A special hyphal branch of order Mucorales, belonging to class Zygomycetes producing copulation branches.

zygospore. 1: The resting spore resulting from the fusion of like gametangia (*e.g.*, in *Spirogyra*) an alga belonging to Class Chlorophyceae.

2: The resting spore formed from the complete fusion of isogametes in Zygomycetes,

zygophyte. A plant produced by sexual reproduction.

zygosis. Union of gametes; conjugation; zygotic.

zygotaxis. The mutual attraction between male and female gametes; zygotactism, zygotactic.

zygote. 1: A diploid cell which is produced by the fusion of two haploid gametes.

2: A fertilized ovum is a zygote. In plants, the zygote develops first into an embryo and than into a sporophyte.

Zygomycetes.. A class of fungi belonging to the subdivison Zygomycotina; cosmopolitan, saprobes or parasites of insects; two orders (1) Mucorales; (2) Entomophthorales.

zygotene. The stage of first meiotic prophase during which homologous chromosomes synapse; visible bodies in the nucleus are now bivalents; amphitene.

zygotic induction. Phage induction following conjugation of a lysogenic bacterium with a nonlysogenic one.

zygotropic reaction. A reaction in which directed growth of zygophores of (+) and (—) mating partners towards each other is observed. This is a chemotropc response to a volatile stimulus, *i.e.*, the prehormones. *See* trisporic acids.

zymogenic. Causing fermentation; zymogenous.

zymase. A mixture of numerous enzymes (at least 14). Zymase was extracted from yeast in 1903 and was then believed to be a single enzyme. Zymase catalyses the breakdown of glucose to ethanol and CO_2 in fermentation, requiring the presence of a number of coenzymes and certain metal ions in order to act completely.

zymologist. One who studies yeasts.

zymology. The study of yeasts.

zymonic acid. A metablic product of *Trichosporon*, *Hansenula* and other yeasts.

zymosterol. $C_{27}H_{43}OH$. An unsaturated sterol obtained from yeast fat; yields cholesterol on hydrogenation.

APPENDIX-A

International System of Units (SI)

Prefixes

Prefix	*Factor*	*Sign*
milli-	$\times 10^{-3}$	m
micro-	$\times 10^{-6}$	μ
nano-	$\times 10^{-9}$	n
pico-	$\times 10^{-12}$	p
kilo-	$\times 10^{3}$	k
mega-	$\times 10^{6}$	M
giga-	$\times 10^{9}$	G
tera-	$\times 10^{12}$	T

Basic Units

Unit	*Symbol*	*Measurement*
metre	m	length
kilogram	kg	mass
second	s	time
ampere	A	electric current
kelvin	K	thermodynamic temperature
mole	mol	amount of substance

Derived Units

Unit	*Symbol*	*Measurement*	*Definition*
newton	N	force	kg. m/s^2
joule	J	energy, work	kg. m^2/s^2
hertz	Hz	frequency	s^{-1}
pascal	Pa	pressure	$kg/m.s^2$
coulomb	C	quantity of electric charge	A.s
volt	V	electrical potential	kg. $m^2/A.s^3$
ohm	Ω	electrical resistance	kg. $m^2/A^2.s^3$

Some Multiples of SI Units

Unit	*Symbol*	*Definition*	*Measurement*
angstrom	Å	$10^{-10}m = 10^{-1}nm$	length
micron	μm	$10^{-6}m$	length
litre	l	$10^{-3}m^3 = dm^3$	volume
tonne	t	10^3kg	mass
dyne	dyn	$10^{-5}N$	force
bar	bar	10^5Pa	pressure

Some Non-SI Units

Unit	*Symbol*	*Definition*	*Measurement*
atm	atm	101325 Pa, 1.01325 bar	pressure
degree Celsius	°C	temp. °C= temp. K—273	temperature
million years	Ma. m.y.	10^6 years	time
billion (US) years	Ga	10^9 years	time

APPENDIX-B

Prefixes Used in Botanical Words

a- without, not, e.g., asexual, not sexual; asymmetrical, without symmetry.

ab- from, away from, e.g., abaxial, the side of a leaf facing away from the stem.

ad- to, towards, on the side of, e.g., adaxial, the side of a leaf facing towards the stem.

aer-air, atmosphere.

allelo- one another, e.g., allelopathy, the chemical inhibition of one organism by another.

allo- different, differing, other, e.g., allopolyploid, a polyploid resulting from the fusion of two different nuclei; allopatric, of species occurring in different regions.

amphi- both, on both sides.

an- same prefix as a-, used before words beginning with *a* vowel or the letter h, e.g., anaerobic, not aerobic.

andro- male, e.g., androecium, the male parts of a flower.

aniso- unequal, e.g., anisogamous, having gametes of unequal size, shape or behaviour.

ante- before, in front of.

anti- against, opposite, e.g., antibiotic, a substance that acts against living organisms (especially bacteria); antipodal, the cells at the opposite end of the embryo sac from the micropyle.

apo- away from, without, e.g., apogamy, reproduction without sexual fusion, apocarpous, of flowers with carpels separate from one another.

arche- first, beginning, primitive, earliest.

archi- first; arche-.

auto- caused by or originating in itself automatic, self operating, e.g., autopolyploid, a polyploid resulting from an increase in the number of sets of chromosomes within a nucleus; autotroph, an organism producing its own food.

benth- sea or river bed, ocean depth, e.g., benthophyte, a plant living at the bottom of a water body or on the bed of a river.

bi- two, twice, double, e.g., binomial, the Latin name of a species, consisting of two words; biennial, a plant with a two-year life cycle.

bio- life, living organism, e.g., biology, the study of living things.

brachy- short, e.g., brachymeiosis, a simplified form of meiosis completed in a single division by suppression of the second division.

brady- slow, e.g., bradygenesis, retarded phylogenetic development.

cata- down, against; **kata-**, e.g., catabatic, used of winds blowing down a slope.

caul(i)- relating to stems, e.g., cauliflorous, having flowers growing directly from the stem.

chromo- colour, coloured, e.g., chromoplast, a plastid containing pigments; chromosomes, so called because they become deeply coloured when stained for microscopy.

chrono- time, e.g., chronometry, the science of precise time measurement.

cis- on the same side as, e.g., cisalpine, one which is situated on the south side of the Alps.

clado- branch, offshoot; **klado-**, e.g., cladoptosis, the periodic shedding of twigs.

cleisto- closed, without an opening, e.g., cleistogamy, self-pollination before the flower opens; **kleisto-**.

co- together, with, sharing, associated, e.g., coenzyme, a substance (not a substrate) that is necessary for the functioning of an enzyme.

coeno- sharing, in common, e.g., coenosis, an assemblage of organisms having similar ecological preferences.

contra- opposite, against, e.g., contranatant, swimming, or migrating against the current.

crypto- hidden, e.g., cryptophyte, a plant whose perennating organs are underground; cryptogam, a plant whose reproductive organs are very small or hidden.

cyto- relating to cells, e.g., cytology, the study of cells; cytoplasm, the parts of the cell outside the nucleus.

di- two, twice, double, e.g., disaccharide, a carbohydrate consisting of two sugar molecules (monosaccharides); dicotyledon, a plant with two cotyledons in the seed.

dia- across, through, e.g., diaphototropism, orientation at right angle to the direction of incident light.

diplo- double, two fold, e.g., diplobiont, an organism exhibiting a regular alternation of haploid and diploid generations during the life cycle.

dys- bad, abnormal, insufficient, malfunction, difficult, e.g., dystrophic, used of fresh water bodies rich in organic matter but having low nutrient content.

ecto- outside, outer, e.g., ectotrophic mycorrhizae grow outside the cells of the host root.

endo- inside, inner, within, inwards, e.g., endocarp, the inner layer of the fruit wall; endotrophic mycorrhiza, with hyphae growing into the cells of the host root.

epi- on the surface of, upon, above, outer, e.g., epicarp, the outer layer of the fruit wall; epiphyte, a plant growing on another plant; epigeal germination, when the cotyledons emerge above the ground.

eu- good, normal, e.g., eutrophic, a habitat well-supplied or rich in nutrients.

eury- wide, e.g., euryadaptive, parasites tolerating a wide range of host species.

ex- without, e.g., exalbuminous, without endosperm; exstipulate, without stipules.

extra- outside, beyond, separate from, e.g., extrafloral, away from the flower.

flavo- yellow, e.g., flavoprotein, one of a group of yellow-coloured proteins.

gam(o)- joining together, fusion, e.g., gamopetalous, having fused petals.

geo- earth, terrestrial, e.g., geocarpy, ripening of fruits underground.

gymno- naked, exposed, e.g., gymnosperm, a plant in which the seed is not enclosed in an ovary.

gyno- female, e.g., gynoecium, the female parts of a flower.

halo- salt, salty, e.g., halophyte, a plant that grows in salty habitats.

hap- once, e.g., hapaxanthic, having a single flowering phase during the life cycle.

haplo- single, simple, e.g., haplobiont, a plant flowering once per season.

heli- sun, e.g., heliad, a plant thriving in full sunlight.

hemi- half, partly, e.g., hemiparasite, a parasite that produces some of its own food.

hetero- different, other than usual, e.g., heterozygous, with different alleles at the same locus on homologous chromosomes; heterotroph, an organism that obtains food other than from itself.

hexa- six, sixfold.

holo- whole, complete, entire, total, e.g., holoparasite, a parasite that cannot survive away from its host.

homo- same, similar, alike, e.g., homologous chromosomes have the same sequence of loci; homosporous plants produce spores all of the same size.

hydro- relating to water, e g., hydrophyte, a plant with perennating organs under water; hydrolysis, a chemical reaction involving the addition of water molecules and the breakdown of organic molecules.

hygro- wet, humid, damp, e.g., hygrotropism, orientation in response to humidity or moisture.

hyl- wood, e.g., hylad, a forest or wood land plant.

hyper- more, above, higher (position), e.g., hypertonic, a more concentrated solution.

hypo- lower, below, under (position), e.g., hypotonic, a less concentrated solution; hypogynous, a flower in which the corolla, calyx and anthers arise below the gynoecium.

infra- below, smaller, e.g., infraspecific, variation below the the level of species.

inter- between, e.g., interspecific, competition between species.

intra- within, e.g., intraspecific, competition within a species.

iso- identical, equal, e.g., isogamy, the fusion of morphologically identical gametes.

kata- down, against; **cata-**.

lepto- thin, slender, e.g., leptotene, the stage in the first meiotic prophase when the chromosomes appear as thin threads.

limn- lake, e.g,, limnad, a lake plant.

limno- marshy, e.g., limnodophyte, a salt-marsh plant.

macro- large, great, long, e.g., macromolecule, a large molecule composed of many smaller molecular units.

mega- 1: great, large, greater than usual, e.g., megaspore, the larger of the two kinds of spore produced by heterosporous plants.

2: used to denote unit one million times ($\times 10^6$).

meio- smaller, less than, e.g., meiospore, a haploid cell produced by meiosis that forms the gametophyte by mitotic division.

mero- part, incomplete, e.g., meroparasite, a partial or facultative parasite that can survive in the absence of the host.

meso- middle, between intermediate, e.g., mesophyll, the layer of tissue between the palisade and lower epidermis in a leaf; mesocarp, the middle layer of the pericarp of a fruit.

meta- after, change, between, among, e.g., metabiosis, a symbiosis between two organisms.

micro- 1: small, short, e.g., microscope, an instrument used for observing very small objects; microspore, the smaller of the two kinds of spore produced by heterosporous plants.

2: used to denote unit $\times 10^{-6}$.

mono- one, once, single, e.g., monocotyledon, a plant with one cotyledon in the seed; monocarpic, a plant that produces fruit once in its lifetime.

morph(o)- shape, form, e.g., morphology, the study of shape.

multi- many.

myco- relating to fungi, e.g., mycology, the study of fungi.

myria- 1: many, innumerable.

2: used to denote unit- $x10^{4}$.

myrmeco- ant; often used to include termites, e.g., myrmecochorous, dispersed by the agency of ants.

nano: 1: dwarf, e.g., nanoplankton, minute planktonic organisms.

2: used to denote unit $\times 10^{-9}$.

neo- new, e,g., neo Darwinism, the science of evolution developed after Darwin, including the more recently-discovered principles of genetics.

octa- eight, eightfold.

oligo- few little, e.g., oligotrophic, a habitat with few nutrients or low fertility, oligosaccharide, a carbohydrate consisting of a few monosaccharide units.

omni- all, universally, e.g., omnicolous, living on a wide variety of substrates.

ortho- upright, straight, at right angles e.g., orthotropic, an upright axis.

pachy- thick, fat, e.g., pachytene, the stage in the first meiotic prophase when the chromosomes become short and thick.

palaeo- old, ancient, e.g., palaeobotany, the study of fossil plants.

pan- all, whole, completely, e.g., pandemic, a disease reaching epidemic proportions in many parts of the world simultaneously.

para- by, beside, along, beyond, e.g., paragynous, having the antheridium at the side of the oogonium.

patho-, pathy- suffering, feeling, associated with disease, e.g., pathogenic, producing or capable of producing disease.

pent(a)- five, e.g., pentose, a monosaccharide with five carbon atoms.

per- through, e.g., percolation, downward movement of water through porous sediments.

peri- around, on the surface, e.g., perianth, the parts of the flower around the reproductive parts; pericarp, the wall of the fruit.

phobo- fear; sometimes used to negate the operative prefix as in phobophototropism that means negative phototropism.

photo- relating to light, e.g., photosynthesis, the production of carbohydrates using energy from light; phototropism, curved growth towards light.

phyco- concerning algae, e.g., phycobiont, the algal partner in a lichen symbiosis.

phyll(o)- relating to leaves, e.g., phyllotaxy, the way in which leaves are arranged.

phyto- concerning plants, e.g., phytochemistry of plants.

piezo- pressure, e.g., piezotropism, a growth movement in response to a compression stimulus.

plagio- oblique, e.g., plagiotropism, an orientation response at an oblique angle to the vertical.

pleio- more, e.g., pleiomorphic, exhibiting polymorphism at different stages of the life cycle.

poly- many, e.g., polypeptide, a molecule with many peptide bonds.

pro- before, forward, in front of, e.g., prognosis, a forecast of the likely course of a disease.

proto- first or original, e.g., protophyte, a plant without well differentiated tissues.

pseudo- false, e.g., pseudoaquatic organisms that thrive in wet ground.

psychro- cold, e.g., psychrophilic, thriving in a low temperature.

quadri- four, fourfold, square, at right angles, e.g., quadriflagellate, one having four flagella.

quin- five, fivefold.

re- again, repeat.

retro- backwards, e.g., retrogressive, pertaining to a process of dedifferentiation to a simpler state.

rheo- current, flowing, e.g., rheophyte, a plant inhabiting running water.

rhiz(o)- relating to roots, rootlike organs or other underground plant parts, e.g., rhizoid, the 'roots' of bryophytes; rhizome, an underground stem.

sapro- concerning decay, e.g., saprophyte, a plant that lives on decaying organic matter.

schiz(o)- splitting, dividing, e.g., schizocarp, a fruit that splits into the separate carpels when ripe.

schler(o)- hard, rigid, e.g., sclerenchyma, a hard, supporting plant tissue.

semi- half, partly, e.g., semipermeable, of membranes that allow the passage of some molecules but not others.

sub- under, below, beneath, somewhat, e.g., subspecies, a taxon below the level of species; subacute, a leaf apex that is somewhat acute.

super- above, greater (in structure or significance), e.g., superoptimal, used of a stimulus that elicits a greater response than the natural stimulus.

sym- together, united, e.g., symbiosis, two different organisms living with and depending on each other.

syn- together, united, e.g., syncarpous, of ovaries in which the carpels are united.

tachy- quick, fast, e.g., tachysporous is a plant that disperses its seeds quickly.

tele- beyond, e.g., telemorphosis, a change in form in response to a remote stimulus.

ter- three, thrice, threefold.

tetra- four, e.g., tetraploid, having four sets of homologous chromosomes.

trans- across, e.g., transalpine, situated on the north side of the Alps.

tri- three, e.g., triose, a monosaccharide with three carbon atoms; triploid, having three sets of homologous chromosomes.

tycho- chance, e.g., tychopelagic, organisms that are normally benthic, but have been carried up into the water column by chance factors.

ultra- beyond, e.g., ultradian rhythm, a biological rhythm with a period exceeding 24h.

uni- one, once, single, e.g., unicellular, an organism consisting of one cell.

xero- dry, e.g., xerophyte, a plant that occurs in dry habitats.